# Introduction

**Total Temporal Gravitic and Faster Than Light Propulsion Vol I: Theory and Principles**

Introduction to Theoretical Basis of Engine Design

## Table of Contents

Total Temporal Gravitic and Faster Than Light Propulsion Vol I: Theory and Principles ............ 1
Introduction to Theoretical Basis of Engine Design ................................................................. 1
Introduction ............................................................................................................................. 3
    Introduction: The Demands of General Relativity, Spatial Geometry and Preferential Frame of Reference on all Scales ........................................................................................ 5
    Introduction: Scale of the Zeno Effect and Related Issues ................................................. 11
**Rubidium** ............................................................................................................................ 15
    Introduction: Scaling the Zeno Effect Up ............................................................................ 29
    Introduction: Artificial Alteration, Real, and Artifacts of Observation ............................... 31
    Introduction: Observation ..................................................................................................... 31
    Introduction: Preferential Frame of Reference and the Demands for Mass*energy ... 33
    Introduction: *Observable* Eigenstate vs *Observed* Eigenvalue ........................................ 35
    Introduction: Equating to the AdS Horizon Surface ............................................................. 37
    Introduction: Scope of the Zeno Effect ................................................................................. 38
    Introduction: Some Technological Problems in the Zeno Literature ................................... 40
    Introduction: Some Technological Issues ............................................................................. 42
    Introduction: Tests of Special and General Relativity .......................................................... 48
Terminologies and Descriptions of Terms Used ...................................................................... 50
    Terminologies: ....................................................................................................................... 59
    The Maric Operator ................................................................................................................ 60
    LIGO and Preferential Frame of Reference ......................................................................... 62
Some rules: as put forth in Quantum Information Dynamics Volume III ................................ 63
    I.   When Preferential Frame of Reference Applies ........................................................... 64
    II.  When Preferential Frame of Reference does not Apply: *For Entangled Systems Only* 64
Zeno Effect Dynamics of the AdS Horizon Surface ................................................................ 67
    Normal Detection Rates Over Normal Light Distance ......................................................... 70

# Introduction

**Normal Light Distance** .................................................................................................72

Zeno Dynamic AdS Horizon Surface Quality .................................................................74

    **Zeno-Effect on the AdS Horizon Surface** ..............................................................80

Anti-Zeno Effect Dynamics of the AdS Horizon Surface .................................................82

    **Anti-Zeno Effect** ....................................................................................................91

Describing Temporal Order of Events on the AdS Horizon Surface ..............................96

**Altering the Rate of Time Evolution on the AdS Horizon Surface** ...............................103

Temporal Progression of the AdS System ....................................................................103

Relativistic-Like Conformally Scale Invariant Changes to Epsilon ..............................104

Observation Rate as the Determinant for Pseudo-Locality with respect to Time and Distance .................................................................................................................104

Preservation of Information Causality .........................................................................106

    **A Note on the Bray Spacetime Manifold** ..............................................................113

Briefly, The Zeno Effect Related to the Bray-Alcubierre Metrics is this: ...................125

Coupling to Mass-energy ............................................................................................133

Detection Principles in Radiation Systems and Associated Electronic Designs ..........139

    **Choice of Zeno Systems to Produce A Stable Zeno Effect** ...................................139

        **Detector and Data Transfer Characteristics** ...................................................141

Details of Zeno Design in the Case of Radioactive Detection ....................................145

Deadtime: ....................................................................................................................146

Pile Up: ........................................................................................................................149

The Zeno and Anti-Zeno Effect as Detector Dead Time and Pile Up .........................152

Summary of Quantifiable Zeno Experiment ...............................................................155

A Faster Than Light Discontinuous 'Jump' Drive Spacetime Manifold with Zero Mass-Energy Requirement and Zero Negative Energy Density: Part I ................................156

The Zero Energy FTL Spacetime Manifold ...................................................................156

Making Sense of the Spacetime Manifold ...................................................................166

The Stress Energy Tensor .............................................................................................170

The Bell Loophole-Free Test ........................................................................................173

Alice, Bob and the Horizon Surface .............................................................................178

Shannon Entropy vs. Conventional ΔS Entropy and the Number of Causal Paths they Define .182

    **Why is Shannon Entropy a Description of Greater than One Causal Path?** .......182

The AdS Horizon Surface and the Delayed Choice Quantum Eraser .........................192

# Introduction

| | |
|---|---|
| Preferential Frame of Reference in Entangled and Non-entangled Systems | 200 |
| What exactly is the 'Gap?' | 205 |
| The Delayed Choice Quantum Eraser and Non-Preferential Frame of Reference | 218 |
|    **Entanglement Swapping** | 226 |
|    DERIVATION OF ℏNQ HUP, The Orthogonal Components of The Electric ⊥ Magnetic EigenVectors | 233 |
|    Back to the DCQE: | 241 |
| Alice and Bob on a Schwarzschild Horizon | 244 |
|    What do Alice and Bob see of each other when connected by >1-causal path? | 244 |
|    Alice and Bob, Multiply Connected | 248 |
| The Lunar Landing Anomaly | 269 |
|    Introduction, The Lunar Landing Anomaly | 269 |
|    The Two Train Scenario | 269 |
|    LIGO and GRACE, Coupling to Mass-energy | 276 |
|    Overview of Lunar Landing Anomaly | 279 |
|    **Details of the Lunar Landing Scenario** | 288 |
|    Lunar Landing: Preliminary Summary | 295 |
|    Destruction of Information | 298 |
| FRACTAL SET ASSIGNMENT | 305 |
| References | 309 |
| Appendix I Defining the Qubit | 312 |
| Appendix II THE CONDITIONS OF THE LOCALLY QUANTIZED METER STICK | 324 |
| Appendix III The Zeno Effect and the single quantized photon | 330 |
| Appendix IV The Maric Operator | 335 |
|    **Contributions of Mileva Maric** | 335 |

## Introduction

This text will summarize the essential characterization of artificial alteration of spatial geometry. The goal is straight forward Gravitic Propulsion, evolving to Faster Than Light Propulsion, *using zero energy*. This is not an esoteric nor transcendent hypothesis but based upon the demands that a change in temporal rate *must be* associated with a change in spatial geometry under General Relativity [Schwarzschild transformation], else, violate General Relativity in some way that has no prior art nor description.

# Introduction

The obvious problem with modern attempts at both Gravitic and/or Faster Than Light is the bizarre mass-energy requirements, even more so 'negative mass-energy' requirements equaling more than a cosmos of such 'exotic matter,' and other such transcendent hypotheses of this nature completely dominating the literature with no end in sight. We had the Alcubierre approach, which started as requiring some greater than one thousand cosmos's of this 'negative mass-energy' just standing still and only got more bizarre as other authors chimed in over the years.

This text Vol I Theory and Principles and the Vol II Technological Approaches will describe both Gravitic Propulsion and Faster Than Light Propulsion utilizing *zero energy requirements*. Meaning explicitly, the Alcubierre approach after three decades significantly *proves that* Gravitic and Faster Than Light Propulsion utilizing *any type of* mass-energy is *impossible*. This in turn then suggests to a compelling degree that Gravitic and Faster Than Light Propulsion can only be achieved utilizing *no mass-energy*.

This series of texts, Vol I Theory and Vol II Technological Approaches will describe Gravitic and Faster Than Light Propulsion in sufficient detail to *construct a prototype*.

This zero-energy requirement is founded on the scaffold of 'the observer effect' in non-ontological industrial applications of the Zeno Effect. For example, Quantum Computing approaches currently employ the Zeno Effect as a means of stabilizing quasi-micro-circuit systems, noise suppression, and so on.

Whereas convention incorrectly describes this 'observer effect' as being limited to quantum scales a micro-circuit is small, but not a 'quantum scale.'

Scale, as conventionally described for the Zeno Effect as being limited to 'quantum scale' phenomenon has never been validated nor has it been derived, but is pure supposition, not even elevated to status of hypothesis. A Schwarzschild transformation in path length must accompany any observed, measured, or even an *artifact of observation* of any change in the rate of progression of unitary time, on all scales, always, regardless of de facto suppositions of scale. And, since no one to date has ever observed a change in path length, e.g., 'length transformation,' any supposition regarding constraints of any sort is non-sequitur if not absurd.

Scale will be a subject in great depth throughout this text. Functionally, any Zeno system is a macroscopic scaled system, wherein *by convention the Zeno system is defined* as the detector being coupled to the observed system, perhaps 1-cm to 1-meter [benchtop] scales. Quantum scale in reference to the Zeno Effect is thus unfounded, and certainly not dutifully derived and is contrary to its own conventions.

This Vol I will show to a compelling degree that the Zeno Effect is in fact the universe's natural clock system, from the Planck scale out to and beyond the cosmological horizon. The Zeno Effect in fact is the very quintessential definition of the AdS Horizon Surface, and that description will dominate this text.

The Quantum Zeno Effect is described in general terms as an observed alteration in the progression of unitary time as the causal result of observation rate of some system, where QZE typically refers to the slowing of observed time by rapid detection and measurement and the Anti-Zeno Effect as

# Introduction

generally a quickening of observed rate of progression of time by a similar but opposite mode of detection and measurement rate. In this key concept, this observation is typically of time invariant natural systems, such as radioactive decay, quantum tunneling, and so on. The observed phenomenon is thus 1-clock regarded as a typically but not always some natural clock system [designated herein as $r$] and the observation rate of the detection measurement [observer system] is the $2^{nd}$-clock of the system [designated herein as $R$]. A Zeno System is a 2-clock system, the observer [$R$] and the observed [$r$] systems. There are no further sequitur elements to a Zeno system. There is no energy description for such a system, that is, no energy dependency that occurs in any prior art. A Zeno system alters the rate of progression of unitary time using no applied energy or mass*energy equivalent. Furthermore, the alteration of temporal rate is several orders of magnitude greater than, for instance, being in close proximity to a main sequence star, further compelling that mass*energy is not part of a Zeno dynamic.

This introduction section will be followed by the hard derivations and definitions in later sections of this text. Throughout the text the natural clock system will be regarded as $^{14}C$ for reasons of simplicity and practicality. There are any number of natural, and even non-natural clock systems that have been demonstrated in the Zeno literature. However, rather than delve into each, the most basic beta emitting radioactive isotope will be used as demonstration.

The introduction is long [~40 pages] and covers a lot of requisite concepts that have to be stated. The maths will be dutifully derived in the later sections.

## Introduction: The Demands of General Relativity, Spatial Geometry and Preferential Frame of Reference on all Scales

A key concept in this text is that the Zeno Effects must be accompanied by an alteration in spatial geometry according to Schwarzschild transformations as path length between the 2-clock system, designated $R$ [observer clock system] and $r$ [observed clock system], else represent a violation in General Relativity that has no sequitur description and certainly no prior art. Another key concept is that although there is no causal relationship of the sort, a change in the progression of time results in a change in spatial geometry or vice versa, if one is changed, *the other component must also change; there is no exception to this*. That is, there is no provision wherein the rate of progression of time can change without an accompanying change in *path length between the 2-clocks of the system*. The interdependency between temporal rate and length transformation has no causal logic, they are interdependently associated.

The notion that this is an observed or artifact of observation of a quantum scale, *only*, is unfounded. There is no prior art wherein the scale of a quantum system vs that which is regarded as macroscopic has been dutifully derived, nor has such ever been observed, or experimentally validated. There is no empirical, *tangible* evidence that suggests General Relativity is subject to scale, for example, in quantum mechanics this [scale] relationship is a dilemma, not a resolved or in fact agreed upon relationship of the sort that General Relativity is subject to scale. It exists as a question, a quandary, not a law of nature. The seemingly odd nature of quantum mechanics on atomic scales is puzzling, not a property or theorem.

# Introduction

However, the exact line where that *divergence* [quantum to macroscopic] occurs is derived clearly herein, in later sections.

To clarify, General Relativity has no Preferential Frame of Reference as modern science assigns, we look at LIGO. The Large *Interferometer* Gravitational Observatory is the same as any classic Michelson-Morley interferometer. A Michelson-Morley interferometer cannot detect its own change of state under a Lorentz transformation. The Large *Interferometer* Gravitational Observatory works by design by detecting its own change of state under a Schwarzschild transformation as per General Relativity. Not only does LIGO respond as an alteration in path length and time as a spacetime ripple which describes a Schwarzschild transformation passes through the interferometer, but other large interferometers around the planet coordinate in this detection as a functioning array of interferometric detectors. Under Special Relativistic conditions all detectors would simply detect nothing, however in this case all of the detectors coordinate signals to such extent that they can be tied as an array of detectors around the planet. Each interferometer and the concert of interferometers are functioning by detecting their own change of state under a Schwarzschild transformation which describes General Relativity. There are three primary such Gravitational Observatories in the global array, whose coordination helps 'triangulate' the locality of the source of the Gravity Wave.

The argument that LIGO is merely detecting an indirect secondary effect of a Newtonian alteration in path length is exactly the point; LIGO cannot, nor can it ever detect its motion orbiting the sun, a Lorentzian state change. Such a state change would also have Newtonian variations in path length associated with it, if and only if Preferential Frame of Reference did not apply to such alteration in path length. Since Preferential Frame of reference applies to Lorentzian changes in path length, LIGO cannot detect its motion orbiting the sun, as is a classic Michelson-Morley type experiment.

> Note that 'frame of reference' is non-specific and incorrectly used in General relativity as such:
>
> Preferential Frame of Reference does not apply to Schwarzschild transformations implies only that a system *that is not coupled to a local mass\*energy* can detect its own change of state under a Schwarzschild transformation. Examples would include the Inflation of the cosmos, which can exceed $v > c$, and LIGO detection of its own Schwarzschild transformation as a Gravity Wave passes through the interferometer. This concept will be discussed in detail.
>
> Under conditions where the spatial geometry is coupled to a local mass\*energy, Preferential Frame of Reference does apply. LIGO, for instance, cannot detect *direct variations* of Earth's gravity well in the sense that, hypothetically if we were to mount the device on an elevator and attempt to measure the shape of Earth's gravity well out into orbit, would not work.
>
> However, GRACE [Gravitational Recovery and Climate Experiment] is a 2-satellite system that does monitor dynamic changes in Earth's gravity map as a result of tectonic, lunar, and other changes in dynamic mass. An example is land tides that result from lunar proximity and orbit. This will be discussed at length later in this text.

With respect to Gravitation:

# Introduction

- The first demand is that the *length transformation is thus real, else could not be detected via an interferometer, or any tangible method.*
- The second demand is that Preferential Frame of Reference does not apply to General relativistic, Schwarzschild transformations, in such cases where mass*energy is not coupled to such spacetime geometry, else could not be detected via an interferometer.
- The third demand is that *local* mass*energy is not a demand of spatial geometry.
  - *local* mass*energy is not a demand of spatial geometry in such cases as a gravity wave, as the spatial geometry is not coupled to the spacetime geometry.
  - *local* mass*energy *is* a demand of spatial geometry in the common case where the spatial geometry is coupled to the spatial geometry.
- The fourth demand is that Minkowski spacetime is a null hypothesis, as aberrantly applied to Schwarzschild type transformations. The complex conjugate of sqrt(-1) has no real world relevance, is true of both Schwarzschild and Lorentz transformations. This demand has been placed on, or rather transferred to, GR over the decades as an aberrant postulate. 'Length contraction' has never been observed nor has it been experimentally validated. No aspect of Minkowski space has been or ever can be validated, it is an intangible, unobservable dimensionality.

The term, 'a Schwarzschild transformation that is not coupled to a local mass*energy' is a key factor in this description. The phenomenon of mass*energy coupling to spatial geometry is a very lengthy subject put forth in prior papers and will be a separate text due to length constraints. Briefly, a Gravity Wave is an example of spatial geometry that answers to a Schwarzschild transformation in the case where there is no mass*energy coupling to the moving wave. The Earth's local gravity well is an example of mass*energy that is coupled to its local source or resulting spatial geometry. Of course, temporal variation is interdependent with spatial geometry, and the same law applies in both cases. There is no sequitur description of scale dependency in this phenomenon.

It will be necessary to go into an in-depth discussion on how the LIGO interferometer actually functions, as the principle is not quite the same as classic interferometry, in later sections. In short, a Michelson-Morley type arrangement is a static measurement. LIGO is a dynamic measurement of pulse timing that differentiates the detection, which is a low frequency [~100 Hz] signal against intermittent pulses of light at the standard wavelength.

The Gravity Wave is travelling at the speed of light. However, the amplitude of the signal is not a single peak, but a 'hum' at very roughly 100 Hz. It takes the light pulse roughly a microsecond to travel the distance of the interferometer and back. At the time the Gravity Wave is passing through the interferometer the entire system is 'stretched,' the 'meter stick' and light are the same. However, the subsequent light pulse is not affected, and thus the two are differentiable.

The key factor in this is that the 'stretch' over the $1/100^{th}$ of a second Gravity Wave [hum] can be differentiated over a time of another $100^{th}$ second of the second pulse. If you convert all of this to Lorentzian terms, as a result of motion, these could not be differentiated. For example, we disregard reality for a moment and say that the Earth can move at its normal 30-meters per

# Introduction

second as it orbits the sun for $100^{th}$ of a second and for the next $100^{th}$ of a second remain stationary. This would mimic the same approach as LIGO with a Schwarzschild transformation. However, since the interferometer does not change path length during its forward motion, as a 'law of nature,' there is no such differentiation between Earth in motion vs Earth as stationary, they are both the same null result. It is this 'not null result' of LIGO that is the key factor in determination and detection of a Schwarzschild type transformation.

All back calculations to the source of the gravity wave are Schwarzschild metrics.

The LIGO approach is taking advantage of the fact that a *real change* does occur during that $100^{th}$ of a second that the Gravity Wave's geometry, as does all of the other components of the Large Interferometer because of the size and scale of the Gravity Wave. If Minkowski spacetime applied to the Schwarzschild metric, there could be no 'real' change in path length, regardless of this dynamic pulse timing type of approach. Both pulses would be a null effect, as this effect does not occur in any tangible, observable dimensionality. Thus, in this sense alone LIGO invalidates Minkowski spacetime as having any possible connection to General Relativity.

At $100^{th}$ of a second is 3-million meters. Consequently, if LIGO were perhaps 10-million meters in length the components of measurement would be out of the window of the Gravity Wave and detection without using the pulse approach would *probably* work. However, that is not an important element in the description and not a subject of this text.

The term, *real change* in path length is critical. If the change in spatial geometry were artifact, no such measurement could occur. This key feature is then translated to Lorentzian conditions, where modern convention regards the time transformation as real, as indicated by the permanent disparagement between clocks, but the length transformation as artifact. The length transformation regarded as artifact is because 'length contraction' must occur in Minkowski spacetime, where the unobservable dimensionality is defined by the sqrt(-1). Minkowski space, as well as the very definition of the sqrt(-1) will be described in depth in later sections of this text. Nonetheless, there is no such condition wherein a system can detect its own change of state under a Lorentzian transformation, this applies to both time and {x, y, z} type dimensionality.

Thus, the notion that LIGO detects its own change of state under a Schwarzschild transformation may seem like a grey area. However, if we consider Preferential Frame of Reference under a Lorentz transformation, no such approach could work via any interferometric method. That is, there can be no Lorentzian change of state that is not *coupled to its source* under local conditions of relativistic velocity.

In addition, the key feature is that if the spatial geometry is the result of, 'coupled to' a local source of mass*energy, this LIGO approach could not work. For example, again, this approach could not work to detect Earth's static spatial geometry. This is not a rhetorical argument, but one that is seated in spatial geometry that is not coupled to local mass*energy as not answering to Preferential Frame of Reference.

# Introduction

In the case of Inflation exceeding v > c, this is regarded in the sense that velocity can be considered a non-sequitur in space that, in over simplistic terms, defines itself with respect to unitary intervals. However, the mass*energy within that system must answer to Lorentzian conditions. Thus, the same principle, Information, in this case in the form of mass*energy, is not coupled to the expanding system or vice versa.

The idea that Information may not be coupled to a system is rather straight forward. We will look at Information and derive an exacting hard definition for such.

To clarify, the statement is that:

- In the case where spatial and temporal geometries are coupled to their local source of mass*energy, Preferential Frame of Reference applies for General Relativity under Schwarzschild transformations.
- In cases where spatial and temporal geometries are *not* coupled to their local source of mass*energy, Preferential Frame of Reference *does not apply* for General Relativity under Schwarzschild transformations.
- Gravity Waves can be expressed as mass*energy and momentum. However, they are not coupled to any local source of mass*energy. Like the photon, they do not possess mass and travel at the speed of light. They are not in themselves subject to Schwarzschild or Lorentz transformations.
- There is no case under Lorentzian transformations where a system in motion *is not* coupled to its local source of mass*energy.
- However, the massless photon does not undergo Lorentzian transformation. It can, however, be expressed as energy and momentum. Field effects under Lorentzian and Schwarzschild transformations have never been experimentally observed nor validated.
- A massless photon, however, can be subject to Schwarzschild transformations. [Shapiro, et al.] E.g., Gravitational Redshift is more than just a change in phase, unlike Lorentzian redshift. Nonetheless, neither case has been validated satisfactorily. The Pound-Rebka approach, for instance, only applies under certain specific conditions and is not generalized as confidently validating the postulate, to date.

In this description I will constrain all systems to purely observable dimensionalities. Invoking and evoking the host of unobservable dimensionalities of our simian ancestors can never have any tangible, hence practical meaning.

The point is, the Zeno Effect is clearly observable; *real transformations*, the observation is the alteration of the rate of progression of time purely as a result of detection and measurement, not mass*energy: the mass*energy requirement with respect to time dilation is an erroneous history that has been invalidated, incontrovertibly, as being interdependent with length transformation [Schwarzschild type] via LIGO, again, else violate General Relativity in some non-sequitur fashion that lacks any prior art or even hypothetical description.

The statement that Zeno systems are real transformations is, like the permanent disparagement between clocks under Lorentzian conditions, are 'real' as indicated by the permanent disparagement between, for example, perhaps an expectation value of numbers of decays vs that

# Introduction

which is measured under Zeno conditions. This is an interesting experimental feature that has not been done in the sense that we can regard a radioactive natural clock source as two separate features:

Using $^{14}C$ as an example throughout this text for reasons of practicality and simplicity, we use 1uCi of C-14 which yields an expected 37K decays per second:

- A $^{14}C$ sample of 1uCi observed from one side under normal conditions.
- The same $^{14}C$ source observed from another side under Zeno conditions.

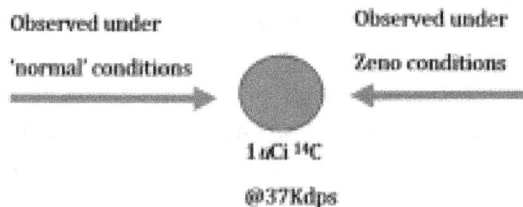

This may seem mundane, however, the information that can be yielded takes the following forms:

- If the two systems observe the same slowing of radioactive decays per second, then the radioactive source has definitively slowed in its rate of temporal progression.
- If the two systems observe different rates of radioactive decay, for example, the Zeno observation is slowed but the normal observation has not, then the radioactive source *has not* slowed in its rate of temporal progression.

As simple as this appears, this approach has not been done.

- The first condition qualifies both that the system is a genuine Zeno Effect and validates the notion that the system has actually slowed in its temporal rate of progression.
- The second condition outcome suggests that the Zeno Effect under the above conditions is an artifact of observation, or as will be explained in detail, the system is not yielding an actual Zeno Effect, but possible detector Dead-Time and Pile-up, data transfer and computational speed issues,* all of which appear exactly like a Zeno Effect. [*These are discussed in great detail in later sections of this text].

In addition, this demands that Preferential Frame of Reference is a null hypothesis as it relates to General Relativity, except in cases where static spatial geometry results from being coupled to a local mass*energy.

Preferential Frame of Reference becomes an important concept in [does not *always* apply to] General Relativity, in particular because prior to LIGO's successful demonstration that this is not the case; despite an entire history of science mythos was constructed and unfortunately, given so much invested in the null hypothesis refuses to go away. This severely limits understanding of General Relativity altogether.

# Introduction

In general, the classic Zeno Effect of slowing of rate of progression of observed unitary time will be described as a dilation of path length and quickening of rate of progression of unitary time as a 'contraction' of path length; between the 2-clock system. This seeming reciprocation comes about in constraining the observable systems to preservation of Information Conservation, which will be described and derived in detail throughout the text.

Invariably the failure to explain the Zeno Effects evokes unobservable dimensionalities, which are non-sequitur as such things can never be proven or even so much as validated to some compelling degree. Furthermore, the alteration of progression of temporal rate is purely observable, thus will be accompanied by purely observable phenomenon that occur constrained to purely observable dimensionalities, in this case, the associated changes in spatial geometry, as both real and observable.

In short, General Relativity demands that the path length between the 2-clocks of the 2-clock Zeno System experiences a classic Schwarzschild transformation logical to the observed change in rate of progression of unitary time. This notion has never occurred to anyone and has thus never been measured. In general, the Quantum Zeno Effect is regarded as a quantum scale, only, phenomenon. In addition, the Zeno Effect has not been scaled up to such macroscopic proportions that such a measurement would be practically achieved.

A Proof of Principle benchtop experiment is described later in the text.

### Introduction: Scale of the Zeno Effect and Related Issues

The notion that the Quantum Zeno Effect is constrained to quantum scale events is not validated by any empirical evidence. Moreover, the scale between the observed system, such as some natural clock system, radioactive source, for example, and observing system, e.g., detector, on a benchtop scale of the classic experimental setup defines the system as macroscopic. That is, the Zeno Effect is a benchtop scale experiment that has the observed system, such as radioactive source, and detection system, perhaps a beta detector, as much as meters apart. Given the Zeno Effect is conventionally regarded as the detection system *coupled to the observed system*, the Zeno Effect cannot be a 'quantum scale' phenomenon, but macroscopic on a meter(s) scale.

The natural clock system being observed and observing clock system cannot be detangled, both are critical elements of any Zeno system, in fact, defines the Zeno Effect. The scale of the system is thus on a classic benchtop, macroscopic proportion. Again, there is no evidence to support the quantum scale constraint, but the empirical evidence of the scale of the 2-clock system itself contradicts such hypothetical quantum scale limitations.

The presumption that the Zeno Effect is a 'quantum scale' phenomenon comes from the aberrant oversight that the observed system is independent of the observing system [detector]. This is the observer effect/measurement problem. For example, we send Alice and Bob to opposite ends of the galaxy in a quantum entanglement experiment. We regard Alice and Bob as an eigenstate of say, spin states as they travel to opposite ends of the galaxy. At some instant in the future a detector

## Introduction

on that side of the galaxy where Bob is detects and determines his spin state and we say that Alice and Bob are then decoupled. No detector is involved in Alice's spin state. The entanglement experiment demonstrates eigenstate precipitating to eigenvalue, Alice's eigenvalue is independent of observation.

This is not how the Zeno Effect works, nor is it any valid part of the description. The Zeno Effect is defined as the observed system [herein $r$] coupled to the observer system [herein $R$]. The Zeno Effect cannot be described by the observed system as independent of the observer system. In the case above of Alice and Bob, they can be regarded as independent of the detection system indefinitely until the moment of detection. However, detection results in the eigenstate precipitating to eigenvalue. That precipitation [I am reluctant to use the term 'collapse'] from state to value is interdependent with the detector. However, although Alice and Bob are interdependent with one another, prior to detection they are not interdependent with the detection system.

The notion that the Zeno Effect is limited to a quantum scale is also due to there having been little scaffold of prior art in rendering hypotheses of a quantum mechanical nature on macroscopic scales. However, that particular issue is changing rapidly, in recent years a host of quantum phenomenon have been performed on macroscopic scales. For example, superfluidity, superconductivity, magnetoresistance, Bose-Einstein condensates, are common examples, the list of less common systems is extensive. It is becoming clearer every year that quantum behaviors being limited to a quantum scale is limited to special cases, with the general case being that quantum scale behaviors as being a generally null hypothesis, all behaviors are macroscopic. Simply, the technological approach has shifted in light of advancing technology and we find the ability to determine the scales of these phenomenon as being universal.

In general, the natural clock source is regarded by convention as a quantum system, such as a radioactive source, for example. In these examples we will use a C-14 source on the micro-Curie scale. There are of course a host of other natural clock systems, however we will use a single, simple example for practicality. For clarity's sake, we will look at radioactivity as counts per second:

- 1Ci = 3.7E10 events/second
- 1mCi = 3.7E7 /s
- 1$u$Ci = 3.7E4/s
- 1nCi = 37/s

Since there seems to be a general lacking knowledge of radioactive sources, as the lack of such specific knowledge in the literature seems to suggest, I will also spend some time clarifying specific issues to this effect. A Becquerel is simply 1-event per second. The Curie and Becquerel are the standard units of emission. 1-Curie is equal to 3.7E10 decays per second, 3.7E10 Becquerel.

Specific Activity is given by:

## Introduction

$$\frac{dps}{g} = \frac{Ln2 \times 6.02E23(\frac{atoms}{mol})}{t_{1/2}sec \times M(\frac{g}{mol})} = \frac{4.17E23/mol}{t_{1/2}sec \times M(\frac{g}{mol})}$$

Where $T_{1/2}$ is in seconds. Thus, half lives in hours or years are converted to seconds.

- Ex: U-238 $T_{1/2} = 1.41E17s \therefore (4.17E23/mol)/(1.41E17 \times 238[g/mol]) = 12,432$ dps/g ~3.4E-7Ci/g
- Ex: C-14 $T_{1/2} = 5730y = 1.807E11s \therefore (4.17E23/mol)/(1.807E11 \times 14[g/mol]) = 1.65E11$ dps/g ~4.5Ci/g

The point of this exercise is that, for example, in the case of $^{238}$U a natural clock would require a macroscopic gram of material to produce a sufficient population of events to monitor. A milligram of $^{238}$U, for instance, would only yield about 12 events per second, barely above background, meaning that it would take days of measurement just to separate from the noise floor, but still be a macroscopic quantity of material. If we move on to $^{14}$C, we have a sweet spot of 0.225-microgram yielding about 37,000 events per second, or 37KHz. This 'sweet spot' will be explained in detail. The case for C-14 is roughly on the microgram scale, which is not a 'quantum scale' mass of material, nor is it a 'quantum scale' physical size, but about a microgram of total mass.

Some issues in the Zeno literature:

- A lot of work seems to focus on Rubidium, probably because it is commercially available as typified for its use in Drug Metabolism and Pharmacokinetics. I will thus use rubidium as 'the bad example of bad science.' However, in the numerous papers where Rubidium is mentioned as the source of radioactive material [natural clock system], the isotope is rarely mentioned, hundreds of papers do not even state the isotope. Only a few [exactly three] state the exact isotope but not quantity of isotope such that expectation values in dps can be back calculated. This isn't unique to rubidium; this is the case across the Zeno literature for all radioactive sources of every type. There are about a dozen papers that describe the activity sufficient to back calculate the expected dps for the experiment. However, insufficient information to reproduce the experiment in other necessary details.
- Given there is no practical reason, such as confidentiality and so on, for not reporting the isotope and expectation values of decays per second sufficient to back calculate the precise values, perhaps *to reproduce the experiment*, I can only surmise that the authors do not know the expected decays per second. Furthermore, this extends to quantum tunneling, along with the host of observed systems that appear throughout the Zeno literature. It seems the authors do not know the expectation value, else, regarded as fundamental good science, would state the numbers clearly.
- In all but a very few cases, perhaps a dozen or so, there is no quantifiable expectation values of expected number of events per second. There is no reporting of actual detector data acquisition, data transfer, and computation rates nor methods, such as algorithms. There is no quantifiable data whatsoever, and consequently both x and y-axes are *always* normalized to some scale of penchant. There is no evidence whatsoever of a 'Zeno Effect,' because there is no quantifiable expectation temporal rate and no quantifiable resultant rate reported. How these

# Introduction

thousands of papers have made it past peer review is an enigma. There is zero information that any good science has occurred in a lab.

- All information that is reported in the literature is also lacking in detector specifics, data transfer rate and type specifics, and computational specifics. These will be discussed at length later in the text. It would seem the authors also do not know these values such as to report them confidently. For example, not stating the isotope, nor the type of detector; is it a beta emitter? Is it a beta detector? No beta, alpha, IT, EC detector can separate events rapidly enough to produce a Zeno effect; assuming we have to acquire data at perhaps a million times that of the temporal rate of the observed system.
- All information that is reported in the literature is invariably normalized to some {x, y, t} set of values. Rather than provide graphs with a, for example, y-axis in expected counts per second and/or x-axis in data acquisition rate, both x and y-axes are generally normalized to some value $\equiv 1$. These are invariably clouded in some penchant, for example, a penchant of y-axis normalized to the hypothesis at hand, rather than a quantifiable value.
- In general, quantum computing experiments tend to provide more specific quantifiable information.
- *In no case* in the Zeno literature is there sufficient information to reproduce the experiment from the paper. Contrasting, in chemistry insufficient information to reproduce the experiment will not pass peer review.

It is not clear to me how these papers pass peer review. I can only surmise that theoretical physics has produced nothing tangible in almost a century [maser, 1954, transistor, 1924, nuclear fission 1944], thus, it is a non-issue to theoretical physics, *reproducibility*. At this time, superfluidity, superconductivity, Bose-Einstein condensates, are still in the infant, discovery stages and cannot be regarded as tangible yet, at this time. In fact, they are not even characterized. The Zeno literature that is not specific to quantum computing, which is a tangible field of applied physics, seems to regard hypotheses, only.

The point is that:

- A stable, robust, and reproducible Zeno Effect has not yet been performed nor has such been reported in the literature. No oner has ever reproduced any Zeno effect from prior art.
- The Zeno Effect seems to be treated ad-hoc, each experiment yielding and reporting no quantifiable values.
- No system design of the Zeno Effect has been reported in sufficient detail to reproduce the experiment, a fundamental of good science.
- No reported data is quantifiable, and consequently, cannot be said to represent a Zeno Effect.
- No expectation values of temporal rate are described in sufficient detail to back calculate nor reproduce the experiments, such as quantifiable radioactive decay rates, and as such cannot be said to represent a Zeno Effect.
- The lacking is specifics of detector types and manufacture suggests the data acquisition rates are not known to the authors, nor are the data transfer rates or computational speeds. In most cases, the likelihood that the authors were able to 1) detect at sufficient rates 2) transfer data at sufficient rate, 3) compute data at sufficient rate is impossible.

# Introduction

This isn't a great sales pitch for a Zeno effect text. However, there are sufficient experiments throughout the half century history of the QZE-QAZE to establish some stabler hypotheses that the Zeno effect is not a null hypothesis.

## Rubidium

Rubidium appears often; thus, we will look at Rubidium. A table, for example, of Rubidium isotopes looks like this, Rubidium being a common source in the Zeno literature:

| Nuclide [n 1] | MW | Isotopic mass (Da) [n 2][n 3] | t1/2 s | Decay mode [n 6] | Daughter isotope [n 7][n 8] | DPS/g | dps/ug |
|---|---|---|---|---|---|---|---|
| $^{72}$Rb | 72 | 71.95908(54)# | 1.50E-06 | p | $^{71}$Kr | 3.86E+27 | 3.86E+21 |
| $^{72m}$Rb | 72 | | 1.00E-06 | p | $^{71}$Kr | 5.79E+27 | 5.79E+21 |
| $^{73}$Rb | 73 | 72.95056(16)# | 3.00E-08 | p | $^{72}$Kr | 1.90E+29 | 1.90E+23 |
| $^{74}$Rb | 74 | 73.944265(4) | 0.648 | β+ | $^{74}$Kr | 8.70E+21 | 8.70E+15 |
| $^{75}$Rb | 75 | 74.938570(8) | 19 | β+ | $^{75}$Kr | 2.93E+20 | 2.93E+14 |
| $^{76}$Rb | 76 | 75.9350722(20) | 35.6 | β+ | $^{76}$Kr | 1.54E+20 | 1.54E+14 |
| | | | | β+, α (3.8×10−7%) | $^{72}$Se | | |
| $^{76m}$Rb | 76 | | 3.00E-06 | | | 1.83E+27 | 1.83E+21 |
| $^{77}$Rb | 77 | 76.930408(8) | 226.2 | β+ | $^{77}$Kr | 2.39E+19 | 2.39E+13 |
| $^{78}$Rb | 78 | 77.928141(8) | 1059.6 | β+ | $^{78}$Kr | 5.05E+18 | 5.05E+12 |
| $^{78m}$Rb | 78 | | 345 | β+ (90%) | $^{78}$Kr | 1.55E+19 | 1.55E+13 |
| | | | | IT (10%) | $^{78}$Rb | | |
| $^{79}$Rb | 79 | 78.923989(6) | 1380 | β+ | $^{79}$Kr | 3.82E+18 | 3.82E+12 |

## Introduction

| Isotope | A | Mass | Half-life | Decay mode | Product | | |
|---|---|---|---|---|---|---|---|
| $^{80}$Rb | 80 | 79.922519(7) | 33.4 | $\beta^+$ | $^{80}$Kr | 1.56E+20 | 1.56E+14 |
| $^{80m}$Rb | 80 | | 1.60E-06 | | | 3.26E+27 | 3.26E+21 |
| $^{81}$Rb | 81 | 80.918996(6) | 16452 | $\beta^+$ | $^{81}$Kr | 3.13E+17 | 3.13E+11 |
| $^{81m}$Rb | 81 | | 1830 | IT (97.6%) | $^{81}$Rb | 2.81E+18 | 2.81E+12 |
| | | | | $\beta^+$ (2.4%) | $^{81}$Kr | | |
| $^{82}$Rb | 82 | 81.9182086(30) | 76.38 | $\beta^+$ | $^{82}$Kr | 6.66E+19 | 6.66E+13 |
| $^{82m}$Rb | 82 | | 23302.8 | $\beta^+$ (99.67%) | $^{82}$Kr | 2.18E+17 | 2.18E+11 |
| | | | | IT (.33%) | $^{82}$Rb | | |
| $^{83}$Rb | 83 | 82.915110(6) | 7447680 | EC | $^{83}$Kr | 6.02E+14 | 6.02E+08 |
| $^{83m}$Rb | 83 | | 0.0078 | IT | $^{83}$Rb | 6.44E+23 | 6.44E+17 |
| $^{84}$Rb | 84 | 83.914385(3) | 2859840 | $\beta^+$ (96.2%) | $^{84}$Kr | 1.74E+15 | 1.74E+09 |
| | | | | $\beta^-$ (3.8%) | $^{84}$Sr | | |
| $^{84m}$Rb | 84 | | 1215.6 | IT (>99.9%) | $^{84}$Rb | 4.08E+18 | 4.08E+12 |
| | | | | $\beta^+$ (<.1%) | $^{84}$Kr | | |
| 85Rb[n 10] | - | 84.911789738(12) | | | | | |
| $^{86}$Rb | 86 | 85.91116742(21) | 1610668.8 | $\beta^-$ (99.9948%) | $^{86}$Sr | 3.01E+15 | 3.01E+09 |
| | | | | EC (.0052%) | $^{86}$Kr | | |
| $^{86m}$Rb | 86 | | 61.02 | IT | $^{86}$Rb | 7.95E+19 | 7.95E+13 |
| $^{87}$Rb[n 11][n 12][n 10] | 87 | 86.909180527(13) | 1.5525E+18 | $\beta^-$ | $^{87}$Sr | 3.09E+03 | 3.09E-03 |
| $^{88}$Rb | 88 | 87.91131559(17) | 1066.38 | $\beta^-$ | $^{88}$Sr | 4.44E+18 | 4.44E+12 |
| $^{89}$Rb | 89 | 88.912278(6) | 909 | $\beta^-$ | $^{89}$Sr | 5.15E+18 | 5.15E+12 |
| $^{90}$Rb | 90 | 89.914802(7) | 158.5 | $\beta^-$ | $^{90}$Sr | 2.92E+19 | 2.92E+13 |

## Introduction

| Isotope | A | Mass | Half-life | Decay mode | Daughter | Col1 | Col2 |
|---|---|---|---|---|---|---|---|
| $^{90m}$Rb | 90 | | 258.4 | β⁻ (97.4%) | $^{90}$Sr | 1.79E+19 | 1.79E+13 |
| | | | | IT (2.6%) | $^{90}$Rb | | |
| $^{91}$Rb | 91 | 90.916537(9) | 58.4 | β⁻ | $^{91}$Sr | 7.85E+19 | 7.85E+13 |
| $^{92}$Rb | 92 | 91.919729(7) | 4.492 | β⁻ (99.98%) | $^{92}$Sr | 1.01E+21 | 1.01E+15 |
| | | | | β⁻, n (.0107%) | $^{91}$Sr | | |
| $^{93}$Rb | 93 | 92.922042(8) | 5.84 | β⁻ (98.65%) | $^{93}$Sr | 7.68E+20 | 7.68E+14 |
| | | | | β⁻, n (1.35%) | $^{92}$Sr | | |
| $^{93m}$Rb | 93 | | 5.72E-05 | | | 7.85E+25 | 7.85E+19 |
| $^{94}$Rb | 94 | 93.926405(9) | 2.7 | β⁻ (89.99%) | $^{94}$Sr | 1.64E+21 | 1.64E+15 |
| | | | | β⁻, n (10.01%) | $^{93}$Sr | | |
| $^{95}$Rb | 95 | 94.929303(23) | 0.3776 | β⁻ (91.27%) | $^{95}$Sr | 1.16E+22 | 1.16E+16 |
| | | | | β⁻, n (8.73%) | $^{94}$Sr | | |
| $^{96}$Rb | 96 | 95.93427(3) | 0.208 | β⁻ (86.6%) | $^{96}$Sr | 2.09E+22 | 2.09E+16 |
| | | | | β⁻, n (13.4%) | $^{95}$Sr | | |
| $^{96m}$Rb | 96 | | 0.2 | β⁻ | $^{96}$Sr | 2.17E+22 | 2.17E+16 |
| | | | | IT | $^{96}$Rb | | |
| | | | | β⁻, n | $^{95}$Sr | | |
| $^{97}$Rb | 97 | 96.93735(3) | 0.17 | β⁻ (74.3%) | $^{97}$Sr | 2.53E+22 | 2.53E+16 |
| | | | | β⁻, n (25.7%) | $^{96}$Sr | | |
| $^{98}$Rb | 98 | 97.94179(5) | 0.114 | β⁻ (86.14%) | $^{98}$Sr | 3.73E+22 | 3.73E+16 |
| | | | | β⁻, n (13.8%) | $^{97}$Sr | | |
| | | | | β⁻, 2n (.051%) | $^{96}$Sr | | |
| $^{98m}$Rb | 98 | | 0.096 | β⁻ | $^{97}$Sr | 4.43E+22 | 4.43E+16 |
| $^{99}$Rb | | 98.94538(13) | 0.05 | β⁻ (84.1%) | $^{99}$Sr | 8.42E+22 | 8.42E+16 |

17

## Introduction

| | | | | | | |
|---|---|---|---|---|---|---|
| | 99 | | | β⁻, n (15.9%) | ⁹⁸Sr | |
| | 100 | | | β⁻ (94.25%) | ¹⁰⁰Sr | 8.02E+22 | 8.02E+16 |
| ¹⁰⁰Rb | | 99.94987(32)# | 0.052 | β⁻, n (5.6%) | ⁹⁹Sr | | |
| | | | | β⁻, 2n (.15%) | ⁹⁸Sr | | |
| ¹⁰¹Rb | 101 | 100.95320(18) | 0.032 | β⁻ (69%) | ¹⁰¹Sr | 1.29E+23 | 1.29E+17 |
| | | | | β⁻, n (31%) | ¹⁰⁰Sr | | |
| ¹⁰²Rb | 102 | 101.95887(54)# | 0.0375 | β⁻ (82%) | ¹⁰²Sr | 1.09E+23 | 1.09E+17 |
| | | | | β⁻, n (18%) | ¹⁰¹Sr | | |
| 103Rb[3] | 103 | | 0.026 | β⁻ | ¹⁰³Sr | 1.56E+23 | 1.56E+17 |
| 104Rb[4] | 104 | | 0.035 | β⁻? | ¹⁰⁴Sr | 1.15E+23 | 1.15E+17 |

Looking at Rb-87, which is a common isotope in the sales of Rb-isotopes, 1-microgram of Rb-87 yields 3E-3 dps, or one decay every 333-seconds. Unless the energy spectrum window is set very narrow, a common researcher will never measure anything above background, and counting has to be done for days. Furthermore, from what I've seen in the literature I seriously doubt any such researcher is even aware that there is an energy spectrum associated with the isotope in question.

There are numerous papers that suggest, but do not state unambiguously, $^{87}$Rb as the natural clock system being used. $^{87}$Rb is often used as a natural clock calibrant in industrial applications, however, at milli-gram quantities, the counting rate is on the order of one decay per second, which is why $^{87}$Rb is ideal. In Zeno Dynamics this is not the case. I do not see how containment of an entire milligram of $^{87}$Rb can be done on a benchtop scale.

On the other end of the scale, we look at Rb-86, another common Rb isotope. Rb-86 has a specific activity of 3E9 dps per microgram. In our Zeno Dynamic description, which involves detection and measurement at some 'phenomenal' rate, we suggest 1MHz detection rate. This puts 1-microgram of Rb-86 at 3E15 Hz, which does not exist. In order to get that detection rate down to something believable, we have to dilute that 1-microgram down 1-million-fold such that we have 3K dps thus requiring a realistic 3GHz detection and processing rate. 1-atto-gram of material.

You are thinking, 'that's fine, dilute the sample.' However, in no such piece of literature on the subject do the authors state the amount of Rubidium being counted, hence the number of expected decays per second and the Zeno result decays measured per second, detection rate, detector type [if it is a beta emitter, are they using a beta detector?], no indication whatsoever

## Introduction

that we can believe the experiment was even performed. Flat out, honest. Simply, there is no indication in the literature on this subject that suggests the authors have sufficient knowledge to design the experiment and every indication that the authors, almost without exception, couldn't possibly have carried the experiment out. This is a big problem and is unique to theoretical physics, as theoretical physics has become a stage of science-fiction and fantasy that simply could not get passed in any other field of science. Frankly, theoretical physics has become simian dope smoking quackery.

The exact characteristics of various detectors are described in detail in later sections of this text. In general, in the Zeno literature we see beta emitters being used, e.g., Rubidium 86 and 87. Keep in mind that there is a Rb-86 isotope that undergoes Internal Transition, however, no benchtop experiment is going to work with that, and the IT isotope is merely on its way to the next step, which is beta decay. I have not seen any gamma emitters mentioned in the literature in regard to the Zeno Effect. Thus, we will focus on beta emitters.

The practical upper limit of a beta detector's capabilities is in the 100MHz range. There are detectors that can go as high as 1GHz, but these are very costly and not likely in an academic setting. The upper limit for fiber optic is 5GHz, without data compression. The current upper limit of a benchtop computer is 5GHz as well.

We can regard 'phenomenal' detection rate as 1-million detections per second. That means we must limit our radioactive sample to perhaps 1 to 50KHz at the very upper end. That means that our Rb-86 sample of 1-microgram has to be diluted 60,000-fold.

If one can understand these numbers, one can begin to understand my misgivings regarding: rather or not these experiments were actually performed and/or are the results merely detector, data transfer, and processor artifacts?

Each isotope of each element has a specific decay mode, in this example beta, with a specific energy spectrum:

> Bernabei, R & Belli, Pierluigi & Cappella, Fabio & Cerulli, Riccardo & Danevich, Fedor & Grinyov, B. & Incicchitti, A. & Kobychev, V. & Mokina, V. & Nagorny, S.S. & Nagornaya, L.L. & Nisi, S. & Nozzoli, F. & Poda, Denys & Prosperi, D & Tretyak, Volodymyr & Yurchenko, S.S.. (2010). Search for double beta decay of zinc and tungsten with low background ZnWO4 crystal scintillators. Journal of Physics: Conference Series. 202. 012038. 10.1088/1742-6596/202/1/012038.

# Introduction

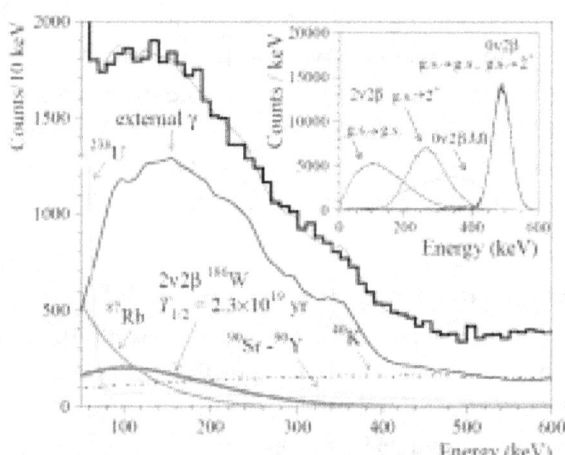

Here, the authors have done a fine job of isolating the beta emission energy spectra of some common elemental isotopes. It's a bit of a shame that $^{14}C$ isn't present so that you would be able to see the degree of overlap with $^{40}K$. The reason I mention this is because Archeologiests and such rely so much on $^{14}C$ dating, when $^{40}K$ is not only more abundant in bone but has a different half-life with almost total overlap in energy spectra.

Meaning, no Archeologist to date has measured $^{14}C$, they are measuring $^{40}K$ completely amiss to this issue.

Below, I have taken the liberty of overlaying $^{14}C$ and $^{40}K$ energy spectra so that you can see how seriously they overlap. $^{40}K$ is more abundant in bone than $^{14}C$ and has a more powerful energy spectrum than $^{14}C$:

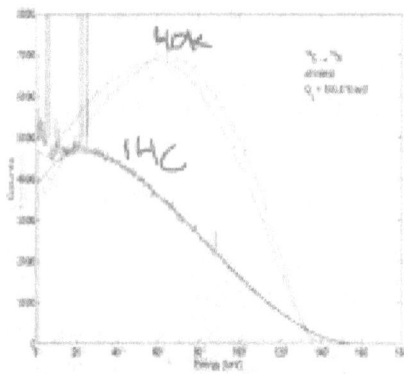

There is no way to physically separate the two *isotopes* in a sample and no part of the spectrum to isolate one from the other. The only possible thing one could do would be to *chemically* separate carbon and potassium in the sample. This obviously has never been done because no Archeologist is knowledgeable enough in RadioChemistry to even be aware that this problem exists.

Thus, all of your dating, the dinosaurs, man, animal, and even ancient astronauts is nonsensical misconception in looking at the wrong decay of the wrong element of the wrong isotope in the wrong type of the wrong sample.

*However*, one can back calculate and correct rather easily for this simply by adjusting the signal strength, which is a factor of about 7/5, and back calculating the radioactive half-life of $^{14}C$ vs. $^{40}K$ 5730 vs. 1.25E9 years. Then of course you have to figure in the dps of $^{14}C$ vs. $^{40}K$ at 1.65E11 dps vs 1.06E7 dps. The next step is to back calculate the natural abundance of each *element* in bone, not atmosphere.

*Then*, you can know how old the dinosaurs are, or Neanderthal, etc.

Typically, Archeologists use Liquid Scintillation Counters, which have no specificity whatsoever, unless you adjust the energy spectrum windows. However, as one can see from the plots, there is no window adjustment that can isolate $^{14}C$ from $^{40}K$. I have never seen any

## Introduction

indication that any Archaeologist is in the least bit knowledgeable or even aware of these issues, regardless.

The Department Of Antiquities, for example, uses Time-Of-Flight Mass Spectrometry to measure the $^{14}C$. TOF MS can isolate a specific mass to 6 or 7-significant figures, on a good day, meaning that the distance between $^{14}C$ and $^{40}K$ is so vast there can be no error. Simply by collecting thousands of scans one can measure the *amount of* $^{14}C$ in the sample and thus the $^{14}C$ dating method is valid.

It should be noted that positron detectors are uncommon. Gamma detectors are common, however, it is very difficult to get licensed to handle gamma samples and highly unlikely in any academic setting.

This is not prohibitive; however, the cost of such units is likely above and beyond that which we would expect to see in an academic lab setting. In addition, the operation of such is likely outside of common academic means. For example, positrons annihilate unless held under near pure vacuum conditions. This is another feature that likely places them outside of academic lab settings. The exact architecture of beta-plus and electron multipliers is described and discussed in great detail later in this text.

Limitations that are discussed in detail in later sections:

- Dead-Time, a commercially available beta or gamma detector cannot respond above 10MHz. Most systems do not extend even into the MHz range.
- Pile-up, a commercially available beta or gamma detector has a Surface area that cannot resolve above 10-million decay particles due to the resolution of the Surface matrix. At best, most commercially available beta [for example] detectors cannot resolve more than a few thousand particles.
- Data transfer, the fastest, fiber optic, commercially available is limited to 10GHz.
- Computation, commercially available computers cannot operate above 5GHz. Invariably will be in the single GHz range at best. The algorithm used to crunch such data will require 32 bit at least, reducing the rate to the 500MHz range at best.
- The data transfer rate cannot exceed the computational speed, but typically are effectively half the computational speed. Thus, a fiber optic is limited to the computer's speed, typically in the single GHz range.

If we go back and look at the table above, for Rubidium we run into serious implications of the sort, at an average decay rate of 1.6E15 decays per second per microgram, in order to place a Rubidium isotope into the GHz range possible for any commercially available system, we have to dilute a standard to the picogram range, to place it at about 1.6GHz. The notion that we are going to observe this dilute under Zeno conditions implies then that we need to observe the sample at perhaps a million times per second. This in turn demands the sample is dilute to the atto-gram level. This is becoming not unlike a Homeopathic dilution, where we only have a few thousand atoms of material present. It seems clear that experimental results from Rubidium sources of radioactive decay are more likely the detection, data transfer, and computation speed problems that will be discussed in later sections of this text. Namely, Dead-Time is the ability

# Introduction

for a detector Surface to clear itself of the incoming [in this example] beta particle, Pile-Up is a phenomenon where the particles literally 'pile up' in the detection Surface, data transfer and computational speeds: All look like a Zeno Effect, indeed even take on a second order curve.

In all cases where the Zeno effect is experiment is not reported in sufficient detail to:

- Back calculate the expected number of decays per second
- Back calculate the quantifiable number of resultant [Zeno effect] events per second
- Verify that the isotope type and amount
- Verify that the correct type of detector was used
- Verify that the type of detector is capable of such acquisition rates
- The data/detector acquisition rates to validate a Zeno effect
- Quantifiable axes, rather than normalized

In all of those cases, which represents about 95% of the reported experiments in the history of the Zeno literature, *are associated with hypotheses that extend into 'unobservable dimensionalities.'* There is no basis to support that an actual experiment was performed at all. It seems the authors have a hypothesis of penchant and write papers that have such lacking information in them to validate, nor reproduce any 'Zeno effect,' as this might invalidate the unobservable dimensionality hypotheses.

Given the Zeno Effect in any sequitur discussion would attempt to measure decays at a phenomenal rate, at least three orders of magnitude above the rate of events, there is no Rubidium isotope on that table that any commercially available system on Earth can possibly detect, as baseline, much less at some phenomenal rate. For example, 1-microgram of Rb-97 decays at about 2E16 Hz, which is far beyond any beta or gamma detector's capabilities, 100-million times the rate of any possible data transfer, and a billion times any commercially available computer's speed. The amount of material is never stated. I honestly believe the authors do not know the amount, hence do not know the expected decays per second, and yet go on to report a 'Zeno Effect.' I do not understand how such things pass peer review.

I think the peer review process is sloppy, lazy, weighted by opinion and absurdity. I honestly do not see any formal journals as accomplishing any positive thing in this world.

In these texts we will use $^{14}C$ as our ideal sample.

For $^{14}C$, the Specific Activity is about 1.65E11 dps/g. This gives 165Kdps/ug. This is falling into the rate limits that are possible using current technology with a bit more dilution.

All of these factors [e.g., Dead-Time, Pile-up, data transfer and computation] are discussed in much detail in later sections of this text.

The point is, a microgram of carbon is not a *quantum scale*, but a macroscopic amount of material, as convention defines quantum scale systems.

*The key is to produce a quantifiable, robust, and reproducible Zeno Effect, is the minimum requirement for good science.*

## Introduction

It should be noted that beta-minus detectors are uncommon, as are positron detectors. This is not prohibitive; however, the cost of such units is likely above and beyond that which we would expect to see in an academic lab setting. In addition, the operation of such is likely outside of common academic means. For example, positrons annihilate unless held under near pure vacuum conditions. This is another feature that likely places them outside of academic lab settings. The exact architecture of beta-plus and electron multipliers is described and discussed in great detail later in this text.

Limitations that are discussed in detail in later sections:

- Dead-Time, a commercially available beta or gamma detector cannot respond above 10MHz. Most systems do not extend even into the MHz range.
- Pile-up, a commercially available beta or gamma detector has a Surface area that cannot resolve above 10-million decay particles due to the resolution of the Surface matrix. At best, most commercially available beta [for example] detectors cannot resolve more than a few thousand particles.
- Data transfer, the fastest, fiber optic, commercially available is limited to 10GHz.
- Computation, commercially available computers cannot operate above 5GHz. Invariably will be in the single GHz range at best. The algorithm used to crunch such data will require 32 bit at least, reducing the rate to the 500MHz range at best.
- The data transfer rate cannot exceed the computational speed, but typically are effectively half the computational speed. Thus, a fiber optic is limited to the computer's speed, typically in the single GHz range.

If we go back and look at the table above, for Rubidium we run into serious implications of the sort, at an average decay rate of 1.6E15 decays per second per microgram, in order to place a Rubidium isotope into the GHz range possible for any commercially available system, we have to dilute a standard to the picogram range, to place it at about 1.6GHz. The notion that we are going to observe this dilute under Zeno conditions implies then that we need to observe the sample at perhaps a million times per second. This in turn demands the sample is dilute to the atto-gram level. This is becoming not unlike a Homeopathic dilution, where we only have a few thousand atoms of material present. It seems clear that experimental results from Rubidium sources of radioactive decay are more likely the detection, data transfer, and computation speed problems that will be discussed in later sections of this text. Namely, Dead-Time is the ability for a detector Surface to clear itself of the incoming [in this example] beta particle, Pile-Up is a phenomenon where the particles literally 'pile up' in the detection Surface, data transfer and computational speeds: All look like a Zeno Effect, indeed even take on a second order curve.

In all cases where the Zeno effect is experiment is not reported in sufficient detail to:

- Back calculate the expected number of decays per second
- Back calculate the quantifiable number of resultant [Zeno effect] events per second
- Verify that the isotope type and amount
- Verify that the correct type of detector was used
- Verify that the type of detector is capable of such acquisition rates

# Introduction

- The data/detector acquisition rates to validate a Zeno effect
- Quantifiable axes, rather than normalized

In all of those cases, which represents about 95% of the reported experiments in the history of the Zeno literature, *are associated with hypotheses that extend into 'unobservable dimensionalities.'* There is no basis to support that an actual experiment was performed at all. It seems the authors have a hypothesis of penchant and write papers that have such lacking information in them to validate, nor reproduce any 'Zeno effect,' as this might invalidate the unobservable dimensionality hypotheses.

Given the Zeno Effect in any sequitur discussion would attempt to measure decays at a phenomenal rate, at least three orders of magnitude above the rate of events, there is no Rubidium isotope on that table that any commercially available system on Earth can possibly detect, as baseline, much less at some phenomenal rate. For example, 1-microgram of Ru-97 [the lowest decay rate on the list of isotopes] decays at about 20GHz, which is far beyond any beta or gamma detector's capabilities, twice the rate of any possible data transfer, and 4-times any commercially available computer's speed. If we dilute the quantity to 1-nanogram, at 20MHz, counting a thousand times per events thus requires data transfer and computation speeds not possible for any commercially available system [20GHz]. There could be some confidence if the authors reported that they diluted the sample to 1-picogram, however, this is never the case. The amount of material is never stated.

So, what Rubidium isotope are these researchers measuring? The isotope and dilution, yielding a quantitative expectation in decays per second does not appear in any paper where Rubidium is described as the isotope in question, *probably because it is unknown to the researcher*. Nonetheless, we must omit all such papers based on the knowledge that there is no system, unless they work for the NSA, that can possibly measure Rubidium isotopes at a sufficient rate to produce a Zeno Effect. Nonetheless, there is no detection system of beta, gamma, what have you, that is capable of operating in the GHz range.

These rates will be important later on, as we look at detector systems, Dead-Time, Pile-up, and other limitations of detector, data transfer [e.g., wired vs fiber optic], and computation speeds. It is important to keep systems within these limitations, as we will address several of these issues, we find that most experimental published results are detector Dead-Time, Pile-up, data transfer limitations, and so on, all of which are nearly indistinguishable from a genuine Zeno Effect. This is rather easy to determine, as in such cases the radioactive material and setup are far beyond any detector system's capability, data transfer via even fiber optic cannot accommodate the event rate, and certainly no commercially available computer can render calculations at such speeds as to suggest a rapid detection leading to a valid Zeno Effect.

In the above example, a gram of C-14 is in the TeraHertz range, which is well above the detection rate of any commercially available beta detector. The result would be Pile-up, which takes on a second order type of curve, and looks like a Zeno Effect, *provided one normalizes the graph* to some y-axis of penchant, which describes all but a very few of the papers in the available literature. Fiber optic is [commercially available] limited to about 10GHz, which causes Pile-up in the data

# Introduction

transfer, which would look exactly like trying to capture a THz image using a GHz camera. No commercially available computer can operate in the THz range of computation. These values are only to match the decay rate [given 1-gram of material]. If we are to assume, we produced some phenomenal effect of detection at extraordinary acquisition rates, we can see that this is obviously out of the question. This example uses a very large amount of material at the given specific activity to make the point clear. However, a dilute sample in the microgram range yields decays in the KHz range, which is entirely practical.

In addition, one can note that the uranium-238, at only 12KHz is well within detectability and computational limitations of any commercially available system. Given a safe margin in the 500MHz range, we can indeed expect some phenomenal result, a Zeno Effect.

The problem is that only a handful of papers [in the single digits] describe the amount of radioactive material or otherwise expectation values of the observed system sufficient to reproduce the experiment. If the experiment cannot be reproduced it simply does not qualify as good science and must be omitted from consideration as not validating a Zeno Effect. Frankly, I do not understand how such papers pass peer review.

For $^{14}C$, the activity is about 0.22g/Ci. Thus, 1uCi = 0.22ug $^{14}C$. The Specific Activity is in Ci/g, for $^{14}C$ is 4.5Ci/g. The 'range of safety,' where we do not tax the limitations of our detection, data transfer, and computation systems will keep us within the micro-Curie range. Thus, the mass in question will be 0.22$ug$ $^{14}C$, yielding 37KHz of events *at the source*. Then we have some distance as well as cross section of the detector surface to deal with, later on. The value, 37KHz is critical because in order to produce a safe *and reproducible*, as well as robust Zeno Effect, we need to constrain our measurement rate to within the 500MHz range. This will not tax our detectors, with respect to Dead-Time and Pile-up, data transfer and computation speeds.

All of these factors [e.g., Dead-Time, Pile-up, data transfer and computation] are discussed in much detail in later sections of this text.

The point is, a microgram of carbon is not a quantum scale, but a macroscopic amount of material, as convention defines quantum scale systems. Furthermore, it is noted that in the literature, the actual activity, for natural clock systems of the radioactive sort, is never discussed. This is a serious problem for a host of reasons. In many cases the type of radioactive material, in terms of specific isotopes is not even mentioned. As our example of rubidium suggests, a macroscopic sample of rubidium is far outside of the range of detection and computation of any commercially available system, which in turn suggest that these papers are rendering aberrant results and associated hypotheses.

*The key is to produce a quantifiable, robust, and reproducible Zeno Effect, is the minimum requirement for good science.*

For example, Uranium-238 has a Specific Activity in the pico-Curie per gram range [3.4 micro-Curie/g] meaning that an entire gram of $^{238}U$ would only produce 12,580 events per second. At the other end of the common isotope scale, we have $^{91}Ru$ at about 5.5E20 dps/g, or 5.5E11 dps/$ng$, or 55GHz. There is no beta detector commercially available that can detect 5.5E11 events per second.

## Introduction

Fiber optic cable in commercial applications is limited to 10GHz. There is no commercially available computer that can perform 5.5E11 calculations per second [5.5GHz]. Thus, the amount of uranium required is about the size of a pea, but Rubidium a trillionth of a gram for the same number of events per second. One would expect the researcher to perform such a 12-order of magnitude dilution and describe such in detail, as a basic lower limit of reporting in good science. There is no such paper in the Zeno literature.

We will focus on radioactive systems for the major portion of this text, rather than other natural clock systems such as quantum tunneling and so on, for practicality's sake.

There are currently over one hundred thousand papers in the literature regarding the Zeno Effect, and I have only been able to find nine [9] that discuss the specific isotope as well as a range of activity, albeit must be back calculated from the available text. Of those experiments that were performed on other natural clock systems, tunneling, traps, and so on, very few, again, describe precise expectation values to such extent that a Zeno Effect can be validated from the information in the paper. They instead only report normalized values, which is not useful information.

It seems most likely that these values, both for radioactive sources as well as other forms of natural clock systems are not known to the authors, which raises the obvious question, *what is it they are writing about?* The answer, invariably, is a hypothesis of penchant, of which there are thousands, without any agreement. The larger problem is that all such hypotheses invoke unobservable dimensionalities of a rather simian line of reasoning, in an attempt to 'explain' a clearly observable phenomenon. In general, experimental procedures indicated for a tangible end result of Quantum Computation have real figures and sufficient information to determine of a Zeno Effect is in fact the case.

We will only look at tangible systems. We will also constrain all of our hypotheses to tangible dimensionalities, rather than simian thinking in unobservable dimensions or 'realms.'

The natural clock system of carbon-14 will be the set example throughout this text, as a matter of practicality. Since we are not necessarily interested in the energy spectrum, only the number of events per second, we can remain focused on the quantity of events. However, for clarity's sake, $^{14}C$ emits beta [-] at about 156KeV.

Thus, we have a macroscopic mass of our natural clock source, with the Zeno System being perhaps a meter in scale, including the detection system. Again, a Zeno system is described as the observed system, such as our radioactive source, as coupled to the observing system, our detector. Thus, the scale of the Zeno Effect is the source to detector, which is non-phenomenally at least 1-centimeter in scale, not a 'quantum scale.' For example, in the case of tunneling, such experiments are performed in a well type of format, similar to a microchip design, which is not a quantum scale system. Mass spec traps are essentially miniature cyclotrons of a sort, most benchtop systems about 10-centimeters in diameter, about the size of a fist.

The key concept is that the Zeno Effect is a *population* of similar, not identical, events. For example, if we were to measure one entangled spin pair, perhaps an electron-positron pair, we only get to measure that particle-antiparticle pair once, at which time it collapses from eigenstate to

## Introduction

eigenvalues, there is no second measurement sequitur to the system. In order to perform a Zeno type experiment, we need to produce a good luminosity of particle-antiparticle pairs, perhaps tens of thousands per second [hence, a population of separate, similar, not identical events]. As we detect each particle-antiparticle pair in order to determine spin states, as a permutation, *not combination*, each detected pair drops out of the system as having collapsed from eigenstate to eigenvalue, and we are left with the remainder of the population, at which point we move on to the next pair. Exactly as any liquid, such as a Bose-Einstein condensate, this population is an indifferentiable population of systems in eigenstate, isolating any one particle-antiparticle pair for detection renders that pair no longer part of that population, they have precipitated from the population eigenstate to eigenvalues: they are an entirely different species of phenomenon.

The point is, there are no examples of a *valid* Zeno Effect being produced on a 'quantum scale.' Given the source, the natural clock system cannot be a quantum scale mass of material, or a system that can be described in terms of Planck lengths [E-35 meters], such as a tunneling setup [microchip scale], ion trap [perhaps 10cm], and so on, there is no source of the phenomenon that can be regarded as 'quantum scale' system to be observed. Then, regarding the detection system as coupled to the observed system, which is a Zeno convention, we extend out from about 1cm to perhaps a meter in scale. These are many orders of magnitude beyond which we can regard a system as 'quantum scaled.' Any observed system is a population of eigenstates. Each element in that population of eigenstates precipitates to some eigenvalue or values at the instant of detection, and as such is no longer part of that population.

There is no acceptable hypothesis that suggests that General Relativity does not apply on all scales, it only exists as a question, not an answer. There is also no empirical evidence of this. In fact, General relativity as a whole is only loosely validated on any scale, and only in special, not general cases. The notion that GR is related to large scale phenomenon and does not apply on other scales in unfounded. Every attempt to generalize GR as summarily correct has failed. In astrophysics there is no issue that is resolved by GR and regarded as a failure to match data to theory, rather than theory to data. In both cases, it is not validated in any general sense, but only in a few specific cases that for the most part only refine Newtonian mechanics. That is not open to debate or opinion, that is the state of science in the year 2020.

The notion that there is some invisible and undefined line between the macroscopic and quantum scale is also dismissed in this text. No definition for such a line exists as anything more than question, pondery. Invariably, classic modern quantum mechanics involves unexplained, *pure observable* phenomenon, which then evolve to acceptance of hypotheses in a complete absence of empirical, tangible, data. That is, the notion that the laws of physics differ on a 'quantum scale,' based on the inability to explain such observations, is rather simian thinking.

A hard definition for these scales is derived and presented in later sections and does not negate the Zeno Effect on any macroscopic scale, based purely upon proximity at 2-Planck lengths from a Schwarzschild Surface. At this point, coalescence occurs with the Surface [ I tend to avoid the term horizon] and conventional $\Delta S$ and Shannon Entropy diverge. That will be derived and discussed in detail.

# Introduction

The problem is that we just demonstrate that typical macroscopic causality can be affected by the Zeno Effect. Even as 'the quantum Zeno effect,' for some presumed quantum scale event(s), such as our classic 'cat paradox:'

- If we do not observe a 'cat' system via a Zeno Effect the radioactive source the cat is superpositioned as alive|dead, as classic.
- If we put Schrodinger's cat in a box [with the radioactive decay activated poison] and observe the system under Zeno conditions, the radioactive source, [slow it to a stop] the cat lives forever.
- *If we produce an Anti-Zeno Effect, such that the radioactive source decays toward infinitely fast the cat is instantly dead.*

This is the:

{Quantum Zeno Cat|Quantum Anti-Zeno Cat}

Causality metaphor. We can effectively scale the Zeno Effect up to any scope and magnitude. All of the tangible experimental evidence is compelling toward the conclusion that 'quantum scale' phenomenon scale up to macroscopic proportions. Indeed, the only notions that exist to the effect that quantum mechanics is limited to very small scales are purely hypothetical and in fact contradict the available evidence.

There is nothing to the Zeno Effect that suggests it is confined to atomic scales, only that it has never been scaled up to vast proportions to date. However, the Proof of Principle detailed later in this text is a rather unsophisticated and simple to construct Surface phenomenon that will be described in detail in that section of this text. Frankly, it is simply too simple to do, not to do it. The Proof of Principle is described in later sections of this text.

- Thus, the notions that the Zeno Effect is somehow characterized in *any way* is subject to great scrutiny, given there has been no reporting of exacting quantifiable values, nor of any reproducible experiments.
- The notions that the Zeno Effect is not subject to General Relativity cannot be supported in light of the facts that there is no quantifiable, robust nor reproducible information describing the phenomenon.
- The notions that the Zeno Effect is does not demand a change in path length as co-mutual associating with temporal progression cannot be supported in a vacuum of data characterizing the phenomenon.
- The notions that the Zeno Effect is not constrained to purely 4-dimensional spacetime is, most of all, rather ludicrous, as this would then demand that the Zeno Effect has been in some way quantifiably validated as occurring in unobservable dimensionalities.
- The notions that the Zeno Effect is constrained to quantum scales is quite opposite of that which answers to what observations have been made, again, the detector system being coupled to the observed system is on a macroscopic, benchtop scale.
- The notions that the Zeno Effect cannot be scaled up indefinitely to any scale has no quantifiable data to support it, all of the data that does exist is macroscopic effect.

## Introduction

- The notions that the Zeno Effect is not a macroscopic change in the rate of temporal progression of the observed system as coupled to the observer system has no support uin any prior art, other than those hypotheses that extend into unobservable dimensionalities, and thus lend themselves to be wholly improvable.

This text will summarize the essential characterization of artificial alteration of spatial geometry via the Zeno and Anti-Zeno Effects, on macroscopic scales, as a means of Gravitic Propulsion. All of the associated maths will be dutifully derived in whole.

### Introduction: Scaling the Zeno Effect Up

This text describes systems to scale the Zeno Effect up to large macroscopic proportions, starting on a benchtop and evolving up to much larger scales. This is essentially described as a surface phenomenon of some recent nanotechnology coming from NIMS in the form of microcircuits being printed as surface materials. A surface of 2-clock nano-systems of microcircuit design creates a highly localized Zeno Effect, altering both the unitary progression of time and thus path length between each of the 2-clock systems of the microcircuit, altering path length on a somewhat macroscopic and measurable scale covering the surface of the microchip, or rather, nanosurface. This scale should be sufficient to validate the alteration in path length by standard interferometry and Gravitational Lensing.

The basic premise of how localized the effect should be is somewhat Newtonian, in that the curvature should fall off with the square of the distance from the *observed* system, $r$. Thus, the microcircuit Surface is millions of 2-clock systems, at the one to ten micrometer scale. In this, the *observer system*, $R$, is a microcircuit detector, in our text example [$^{14}$C], beta detector. Microcircuit scale beta detectors have already been produced. The basic design is sold state scintillation on a 1-micron scale. The observed system, $r$, is a micron scale sample of, in this case, $1\mu$Ci-$^{14}$C, that produces 37KHz decays per second.

For clarity's sake, we will go over the basic design of micron scale scintillation [beta] detection:

- A Silicon photomultiplier, SiPM, is based on the Single-photon avalanche diode [SPAD]. These range in scale, commonly, down to about 10-microns. Surface nanotechnology reduces this to a 1-micron scale, with a conventional density of 10K/mm$^2$, works out to 1-million/mm$^2$ as a Surface nanotech device.
- A SPAD is a basic p-n junction sensitive to, among other things, various forms of ionizing and non-ionizing radiation, in this example, beta particles as beta waves.
- The avalanche diode does not require a photomultiplier stage. The driving phenomenon is avalanche breakdown. This phenomenon is dependent on the number of atoms present as the diode structure, rather than a macroscopic multiplier.

## Introduction

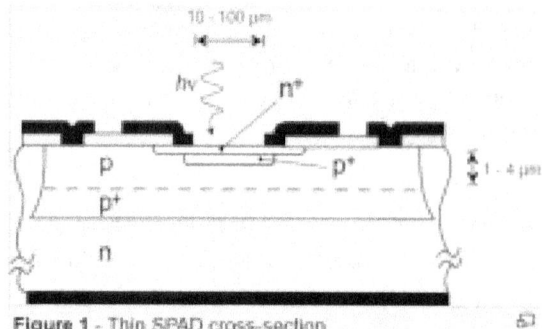

Figure 1 - Thin SPAD cross-section.

Essentially, the entire beta detector is this 1 to 4-micron scale SPAD. The nanotech that describes the 1-micron scale Surface materials by NIMS is summarized in this image: [credit, NIMS, Aug 21, 2016]

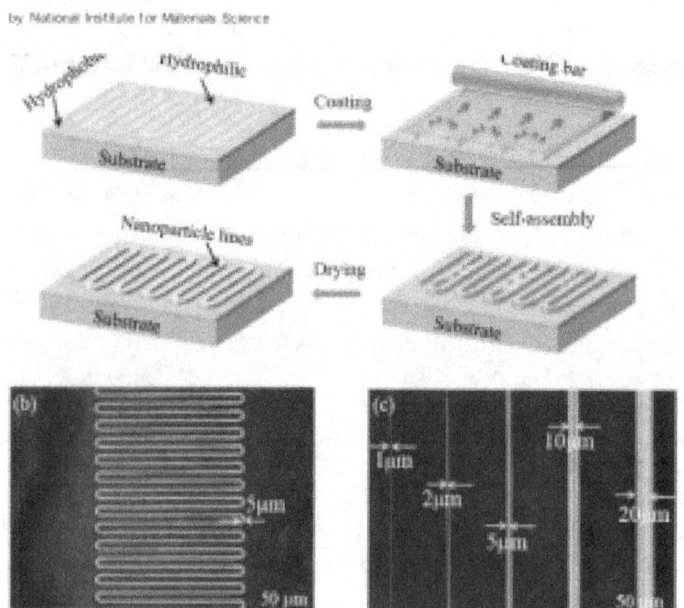

Formation of microcircuit lines using a selective coating technique. (a) Schematic of selective coating technique. Only a hydrophilic region created through irradiation of parallel vacuum ultraviolet (PVUV) is coated with metal ink. (b) Electronic circuit with a line width of 5 μm formed through selective coating. (c) Electrode lines with different widths. Lines as narrow as 1 μm can be formed. Credit: NIMS

The intent discussed by NIMS is fabrication of simple things such as 'Beyond Ultra Resolution' TV screens and such. These are intended as rather large Surface areas such as 50-inch screens and such, just as an idea on scale. The Surface is a flexible substrate.

Given that each 1-micron scale 2-clock system will have a sufficient Zeno Effect, regarding such as a t-prime value, that effect should be localized by Newtonian constraints and fall off with the

# Introduction

square of the distance from the Surface. The localized effect on path length would thus produce a Gravitational Lensing effect that can be measured with great precision by conventional means.

Eventually, the model expands up to space-filling via a Phased Array approach. The natural clock system will consist of electron-positron pairs via a standard gold foil approach. The electron positron pairs are cattle herded via simple magnetic field into a space-filling volume. The Phased Array uses the Near Field bosons to impinge upon the, for instance, electron paths, determining their spin states. This is a computational speed and power issue, only. There is some necessary development to scale a Phased Array system to accommodate such an approach. This will take some time in development of the technological limitations of such. However, this Phased Array approach is not necessary for the simple surface phenomenon described.

I am going to forego the term Quantum Zeno Effect as relevant only to a special case where the description refers specifically to some prior art in explanation and on a scale consistent with the terminology. Zeno Effect will be a term that extrapolates out to cover the entire AdS Horizon Surface, as well as the domain figuratively 'below' the Horizon Surface, for lack of a better term, where the domain 'below' the surface is thus representative of the normal light distance between points, and the Horizon Surface is superluminally isolated, quanta for quanta. A description of the AdS Horizon Surface consistent with the Zeno Effect is a subject throughout this text.

### Introduction: Artificial Alteration, Real, and Artifacts of Observation

In addition, the Zeno Effect is summarized as an *artificial* alteration in the progression of unitary time in any two-clock system by adjusting the observation rate [e.g. 'frame rate'] of detection, measurement, *observation*. There is no rational description that the Zeno Effect can be regarded as *natural*. Thus, the Zeno Effect is 1) artificial, 2) does not involve mass-energy 3) alters the rate of progression of time [disregarding scale momentarily] 4) alters spatial geometry as path length between the 2-clock system, as a consequence of an alteration in the rate of progression of time, *regardless of scale*.

### Introduction: Observation

Observation is [in this text] described as the *forcing* of Information to 're-emerge' [for lack of a better term] at the AdS Horizon Surface, in time, and locality; wherein locality is redefined as:

A) On the AdS Horizon Surface [Ryu-Takayanagi Path L] regarded as superluminal, thus superluminally isolated for each point as defined by the unitary interval epsilon, which is then quantized and equated with the unitary Planck values for length [Lp] and time [tp] and/or

B) *Below* the AdS Horizon Surface Path *l* which defines the normal light distance between points.

## Introduction

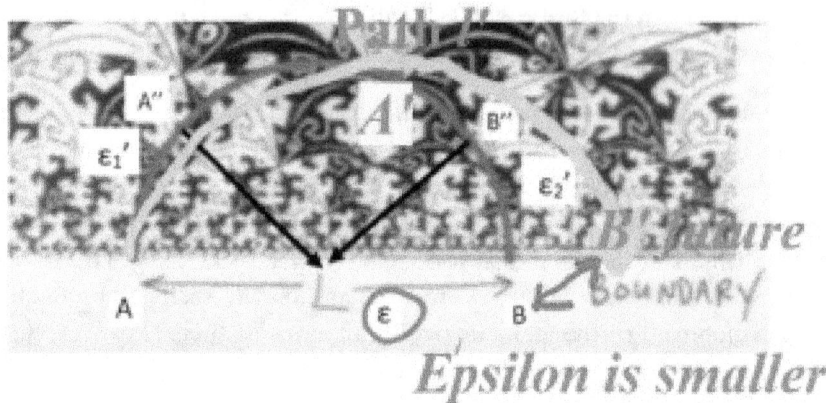

$$l = \frac{e}{3} Ln \frac{L}{\epsilon}$$

Information leaves Alice at point A and travels Path $l$, the red arc. Path $l$ represents the normal light distance between points. As the Information is en route, epsilon [$\epsilon$] is diminishing in value at a scope and rate defined by $c \equiv 1Lp/1tp$, thus, the yellow arc Path $l'$ points toward the point B', with the 'Gap,' represented by the double-red arrow as being equal in magnitude to the red arc, always. This 'Gap' results in many seeming causal violations described in more detail later. It is *Bob* at B and B' who defines the 're-emergence' of Information *onto the Horizon Surface*, and no other factor. This is *observation;* forcing Information to 're-emerge' at the Horizon Surface. Observation can be any type of interaction. It does not presuppose 'consciousness' or other esoteric nor transcendental hypotheses.

It is this metaphoric 're-emergence' that will be described as the universe's clock, the rate of progression of unitary time is defined by the above phenomena.

The artificial alteration, modification, and manipulation of the *spatial* geometry of spacetime is not a transcendent, but mundane hypothesis in this set of postulates and exhaustive derivations based on the Zeno and Anti-Zeno Effects and the demand of General Relativity that:

> 1: an alteration in the rate of the progression of time [via the QZE-QAZE] is *associated with* an accompanying Schwarzschild transformation of path length between the two clock systems [e.g., path length between clock#1 and clock#2],
>
> 2: *else the QZE-QAZE represents some violation of General Relativity that has no prior art nor explanation.* There is no logical description of temporal progression that does not obey the laws of Special and General Relativity. Thus, the QZE-QAZE can and is described by Special and General Relativity.

In addition, *if* the Zeno Effect is regarded as an *artifact of observation*, [e.g. obeys some transcendent hypothesis] then, an artifact of observation in change of path length between the two clock systems *must also be observed*; else again, represent some violation of General Relativity that has no prior art nor explanation. Observation must be slave to [obey] General Relativity, there

# Introduction

is no prior art of any hypothetical nor theoretical work otherwise. If one parameter changes, such as time, length must conform according to General Relativistic transformation [Schwarzschild].

Special Relativistic [Lorentz] transformations do not appley because:

- There is no Preferential Frame of Reference.
- There is no state regarded as relative motion, the systems are fixed and static.

Observation and detection, measurement] are in fact the key components to all of current Quantum Theory. This is true in disregard to transcendental hypotheses, such as regarding *sentient consciousness*, and so on. Observation in this text will be limited to the obvious, detection and measurement, because that is what the benchtop experiments define, only.

Thus, rather the Zeno Effect is regarded as artifact of observation or real change in the progression of unitary time, in both cases must obey General Relativity, else, again, violate General Relativity in some way that has no prior art nor description. A Schwarzschild transformation in path length must accompany any observed, measured, or artifact of observation of a change in the rate of progression of unitary time.

### Introduction: Preferential Frame of Reference and the Demands for Mass*energy

Conventionally the alteration of the progression of unitary time via the QZE-QAZE is regarded as phenomenal thus explained off in unobservable dimensionalities [e.g. 'superspace,' infinite fields, and so on] and not associated with normal temporal progression because of the association of some demand that mass*energy or mass*energy-equivalent must be the *causal component* of temporal progression and spatial geometry. This is not correct, nor is it any sound set of postulates that have ever been historically presented.

Again, there is a divergence that occurs where spatial geometry may or may not be coupled to its source of mass-energy. In Gravity Waves the spatial geometry is not coupled to its source of mas-energy. In Earth's gravity well the spatial geometry is coupled to its causal source of mas-energy. A Gravity Wave has a causal source of mass-energy, however that causal source is not coupled to the moving spatial geometry.

To clarify: the LIGO [Large *Interferometer* Gravitational Observatory] detection of Gravity Waves dismisses the hypothesis that any *local presence* of mass*energy must be directly associated with changes in spatial geometry. This was Wheeler's syntagm, 'A Gravity Wave is Gravitation Without Mass.' It was Wheeler's work that spawned building LIGO altogether. Albeit one can trace a causal path back to some cataclysmic event such as black hole coalescence perhaps a billion lightyears away and hence a billion years ago; the detection of *local changes* in spatial geometry [e.g. Earth's gravity well] via a conventional interferometer [LIGO] of the Michelson-Morley type dismisses aberrant hypotheses that mass*energy, local, must be present to associate with a change in spatial geometry.

# Introduction

Furthermore, the notion that one cannot detect one's own change of state under General Relativistic transformations is dismissed by LIGO's detection of its own change of state under a Schwarzschild transformation. However, this is a special case as was discussed earlier.

Two assumptions must be regarded as aberrant hypotheses:

1) in the demand of local mass as a causal component of local spatial geometry, this is only true when a static spatial geometry is coupled to its causal source.

2) that Preferential Frame of Reference applies to General Relativity. This is only true in the case of static mass*energy.

*None of these proposed systems in this text utilize any energy*, nor mass*energy equivalent, only 2-clock systems, metering the flow of time, *with an associated alteration in spatial geometry* as, again, meeting the demands of General Relativity. This is not transcendent but mundane, again, in the simple demands that both time and space must conform to one another, else represent some violation of General Relativity that has no prior art nor description. That is, this text describes the *artificial alteration of spatial geometry* without any application of energy whatsoever.

*This is new science, not new technology.* The artificial manipulation of spatial geometry and unitary progression of time in the absence of applying any mass*energy or equivalent has no prior art, is not a concept that has been stated. Thus, it is a bit bold to present. However, given the simplicity, 2-clocks, metering one another, alters spacetime, its temporal rate and spatial geometry: this is the 'observer effect,' the 'measurement problem,' applied in a useful way.

LIGO detection of Gravity Waves determines that the Schwarzschild transformations *are real*, not artifact of observation or some other unobservable principle; else, the interferometer would detect nothing. A *real* transformation of spatial geometry is key and at this time still a rather novel concept, oddly. Convention typically treats the time transformations as real but the length transformations as artifact. This is a type of engrained thinking that is artifact of thinking in terms of Lorentz transformations, interferometric detection, and the aberrant assignment of Preferential Frame of Reference as it applies to Special Relativity therefore somehow applies to General Relativity. The treatment of what is a real transformation and what is artifact of observation has been done in detail in prior [Bray] papers, I intend to summarize some of that in later sections.

In short, a transformation that is not *real* cannot be detected, much less detected by an interferometer. For instance, Minkowski spacetime *is not real, but imaginary*, both mathematically as well as literally. No detection of any Schwarzschild transformation can occur in Minkowski spacetime; *which brings us to the next conclusion:*

*Minkowski spacetime, as it is thought to apply in General Relativistic terms, cannot possibly be correct, because the length transformation can be detected by interferometry.*

This demands that spatial geometry via Schwarzschild transformations must be real. These *real changes in spatial geometry* have been detected and quantified in the absence of mass*energy, e.g., Gravity Wave. Again, there is no *real* transformation being detected by the LIGO interferometer that is not the direct causal result of a Schwarzschild transformation. The only *real*

## Introduction

change not associated with a Schwarzschild transformation would be to physically alter the shape of the interferometer without be affected by gravimetrics, e.g., physically move the mirrors around. When we compare this to the Gravity Recovery And Climate Experiment [GRACE; NASA] we see a less direct, or more indirect measure of Schwarzschild transformations; perhaps can be described as 'twice removed.' It should be clear, again, that in *no case* can a Lorentz transformation be detected via interferometry for a system in its own Preferential Frame of Reference, but only via some remote detection in some other frame of reference. LIGO and GRACE, directly or indirectly detect *their own changes of state*. This is an important concept because of the engraining of Preferential Frame of Reference incorrectly applied to all cases of General Relativity.

I will thus state explicitly that the concepts of Temporal Engineering and metrics via the Zeno Effect:

1. Preferential Frame of Reference does not apply to General Relativity, thus Schwarzschild transformations, in all cases where the spatial and temporal geometry are not coupled to their causal source. If Preferential Frame of Reference applied to General Relativistic transformations, LIGO, an interferometer, would not have detected anything in the presence and causal result of a Gravity Wave. LIGO can detect the intermittent Gravity Wave, which is not coupled to its source of mass*energy, among the static background of spatial geometry.
2. Mass*energy is not *required* as the causal component of alterations in the rate of temporal progression of unitary time thus spatial geometry. If mass*energy was required to alter spatial geometry, LIGO would not have detected a Gravity Wave.*Note 1

NOTE 1: A Gravity Wave *can be expressed as* mass*energy, however, the causal relationship, mass*energy is *required* in this discrete locality as the causal component of the change in spatial geometry is dismissed by the absence of mass*energy *that describes* the Gravity Wave. Thus, a Gravity Wave has a causal component, but the Gravity Wave is not coupled to that causal source.

In fact, later we will discuss the fact that the causal sources of the Gravity Wave, labeled black holes Alice and Bob, have no causal paths leading back to them, but only the product black hole Victor has a causal path leading back to it [him].

### Introduction: *Observable* Eigenstate vs *Observed* Eigenvalue

The demand that an alteration in spatial geometry is real and not artifact of observation brings us to eigenstate vs eigenvalue. An eigenstate is not {detected, observed} but regarded as {detectable, observable}. Any {detected, observed} and measured phenomenon must be in eigenvalue, cannot be in eigenstate. If a system is {detected, observed} it is therefore {observable}, it is now {observed} and it is an eigenvalue, not an eigenstate. Entangled systems 'collapse' or as I prefer, *precipitate,** to non-entangled eigenvalues at the moment of detection. Thus, entangled systems are also slave to detection and measurement. [*The term 'collapse' has too many conflicting and even ontological descriptions and will thus be avoided].

This in turn will yield the non-ontological logical sets, such as

# Introduction

{ observer, observed, observable, unobservable }

later discussed in detail as purely proximal to a *Schwarzschild Surface*; here, 'horizon' also has too many conflicting descriptions, Surface is more specifically a 2-dimensional description and is better applied.

In essence, it is this precipitation from eigenstate to eigenvalue that is the universes clock. That is, the metering of time in systems by nature is by, *precipitation from eigenstate to eigenvalue*. This will be discussed at length.

For example, Alice and Bob, are at opposite ends of the universe and so long as they remain in some entangled eigenstate, time is a non-sequitur in their description. At the moment of detection their properties, locality, and eigenvalues are well defined. *Detection has defined* the temporal and spatial relationships between them, and as such, defines the universes clock rate and spatial geometry. It is the history of first establishing a non-sequitur of space and time and *then* introducing Alice and Bob that results in the term, non-locality, and interesting but non-sequitur 'paradoxes.' That is rather backward thinking.

That is, it is because we first establish a scaffold of spacetime geometry, simple distance, that the observations appear non-sequitur. There is no spatial geometry for a thing that is in an eigenstate. We can claim, for instance, to know the predicted whereabouts of a native photon wave function. However, this is not logical because by definition the native photon wave function [which momentarily disregards the Electric and Magnetic bosons associated with it], is infinitely distributed. We conjure fancy maths to describe this distribution with a weighted value somewhat localized to our laboratory benchtop. However, this is a penchant, and a violation of the fundamental Theorem of the Limits at Infinity, which is the scaffold for all non-Platonic math. There is no localization of the native photon wave function, none at all.

This will be described at length along with the estuary derivations.

It is not sequitur to regard Alice and Bob as mere subsystems to the domain of the universe, treat the universe as set, and Alice and Bob as variables; *the observable universe* is defined as and by the sum of its components, Alice and Bob. For example, we define the temporal and spatial relationships between Alice and Bob as still a bit of a not well resolved dilemma in this transition from eigenstate to eigenvalue; if non-locality treats Alice and Bob as infinitely distributed, then the universe is according to Alice and Bob treated as *infinitesimal*. Detection and consequent eigenvalue precipitation thus define the finite characteristics of Alice, Bob, and the scope and disposition of the *observable* universe.

This is actually the heart of Einstein-Maric's approach to the 1905 Electrodynamics [Special Relativity] paper. All systems in the 1905 SR paper are measured against $k$, the rod in motion. The rod in motion is derived as existing at a constant velocity $v = c$, and as such is the single quantized photon, which Einstein had published just five months prior. Against $k$, the single quantized photon, all systems are at rest, regardless of being defined as in motion relative to one another. There is no portion of the 1905 SR paper wherein K [stationary] and K' [speeding] determine one another's states. They in fact set several scenarios up and determine by derivation that there is no

# Introduction

possibility of K and K' determining one another's states, because *time* does not exist in the system(s). This also will be described at great length.

### Introduction: Equating to the AdS Horizon Surface

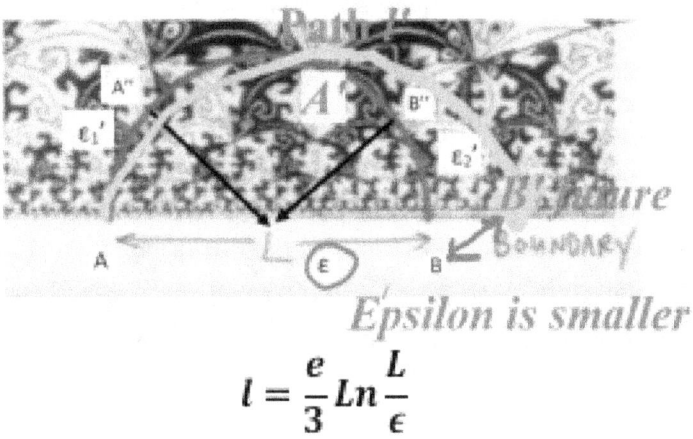

$$l = \frac{e}{3} Ln \frac{L}{\epsilon}$$

On the Anti-DeSitter Horizon, Alice and Bob are on the Horizon Surface which is defined as superluminal Path L in the Ryu-Takayanagi derivation and Path *l* is the normal light distance between points, indicated as the rigid red arc. The Information leaves point A [Alice, or Alice] on the Horizon Surface and for lack of a better term, 'submerges' below the Horizon Surface taking Path *l*, which changes Conformal Scale continually during its Time of Flight, relative to that of the Horizon Surface, which is continually diminishing as the value epsilon, $\epsilon$. Where typically $\epsilon$ is regarded in terms of cosmological time scales, later will be described as being defined by $c \equiv Lp/tp$, results in significant scope and rate of change on a second-to-second basis. Thus, that region 'below' the Horizon Surface must, relative to the Horizon Surface value epsilon be regarded as *fluid*, not rigid. The values are in fact fixed; however, this is purely in terms as relative to the Horizon Surface.

That is, because the Horizon value epsilon is changing at a considerable scope and rate, this changes the vectors that describe Path *l*, the normal light distance between points, such that it must change [shape] during Time Of Flight of Information along the red arc; the red arc is then *fluid*, not rigid. This results in the many anomalies we regard as seeming causality violations, which is also discussed in much detail later, particularly in the history of 2-slit experiments. That is indicated above by the double-headed red arrow, between B and B'. The actual distance, because $c \equiv 1Lp/1tp$ [Planck length per unit of Planck time] is exactly equal to the normal light distance in seconds.

Meaning, the end of the *fluid arc*, indicated in yellow, is B'. That distance between the original pseudo locality of Bob [Bob] as being fixed, as B, and the point indicated B', is equal in temporal distance to the magnitude of the rigid red arc, Path *l*. This will be fully derived throughout later sections. In a later chapter of this text the Lunar Landing Anomaly will describe this very well.

At the moment of detection, the Information 're-emerges' at the Horizon Surface for lack of better term, but I will continually use it metaphorically. The normal light distance and hence the temporal

# Introduction

description has been described by the detector and *only as and by detection*. There can be no *observable* case where re-emergence at the Horizon Surface is not determined by detection, which can be any electromagnetic interaction. There is no dualism between determinism and non-deterministic orientation in this, only the causal series of events with no assigned order of preference, detection defines and is defined by the Horizon Surface. Information leaves Alice at A, and the 're-emergence' of that Information is defined by Bob, who detects it at B. If Bob does not detect it, there is no sequitur description what has become of it. There is no *observable evidence* of what has become of the Information, only presumptions.

Therefore, this text will constrain all systems to purely {observable} dimensionalities and estuary relationships; no {unobservable} and therefore purely hypothetical dimensionalities nor processes will be considered. This includes infinite fields and their descriptions.

The non-ontological relationships of observability were discussed in prior [Bray] papers and texts as purely a proximal relationship to a Schwarzschild Surface. It will be summarized here in terms of this proximal relationship being the point of divergence between conventional $\Delta S$ entropy; limited to 1-causal path, and Shannon entropy, defined as >1-causal path.

## Introduction: Scope of the Zeno Effect

These facts, along with the artificial manipulation of temporal rate without mass*energy as being well demonstrated in the long history of the Quantum Zeno [slowing of temporal rate] and Anti-Zeno [speeding of temporal rate] Effects, dismisses the aberrant hypothesis that mass*energy is obligatory in any alteration in local temporal rate of progression of unitary time and thus spacetime geometry; again, as meeting the demands of General Relativity. Thus, the mass*energy requirement being a non-sequitur regards the QZE-QAZE as a demonstrable alteration in the rate of progression of unitary time and associated with an alteration in path length between any 2-clock system, *in the absence of any mass\*energy requirement*. The energy requirement for the artificial alteration of spatial geometry *is zero*. Only detection and measurement are required to alter the unitary progression of time and hence the local spatial geometry via the phenomenon. This is the heart of the observer effect, as described on the AdS Horizon Surface.

In the simplest sense, 'warping' local spatial geometry is achieved via detection and measurement as the Zeno and Ant-Zeno Effects. Any two-clock system capable of the appropriate data acquisition rates can and does alter local temporal rate and thus spatial geometry. Extending this phenomenon out to a space-filling effect is merely a matter of applying numerous two-clock systems and remote detection of non-local systems. This is also discussed at length and described in detail in later sections as Proof of Principle.

The proposed approach takes the QZE-QAZE, and using Alcubierre's metric* as a fundamental scaffold for mapping spatial geometry, an algorithm for treating the alteration of spatial geometry, does so in a useful fashion. The uses are obvious. A lifting body using such spatial geometry in a highly localized fashion has no regard for the amount of mass it is moving, lifting, etc. The local geometry, for example, of Earth's gravity well is so subtle as to be measured only by the most modern of methods, such as GRACE. This is true and directly associated with the rate of progression of time at sea level vs in orbit, the change is very subtle. The QZE-QAZE benchtop

## Introduction

experiments are altering the rate of progression of time by orders of magnitude greater than this, and as such the spatial geometry. However, because the effect is so localized and the systems of the smallest possible scales, has not been obviated; it has not however been measured. Again, such changes [ Schwarzschild transformations] are real, else could not be directly detected via interferometry.

For instance, in an example classic benchtop experiment [Zhang Man-Chao, Wu Wei, He Lin Ze, Xie Yi, Wu Chun-Wang, Li Quan, Chen Ping-Xing. Demonstration of quantum anti-Zeno effect with a single trapped ion. Chinese Physics B, 2018, 27 (9): 090305] the authors used a Paul trap to trap $^{40}Ca$ ions and obtained temporal results on the order of 1-magnitude temporal variance. Other papers have achieved greater variance; however, this is a classic benchtop experiment that can be done in any academic laboratory.

To achieve a temporal variance of an order of magnitude, the mass*energy of our sun cannot achieve this in any Schwarzschild transformation. A Neutron Star has a minimal mass of about 1.4 solar masses [debatable] with a max of about 2.1 $M_0$ [Tolman–Oppenheimer–Volkoff limit] This is also insufficient to alter temporal rate of progression by an order of magnitude, regardless of distance. If you work the Schwarzschild relationship backwards you find yourself inside the Neutron Star to achieve these temporal variances.

The minimum mass of a black hole is a sticky issue. In principle, 1-Planck mass [~2.2E-8 Kg] can become a 'micro-black hole.' *Stellar* black holes, last year a black hole of only 3.3 $M^0$ was discovered [*Thompson, Todd (1 November 2019)*. *"A noninteracting low-mass black hole–giant star binary system"*, *Science (366): 637. arXiv:1806.02751, doi:10.1126/science.aau4005.*] of about 19.5 Km radius.

However, the *conditions* that bring about a stellar black hole are 'complicated,' and a bit still in discovery mode.

Point is, a temporal variance of an order of magnitude requires a black hole, no less, micro or stellar. A temporal variance of 1-order of magnitude has a radial distance that is of course dependent on the physical size of the black hole, e.g., a micro black hole vs stellar. Nonetheless, one must be only slightly above the Horizon of the black hole in order to achieve a temporal variance of 1-order of magnitude.

*In the benchtop Zeno experiment*, there is obviously no such mass*energy condition present. If we argue the thing away by claiming that it is a 'quantum level' phenomenon and so on, 1) there is no direct evidence of this limitation, this is only supposition 2) a Planck scale [1-Planck mass] 'micro' black hole fulfills this 'quantum scale' limitation claim.

There is no differentiating the Zeno Effect from 'time dilation' under a General Relativistic [Schwarzschild] transformation on, for instance, a Planck scale. All of the systems that have been investigated have a great deal of classic as well as esoteric [transcendental] math, however, there is no answer, agreed upon or otherwise. Again, the notion that this is an issue of scale has no direct evidence that this is the case, it merely has not been scaled up in size. However, again, the Zeno Effect is defined as *observation, detection, and measure*, in some way interdependent with the

# Introduction

*observed* system. thus, the mere distance from the observed system, whatever it may be, perhaps in the above paper a $^{40}Ca+$ ion in a Paul trap, defines the scale of the system [detector and $^{40}Ca$ ion in this case], as perhaps 100 cm in size:

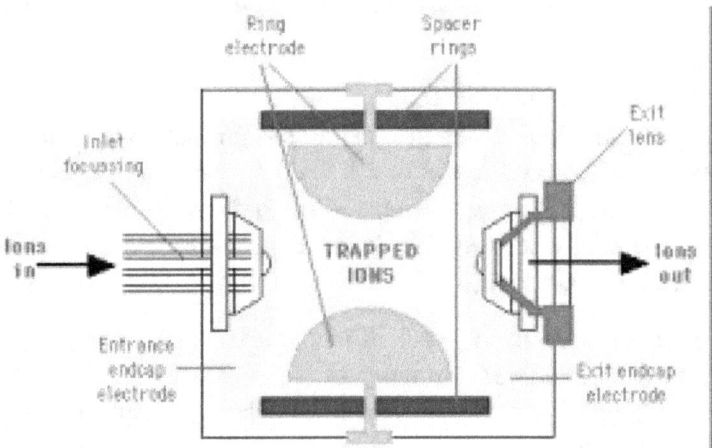

Furthermore, we have to consider that the computer control of the observation and detection rate cannot be excluded as part of the system but is in fact the most vital component that is defining the Zeno Effect and thus the Zeno System. Practically speaking, every benchtop Zeno effect performed to date is on a *room sized scale, not quantum scaled,* because the detector and most importantly, the control arrangement, such as laptops, wiring, etc., that define the system as a Zeno architecture, is room filling. Without the esoterics, this must include the operator(s). Without the operator, no Zeno Effect is happening in that lab today. This is not an ontological issue; this will be very non-ontologically defined and derived as the structure of the AdS Horizon Surface. This is of course defined in great detail throughout this text. The 'observer effect' is explained [later in this text] as *forcing* Information to the AdS Horizon Surface via detection, which is regarded as any electromagnetic interaction.

In brief it is the detector and all of its components, laptops, fiber optic cables, *and personnel,* that *cause* the Information taking the rigid red arc Path *l*, the normal light distance between points, to 're-emerge' at the Horizon Surface. Again, there is no and can be no direct evidence otherwise, only detected systems can be validated by direct evidence, which means every non-detected thing can only be supposition as to its disposition. As the detector, any electromagnetic phenomenon qualifies, is the causal component in the 're-emergence' at the AdS Horizon Surface. This is completely in line with Quantum Mechanics in the agreeable context, such as the history of the 2-slit, and so on.

Scaling the Zeno Effect up to much greater macroscopic proportions is described in great detail throughout this text.

### Introduction: Some Technological Problems in the Zeno Literature

Sadly, we may look at a few examples of such claims of high detection rate in the Zeno literature that turn out to be such issues of detection and data transfer rates, dead-time, and pile-up. Because

## Introduction

these issues take on second order curve characteristics, they look very much like a Zeno Effect, but are electronic issues.

Having various flavors of transistors and gates that operate in the hundreds of gigahertz range is exceptional news. However, we will look a bit later at the data [information] transfer rates in wire and fiber optic cable as being limited to the single gigahertz range. For example, I may have developed a miracle gate that operates in the 200-GHz range, however, the micro-wiring that connects that gate to the other gates on a silicon chip is severely limited to the single GHz range. That chip, regardless of the miraculous speeds of the individual gates will be limited to its slowest component, which is in this case the micro-wiring that connects millions of such high-speed gates together. If the millions of gates were connected by micro-fiber optic, we would still be limited to the 10-GHz range, because that is the max information transfer rate in fiber optic.

Fiber optic is limited by refractive index, most of common types are roughly ~1.5, limiting the photon velocity to 0.6c. For 1-meter, that means the one-way trip takes 3.3-nanoseconds at $v = c$ and 5.5-nanoseconds in fiber optic. The total sum of fiber optic from chip interconnects to detector and back ads up to hundreds of meters. The pile-up time is thus in the single megahertz range.

A micro-chip is wired with typically 20-nanometer interconnects. A billion such circuits ads up to 20-meters of microscopic wiring. However, a typical chip has about 100-Km of wiring in it. Fiber optic has about a tenfold advantage over wire. This extensive interconnecting of billions of individual circuits is a much greater limiting factor than how fast a single gate can operate.

When all of this is summed together, we have to limit our detections for the Zeno Effect down in the single digit mega-Hertz range, if not kilohertz.

Unfortunately, upon reviewing hundreds of experimental setups over the years there are a few things that are apparent:

- Experimenters are not aware or knowledgeable of the limitations of the individual and thus collective hard and software they are using.
- The data transfer rates [wire and fiber optic] are significantly below any reasonable expectation of, for example, a radioactive source:
- The radioactive source with respect to the number of Curies [typically milli or micro] are not stated, are likely unknown, as this would require scintillation counting the source. It requires an in-depth knowledge and experience in radio-analytical skills, which is very rare. Pharmaceutical Drug Metabolism analytical chemists have this skill set, which requires a dedicated doctorate. The unstated amount of radioactivity in [milli or micro] Curies is indicative on a lack of knowledge of that, and the distance and cross section, and efficiency of whatever type of detector is being used. Scintillation counters do not operate in the megahertz but limited to the kilohertz range.
- The overall distance from source to detector is never stated, also likely unknown
- The specifics of detector design are not stated
- The processor speeds are never stated
- The algorithm response time, speed, is never stated, likely unknown
- There is altogether insufficient information to reproduce the experiment: always.

# Introduction

The Law of Thumb is that, unless proprietary or confidential, an experiment should include sufficient information that someone skilled in the art can reproduce the experimental setup. This does not appear in any case with the Zeno Effect in the literature.

Consequently, the greatest 'paradox' is tens of thousands of papers on the subject without two that represent reproducibility, the very backbone of what is called *science*.

And most unfortunately, of those papers I have been able to reverse engineer based on several assumptions, a significant portion of the Zeno literature of reported experimental findings are data and computational dead-time and pile-up, which looks exactly like a Zeno Effect. This will be explained in detail.

The most embarrassing feature is detector and computer dead-time and pile-up being explained via unobservable dimensionalities. Minkowski apace is an unobservable dimensionality. The sqrt(-1) does not have any real world relevance. It is a null hypothesis.

### Introduction: Some Technological Issues

The details are clearly expressed in a later section of this text.

The use and treatment of the Alcubierre metric is a later section of this text. In short, Alcubierre's Metric, after some corrections and quantization, provides a scaffold for treating temporal rate of progression and translating that to spatial geometries as one sees fit. I use the term *Paint* a spacetime manifold. The term relates a large number of surface or volume filling 2-clock systems and coincident spatial geometry, via the metric.

The first step is merely to produce a Zeno Effect as a surface population of events, via some recent developments in Nanoarchitectonics [nano technology taken from biological self-assembly, particularly in the self-assembly of the extremophile archaea] have produced circuits on a 1-micron scale, as a pure surface chemistry:

## Introduction

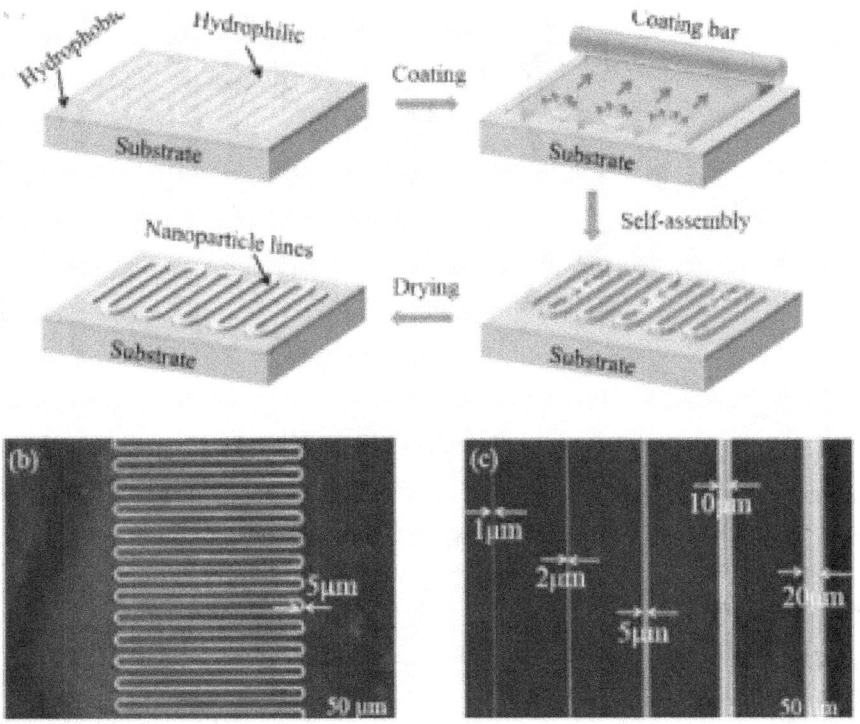

Formation of microcircuit lines using a selective coating technique. (a) Schematic of selective coating technique. Only a hydrophilic region created through irradiation of parallel vacuum ultraviolet (PVUV) is coated with metal ink. (b) Electronic circuit with a line width of 5 µm formed through selective coating. (c) Electrode lines with different widths. Lines as narrow as 1 µm can be formed. Credit: NIMS

The idea is that rather than a serpentine structure as shown above, the structure would be of individual cells, identical to above, but not interconnected as a serpentine structure:

# Introduction

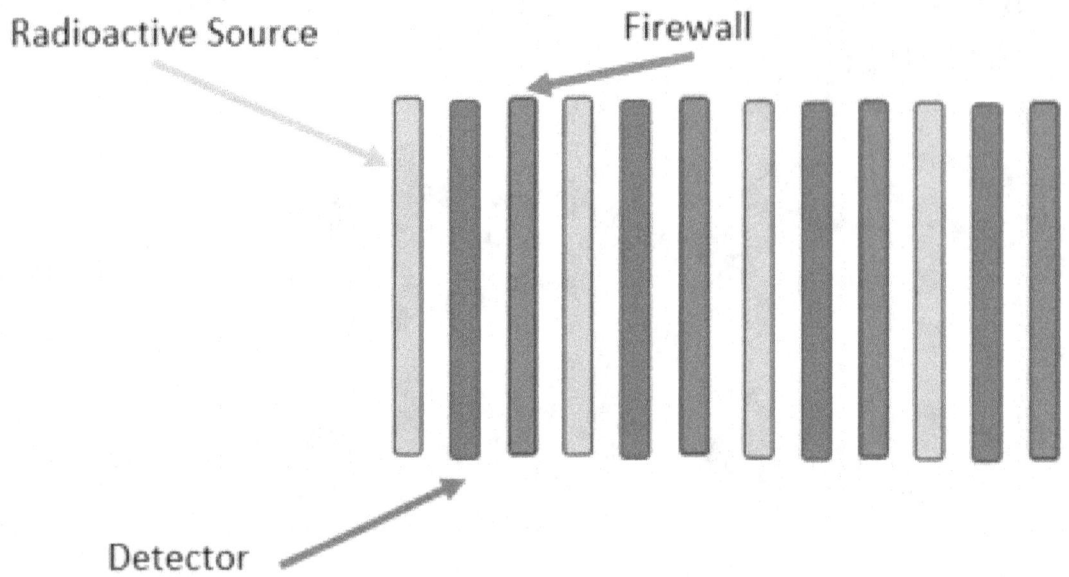

The first Proof of Principle will first involve creating a stable, reproducible, and somewhat predictable, and certainly quantifiable Zeno Effect, described in detail in later chapters of this text. The Zeno effect will explore radioactive sources, quantum tunneling, ion trapping, and numerous other systems described throughout the available literature on the Quantum Zeno Effect.

The system above uses a classic radioactive source indicated in Orange. The detector will be an appropriate adsorptive material that yields an electron [e.g., classic solid-state scintillation] when struck by, in this case a beta particle, indicated in Blue. A firewall isolates each system, and can be any substance, including substrate, as beta particles have insufficient energy to penetrate a silicon barrier.

Millions of such circuits painted on substrate produce a surface Zeno Effect. The temporal rate of progression for a large population of such events, all equal, presumably, in rate change, will thus change over the entire surface of the Thin Film Transistors. [TFT]. Keep in mind that this technology coming from NIMS is suggestive of, for instance, making large screen TVs, as an example of scale. Rather than a host of conventional microcircuits the idea is to produce a single Thin Film Transistor surface, of any size suitable for rendering, in this example, a large screen TV.

On such a large scale, perhaps thinking on the lines of a 40" television set, a large surface area of a large population of Zeno Systems, all active in producing a stable and quantifiable Zeno Effect, will first; alter the rate of progression of time on a large scale, localized to the surface. In this, there is certainly no sequitur counter argument referring off to single quantum systems as some arbitrary limit of penchant with respect to scale and scope.

According to the Theory, there *must be* an associated change in path length interdependent with the change in temporal rate. This should answer to a Schwarzschild metric. The best suited metric to reverse engineer the observed phenomenon is given by the Alcubierre Metric, after some tweaking:

## Introduction

$$f_{R:r}(\pm k) \mapsto \frac{\left(\frac{Lp^4}{tp^4}\right)}{8\pi G'} \frac{\left(\frac{\frac{nLp}{xtp}}{\frac{1Lp}{1tp}}\right)^2 (Lp^2{}_y + Lp^2{}_z)}{4g'^2 \left[(Lp_x - Lp_{xs}(t'))^2 + (Lp^2{}_y + Lp^2{}_z)\right]} \left(\frac{df}{dr}\right)^2$$

$R$ is the rate of observation of the detector, $r$ is the natural clock system being observed. NOTE: the $dr$ term at the far right is not the rate of the natural clock system being observed, it is the radius of the Alcubierre Metric, often referred to as the 'top-hat' term.

In this example, we will call the radioactive source $^{14}C$. The exact math that determines the number of micro-Curies is derived and detailed later. In short, is 1$u$Curie, which produces 37KHz of events.

Only the detectors are wired into the system when using this approach. Their detection rate is suited with concerns for Dead Time and Pile Up, which is parvasive throughout the literature. Essentially, upon reviewing thousands of Zeno experiments in the literature, a very large portion of them are the result of detector and associated transfer line and computer Dead Time and Pile Up, not stable [nor true] Zeno Effects, which accounts for a lot of the aberrant response curves that do not coincide with rational models.

The choice of 1$u$Ci for carbon-14 yielding 37 KHz events is based on detector, transfer lines [wire, fiber optic], computer scope and speed, and algorithmic response. We can assume that a safe margin for detection is given at 500 MHz. This number is based on the many parameters discussed in that section. This is universal across all of the detection methods, for al forms of radioactive detection, as well as extending into counting electron tunneling events, ion trapping, and just about everything else. Simply, too many authors have ignored the issues associated with the real imitations of detection, transfer of information from detector to computer, and computational speed.

This is an example of detector Pile Up for a gamma detector:

## Introduction

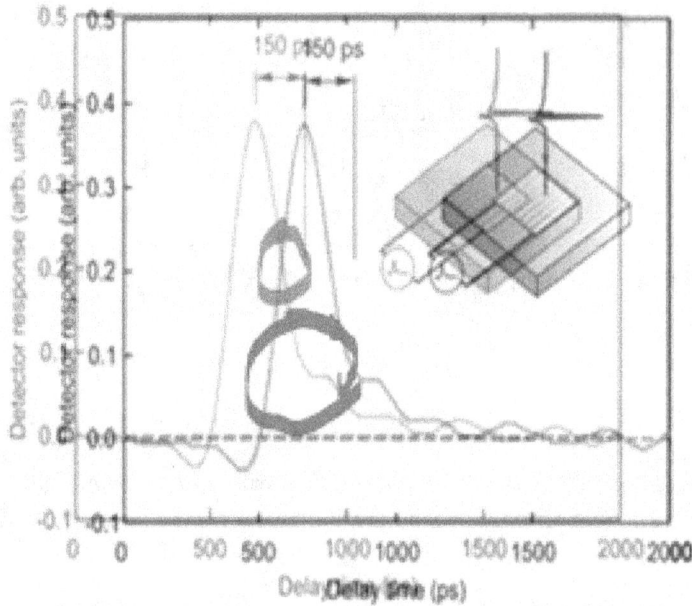

The Peak-to-Peak response, even though the manufacturer's claim is a 150 ps response time, [THz range], is not instantaneous, but as with all electronics, a weighted Gaussian response. If a second signal is detected prior to the peak falling back to baseline, it is not differentiated from the first signal. the purple circle indicates 100% overlap of signal. The red circle indicates that depending on the baseline settings, the response will appear to have 'frozen time,' because the detector is sending a constant 'yes,' a constant response. As the events move slightly away from this 100% overlap, we see a pseudo-Zeno Effect:

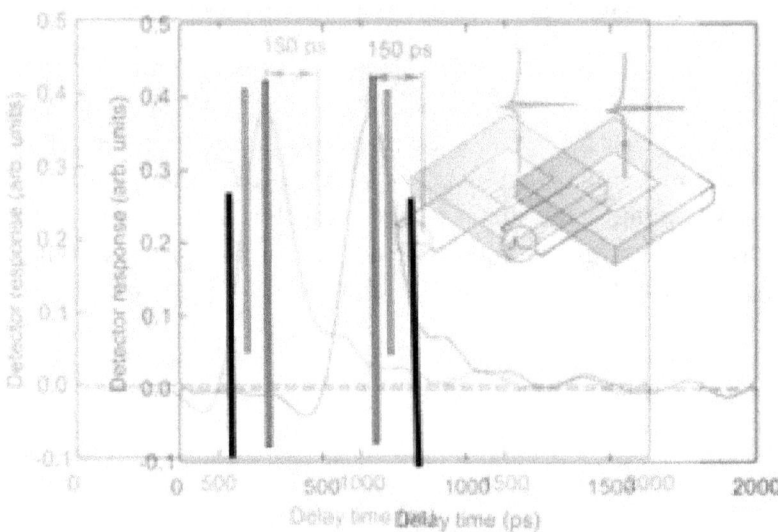

If the detector zero baseline is set at, in this example, 0.1, it appears as though the system has *slowed down*. The red lines indicate the real events. The black lines indicate the detector's ability to differentiate the two events with the zero-baseline set at 0.1. The pseudo-Zeno Effect then

## Introduction

becomes arbitrary. If one wants to 'tune' the experiment in, in such a way as to meet some penchant of hypothetical math the authors are trying to validate, one can move the zero baseline around, to obtain any result of penchant. The green lines indicate the Pile Up [pseudo-Zeno Effect] of setting the zero-baseline at 0.3.

Because the slopes of the peaks are Gaussian, the response will appear pseudo-Gaussian, because we can move the zero baseline around at will and obtain any 'Zeno response' we are trying to validate by experiment.

The same is true for transfer of information via wiring and fiber optic cables. There may be analog to digital converters at the end of the detector, for example, an electron multiplier yields an analog, not digital response. There is delay in signal conversion, and the same Pile up issues can occur. The computer is another limitation, given the algorithm in use and computer speed, which are never reported in the literature, a conventional laptop running at 2GHz is essentially reduced to 1GHz as a $\{1, 0\}$ choice, then like a CD, there is oversampling, to obscure the zero errors, and the practical upper limit will be 500-MHz.

Then of course, no authors report any of these settings and speed limitations *because they do not know them*. In only rare cases are the actual detector speeds reported in real numbers, rather the choice is to use arbitrary units, again, because the authors do not know the actual real detector speeds.

*Most of the Zeno literature is Dead Time and Pile Up, not real Zeno Effects*. They are easily characterized and fill the literature. These will look, not only just like a Zeno Effect, *but will invariably pseudo-validate the authors' hypothetical math. That in itself is a dead giveaway, because no two mathematical penchants of hypothesis are the same*. So how then can everyone's math, all unique, novel, and different, be correct? Quantum systems have a reputation for being *weird*, but not that weird.

These issues are discussed at length in later sections.

The proposed *technology* to achieve a gravimetric via the QZE-QAZE involves *first* the design of a prototype microchip. The chip is many millions of 2-clock systems that are finely tuned according to the mathematical algorithms, which are derived in this paper and prior papers. The later prototype version uses a Phased Array type of technology to measure the spin states of large populations of electron-positron pairs as a means for establishing a volumetric, isolated spacetime manifold as an approach to FTL physics under General Relativistic conditions. Where FTL remains hypothetical a more practical gravimetric drive system has immediate utility and easy to achieve.

There can be no unobservables, dimensions, fields, and so on in any postulate designed to yield an observable, real result.

There will be a rather lengthy appendix that will include a lot of estuary material that was derived and discussed in prior papers.

# Introduction

- The Zeno Effect is the causal result of a 2-clock system, the observing clock, designated $R$ is observing system $r$.
- The Zeno Effect is an *artificial alteration* in the rate of progression of unitary time in $r$ as the causal result of alteration in $R$.
- An alteration in the system $R{:}r$ must be accompanied by an alteration in path length between $R$ and $r$.
- The alteration in rate $R{:}r$ and associated change in path length between them must be *real*.
- Preferential Frame of Reference does not apply to General Relativity and associated Schwarzschild transformations. Thus, the system both causes and detects its own change of state. In this case, there is no mass*energy causal component of the change in spatial geometry.
- Mass-energy is not required in these sets of transformations, only detection and measurement.

### Introduction: Tests of Special and General Relativity

This is where you the reader will probably stop reading this text, because it addresses some issues of convention with a negative outcome. Thus, I will put this in the first page. Some critical papers that must be reviewed:

1. Experimental Basis for Special Relativity in the Photon Sector. Daniel Y. Gezari: arXiv:0912.3818 :

    A search of the literature reveals that none of the five new optical effects predicted by the special theory of relativity have ever been observed to occur in nature. In particular, the speed of light (c) has never been measured directly with a moving detector to validate the invariance of c to motion of the observer, a necessary condition for the Lorentz invariance of c. The invariance of c can now only be inferred from indirect experimental evidence. It is also not widely recognized that essentially all of the experimental support for special relativity in the photon sector consists of null results. The experimental basis for special relativity in the photon sector is summarized, and concerns about the completeness, integrity and interpretation of the present body of experimental evidence are discussed.

2. Theory and Experiment in the Quantum-Relativity Revolution, expanded version of lecture presented at American Physical Society meeting, 2/14/10 (Abraham Pais History of Physics Prize for 2009) by Stephen G. Brush
3. What is the experimental basis of Special Relativity? 1999 available at: https://math.ucr.edu/home/baez/physics/Relativity/SR/experiments.html#Tests_of_time_dilation
4. Z. Zhang, Special Relativity and its Experimental Foundations, World Scientific (1997).

Contrary to modern science mythos or perhaps misunderstanding, Special Relativity has not been validated on several key points:

## Introduction

1. invariance of the speed of light (c) to motion of the observer
2. Lorentz-FitzGerald length contraction
3. relativistic Doppler effect
4. relativistic stellar aberration
5. relativistic source brightening

In fact, the above have been tested on numerous occasions, as {3, 4} compilation of the history of experiments indicate, with null results. I am not suggesting by any means that Special Relativity is a Theory that is not correct, just that line items 2) Lorentz contraction and 3) relativistic Doppler Effect are incorrectly interpreted by modern convention. There is an entire section of this text based on prior [Bray] papers that follow the Einstein-Maric 1905 derivation, 'On the electrodynamics of moving bodies.' Briefly, there is misconception that is based on reciprocation of what Einstein published in 1905 [e.g., original German text] as translated by Princeton later that year into English. This may seem like a puerile rant, as there is no shortage of such in the literature. However, I am providing snapshots of the original [1905] text, as well as snapshots from a second paper in 1907 where Einstein rederived the entire thing in response to Bucherer's viral publishing of his length transformation reciprocated [originally dilating, translated to 'contraction']. These are explicit, clear, and decisive, that Einstein was infuriated by Bucherer's [a science journalist] viral publication of his work, incorrectly.

The length transformation is key to unwrapping the Zeno Effect on a macroscopic scale, applying a modified [corrected and quantized] version of Alcubierre's Metric, which provides a scaffold for relating temporal changes to length transformations [Schwarzschild type]. In short, *all transformations must be real, not artifact, as indicated by the permanent disparagement between clocks.* Thus, 'Minkowski spacetime,' which is an unobservable dimensionality that must be treated as artifact to accommodate the 'length contraction' hypothesis, cannot be regarded as anything other than a null hypothesis. That is, Minkowski assigns real meaning to the complex conjugate of sqrt(-1), based on 'length contraction,' which Einstein published on numerous occasions [all referenced] as Length Dilation: *which is the very principle of the constancy of the speed of light. [e.g., $c \equiv l/t$ ]*

Terminologies and Descriptions of Terms Used

## Terminologies and Descriptions of Terms Used

W Bray DoD:

In 1994 Alcubierre published a Faster than Light spacetime manifold that because of a negative energy density Tensor invoked the notion of 'exotic matter' to manifest 'negative energy' so as to fulfill this negative mass-energy requirement, on the order of 10E38 solar masses [10,000-trillion universes] of non-existent "negative universes." To date, no one has established what this negative energy is, at least not in any agreed upon sense. Because of these issues the manifold fell into a sort of novelty category, suitable for amateur sci-fi documentaries that persist to this day but can never have any real-world tangible use. Part of the problem was trying to establish a type of 'propulsion' that like Star Trek, was a conventional continuous motion, but at non-conventional speeds.

A novel spacetime manifold is derived and presented herein. The spacetime manifold has *zero mass-energy requirement*. The spacetime manifold has zero energy density, thus, no negative energy density. Energy is non-sequitur to the description in every sense. Mass-energy is not required for Faster than Light 'propulsion.' Instead, the system uses only rapid detection and measurement, via the Zeno Effect, creating an alteration in the progression of unitary time on a macroscopic scale, thus altering the geometry of spacetime in a highly localized fashion. There is no sense nor logical description of motion, it is a 'jump drive,' that works by extending the unitary Planck intervals of time and consequently length out to great distances, repositioning the manifold at some distant locality. Because this change in Conformal Scale is a fractal relationship based on artifacts of Preferential Frame of Reference *of that of being inside the manifold looking out at flat spacetime*, it is for this reason there is no energy description that is sequitur to the drive mechanism. The 'zero energy' condition is thus a natural consequence of the fact that *no motion is taking place, in any frame of reference.*

Meaning, it is not the *quantity of* unitary Planck intervals that is changing, it is the *magnitude of* each unitary Planck interval that is changing. The Planck intervals of length [Lp] and of time [tp] represent energy, in the Wheeler derivation, related directly to $h$:

$$lp, tp = sqrt\left(\frac{hG}{2\pi c^x}\right)$$

Thus, any change in the *quantity of* Planck intervals; demands mass-energy conversion [e.g., Alcubierre], else, is a *violation of Preservation of Information Conservation*. The means and methods by which we change the magnitude of the unitary Planck intervals is described in several prior Bray texts and papers and summarized here.

Oddly, convention never seems to address this issue, but seems to regard the *quantity of* unitary Planck intervals as changing in the case of a Lorentz and/or Schwarzschild transformation. Every paradox of both Special and General Relativity goes without empirical validation to this day

Terminologies and Descriptions of Terms Used

because they simply cannot happen as described in said, 'paradox,' as these 'paradoxes' require, by convention, a shifting in the *quantity of* unitary Planck intervals, holding the *magnitude of* the unitary Planck intervals constant. Convention actually regards the Planck interval as fixed, neglecting the fact that they are described as mass-energy, $h$. However, in any Lorentz and/or Schwarzschild transformation, changing the *quantity of* Planck intervals of length [Lp] and time [tp] is a violation of every conservation law that exists.

*Therefore, we need a more viable solution than changing the 'quantity of' Planck intervals of length [Lp] and time [tp] in a Lorentz and/or Schwarzschild transformation.* The only alternative is thus to change *the magnitude of* the Planck intervals. The regard then, is to find a 'constant' in the description of the Planck interval that *may be regarded as variable under certain discrete conditions.*

Keeping in mind that the primary failure of the Alcubierre approach is to regard the Eulerian or Lagrangian arbitrarily distant observer. There is no logic to this approach at all. The only Preferential Frame of Reference that is relevant to any isolated region of spacetime regarded as a 'spacetime manifold' is that of *being inside the manifold, looking out at flat spacetime.* All other approaches are non-sequitur. As such, we can regard:

$$G' = 6.67384(80) \times 10^{-11} \text{ m}^3/\text{Kg (s)}^2$$

As:

$$G' \rightleftharpoons 6.67385E-11 \frac{Lp^3}{Kg * \left(\frac{t_0}{t}\right)_a^2}$$

Then,

$$\varphi(\pm L'_p) \rightleftharpoons |\sqrt{\frac{hG'}{2\pi c^3}}|; \quad \varphi(\pm tp') \rightleftharpoons |\sqrt{\frac{hG'}{2\pi c^5}}|$$

Where, in quantized form:

$$v = nLp/xtp$$
$$c = 1Lp/1tp$$

Terminologies and Descriptions of Terms Used

$$\theta_{Sc} \rightleftharpoons \sqrt{1 - \frac{2G'M}{4\pi L'p \left(\frac{1Lp}{1tp}\right)^2}}$$

And/or

$$\theta_{Lo} \rightleftharpoons \sqrt{1 - \left(\frac{\frac{nLp'}{xtp'}}{\frac{1Lp}{1tp}}\right)^2}$$

Note that these are fractal forms and not equalities.

The spacetime manifold is unlike Alcubierre's work in most every sense, just referred to as the 'Bray Spacetime Manifold,' so as not to be confused with Alcubierre's work. There is no 'length contraction,' as this is an erroneous approach that regards Eulerian observation, when the Preferential Frame of Reference must be that of being *inside the manifold, looking out at flat spacetime*. It is this Eulerian and even Lagrangian approach that is inherently incorrect because it does not regard the Preferential Frame of Reference of the system that is supposedly in motion. Both Eulerian and Lagrangian approaches share this bizarre 10E38M☉ 10,000-trillion universes of 'negative mass-energy,' via 10,000-trillion non-existent 'negative universes.' Obviously, therefore, they cannot be regarded as correct in any sense at all.

Unlike the conventional spacetime manifold Alcubierre presented the manifold is, rather than a type of continuous 'motion,' is a discontinuous 'jump drive.' Hence, 'Bray Manifold.' I refer to this as 'pseudo-motion,' rather it is more of a discontinuous 'jump.' The actual detailed description of the physical design of such an engine is in:

Faster Than Light Propulsion via the Zeno Effect. W. Bray, March 2021 DOI: 10.13140/RG.2.2.28575.48809

A text describing the *physical engine technology and design* in explicit detail is forthcoming, likely toward the end of 2022. The actual physical architecture of the engine design encompasses a variety of conventional 21st century technology used in a non-conventional way. There is a particle fountain in a central nacelle that emits beta radiation at very high energies and luminosity. These are steered and 'wiggled' by Near Field Phased Array systems in nacelles two and three. The 'wiggling' of the beta particles in flight produces magneto-Bremsstrahlung radiation, which allows for *non-destructive detection* via RF, whose fractal Koch antenna arrays are also housed in nacelles one through three. The non-destructive detection via magneto-Bremsstrahlung radiation at phenomenal rates is a Zeno Effect, which, as described herein, is the key to the working principle of the engine.

Terminologies and Descriptions of Terms Used

Thus, it is both the theory and physical working design of the 'engine,' spread across this and a later text.

There seems to be this mystique that is rather illogical regarding the nature of a spacetime manifold that is localized in any sense: All work to date seems to be from the Preferential Frame of Reference of any type or number of arbitrarily distant observers, observing the manifold. There is no logic to this approach at all. Meaning, a single quantized photon neither experiences time nor distance. And, because we regard the Preferential Frame of Reference of the Eulerian or Lagrangian observer, by convention only and always, we cannot understand such simple behaviors as the collective history of the double slit experiment, which is the root to the invoking of unobservable dimensions in an attempt to understand the single quantized photon. For the single quantized photon, all distances in the cosmos are zero, as is the travel time. It is from this Preferential Frame of Reference that Albert Einstein and Mileva Maric derived and described "Special Relativity," which appears to escape convention to this day.

The key feature to the Einstein-Maric 1905 derivation of "Special Relativity" is that: Having set the Stationary Twin [$K$] and the Speeding Twin [$K'$] in motion relative to each other, *immediately prior to the length transformation and therefore not a consequence of it; they state explicitly* that all systems [throughout the entire paper, not just this instance] are measured against $k$, not each other, which is "a rod in motion in the x direction." Explicitly, there is no instance in the 1905 Einstein-Maric "On the electrodynamics of moving bodies" wherein the stationary [K] and speeding [K'] measure one another's states, all systems are measured against [$k$] the "rod in motion." Given *the original paper* only has 100-citations to it in over a century, it is the most unread paper in science history, students read *about it, instead.*

The "rod in motion" is being derived as the single quantized photon, which Einstein had published just five months prior. They are methodically deriving the single quantized photon, $k$, as being at $v = c$, which *as a consequence of being at the lower [contraction] and upper [dilation] boundaries of the Limits at Infinity, is thus constant*. Keep in mind that this description is novel, *there is no conventional derivation of the constancy of c that is not hypothetical, none bare any empirical evidence, and no two are complimentary*. It is the Limits at Infinity that define the constancy of the speed of light, as per the 1905 Einstein-Maric derivation. This seems to have eluded convention for a century at this point. This key is not off-topic, but critical to understand.

They determine that the stationary twin $K$ and the speeding twin $K'$ as measured against the single quantized photon $k$ "are at rest relative to each other," regardless of having been set in relative motion. This is a natural consequence of being measured against the photon's property as existing on both the lower boundary [infinite contraction] and upper boundary [infinite dilation] at the Limits at Infinity. Meaning, against either limit at infinity, lower [infinitesimal] or upper [infinite] any system measured against either limit is of infinitesimal or otherwise of infinite distance, thus, stationary, regardless of our ideological notions of relative motion. This Einstein-Maric approach is in the Appendixes.

The idea of continuous motion is thus non sequitur. As strange as this might sound, we have no incontrovertible evidence of continuous motion beyond Mars, which is only 1-AU distance, on average. I am not stating that continuous motion does not exist, at this time, only that it is

Terminologies and Descriptions of Terms Used

undefendable in terms of empirical data. There is a scale limitation, however, that is deeply tied to the AdS Horizon Surface, which is a subject in prior [29, 30] and forthcoming works.

Oddly, any claim of empirical measurement beyond this distance of only 1-AU is conjecture and are not compelling to the extent that it has *only one theoretical description, but instead, many competing hypotheses*. Man has only been as distant as the moon, interestingly, and our instruments have only been *successfully* to Mars. We have two probes extending beyond this; however, the 'Pioneer Anomaly' is a riddle wherein we cannot explain where these probes are and why they are not where they are supposed to be. Thus, we cannot regard this as compelling evidence of continuous motion that has a correct mathematical description. It is in fact General Relativity that has failed to explain their positions and vectors successfully.

However, upon close examination of this, in turns out that JPL used a different NIST value for G [1998 $6.6745 m^3/Kgs^2$] than used at launch [1973 $6.6720 m^3/Kgs^2$] yields $8.7E\text{-}8 \pm 1.5$ $cm/s^2$, which is *exactly* the velocity difference in question. The point is that there are several hundred papers that attempt to explain this via unobservable dimensions, unobservable forces, and unobservable 'negative mass energy,' such as Exotic Dark Matter, when the correct answer is utterly mundane. Key: simple motion cannot be explained via constrained to purely observable forces, dimensionalities, and substance.

We have to swallow our presumptions and in light of requiring 10E20 nonexistent "negative universes" concede that the idea of continuous motion of any type at any velocity is in every sense illogical. Faster than Light "propulsion" will thus take the form of a discontinuous "jump." Furthermore, the Star Trekian notion of vast amounts of brute energy to engage FTL is also illogical, the logical approach will require *no energy nor mass-energy equivalent. This is a logical consequence of a system that is not in motion.*

Meaning, the logical Preferential Frame of Reference for an isolated, localized spacetime manifold must be *purely* from *inside the manifold, looking out at flat spacetime, and what will the manifold thus do in order to "reach out" to some great distance without any type of motion in its description, thus, no energy*. It does not appear anyone has done this in any pure sense, such manifolds are invariably described by the classic flip flopping between "frames of reference," most horrifically with "Eulerian" observers. The Eulerian observer has to be the least correct approach to an isolated spacetime manifold. The Lagrangian approach is only half as futile, but futile nonetheless.

The correct Preferential Frame of Reference must be that of inside the manifold looking out at flat spacetime. All other frames of reference must be negated, without exception. I refer to this as the Dead Twin Paradox, where the speeding twin kills his stationary twin and is thus forced to meter his own progress, second-for-second, meter-for-meter, along his journey. No one has ever considered that scenario, oddly, measuring his own state, as convention regards that he cannot do so, which is erroneous in every possible sense and has never been empirically observed and certainly not validated in General Relativity.

LIGO, for example, cannot detect its own change of state as a result of its motion around the sun, e.g., Lorentzian. However, LIGO *can detect its own change of state under a Schwarzschild transform, either directly or indirectly.* This Preferential Frame of Reference issue is key to

Terminologies and Descriptions of Terms Used

manifesting a spacetime manifold: *the engine has to produce a Schwarzschild transform of its own state.* That requirement is never discussed in the literature. Meaning, convention *always regards* detection of one's own state under Lorentz transform as impossible, and *usually* under Schwarzschild transform as impossible. However, somehow one is supposed to change one's both Lorentzian and Schwarzschild state in order to manifest this localized spacetime manifold. Thus, the abandonment of one's own principles and logic is required to engage in this fantasy work.

Just as a practical argument, In Special Relativity, a zero-fringe result is a valid result [e.g. Earth around the sun], when you can measure a non-zero value for any other system [e.g., Mars around the sun]: is detecting your own state under Lorentz transform. Meaning, if we have a fringe result of some observed system of say, 0.2, what does that mean if we do not have a zero baseline? It is a meaningless number.

In several prior papers I described that the reason for this ability to detect one's own change of state under Schwarzschild transform is rather a local system of spacetime, such as a Gravity Wave, is coupled to a source of mass-energy. This is the failing point of the Alcubierre and all subsequent attempts at localized spacetime manifolds. Regardless of the energy requirements, because they employ mass-energy in an attempt to alter spatial geometry, the spatial geometry is thus coupled to the mass-energy. As a result, no such system can alter its change of state under a Schwarzschild transform.

It is this flip flopping between frames of reference that creates these bizarre non-existent negative mass-energy requirements on the order of 10E38 solar masses of "exotic universes," which obviously do not exist. Requiring unobservables from the Realms of the Gods is as desperate as it is primitive as it is absurd. The mathematical description of the energy tensor, which is quantized, is defined herein, derived in prior [Bray] papers. As for the shaping functions these are vaguely defined herein and forthcoming in a later paper, as the notions of symmetry on an isolated local scale must be negated. The actual physical description of the engine design in technological terms is nearly complete and will follow this paper.

Next, we need to address the ontology of observation and *conscious observation*, suitable within the framework of quantum mechanics with the least number of unagreeable elements. Since any "conscious" sense of observation begins with any electromagnetic interaction, which is a definition that back engineers to the brain, not limited to any particular sensory quality, nor any unobservable quality of 'self,' we thus can regard detection and/or observation as any electromagnetic interaction. All other forces of nature are excluded because there is zero evidence that these play any role in detection, in any immediate sense. For example, gravitation *between* distant objects is a misnomer: the interaction is with the *immediately local spacetime geometry, not the distant object.* As for the weak forces and the strong force, there is zero evidence that these can be regarded as interactive, any counter argument is sub-speculative not even worthy of conjecture.

In order to separate "conscious observation" out, we can regard the Zeno Effect as providing the first and only definition for "conscious observation" universally acceptable within the framework of quantum mechanics:

1. Only humans can change the rate at which we observe and detect phenomena.

Terminologies and Descriptions of Terms Used

2. All systems in nature are set with respect to how rapidly they interact with other systems. If this were not so, this universe could not exist, e.g., an Anthropic Principle.
3. Furthermore, it is these natural clock systems that we regard as the golden standard by which we meter time.
4. That golden standard of temporal rate of progression of natural clock systems *changes when we change our observation rate of it.*

It is this unique ability for humans, and only humans [thus far], to change the rate at which we electromagnetically interact with systems, e.g., detection, which can be regarded as: *Observation*.

The second principle of *observation* is that, as is true with all systems which electromagnetically interact, observation 'forces' a system to precipitate from an eigenstate to some eigenvalue. This is the key, and is by convention, regarding the 'quantum' Zeno Effect. However, it is a misnomer that the Zeno Effect is limited to quantum scales and there is zero evidence that this is the case. By convention, the Zeno Effect is defined as the observed system coupled to the detector, which puts the Zeno Effect on a centimeter and in some cases a meter scale. As will be discussed and was derived fully in prior [Bray] papers and texts, the Zeno Effect, by being the causal component of precipitating systems from eigenstates to eigenvalues, changes their Conformal Position on the AdS Horizon Surface. That is the key factor that defines the Zeno Effect.

The Zeno Effect is the universe's natural clock system. The Zeno effect is simply, *two clocks, observing one another.*

The manifold extends spacetime out to very large distances in a fraction of a second, repositioning itself at some distant locality. This isn't exactly a type of superpositioning, but an alteration in spatial geometry in a highly localized fashion but extending the unitary Planck intervals out to cosmological, or at least very large distances, by altering them in a fractal relationship, equivalent in concept to altering their Conformal Scale, *not their quantity.*

It is by regarding the need to change *the quantity* of Planck intervals that creates this bizarre mass-energy requirement, as the Planck intervals are defined by $h$, $G$, and $c$, all of which represent mass-energy. Altering the *quantity of* Planck intervals is a violation of Preservation of Information Conservation in the worst possible sense. However, no one seems to have taken this into account when regarding both Lorentz and Schwarzschild transformations. Convention simply regards time dilation as increasing the *quantity of unitary Planck intervals of time*, and the same for length transformations. Even worse, convention regards time dilations as 'real' based on the permanent disparagement between clocks, but length transformation as 'artifact of observation,' thinking that no permanent record of such exists; with the unitary time having an explicit accounting but space as a zero budget, is the truly worst case of violation of Information Conservation possible.

In this description, all transformations are *real*, there are no artifacts. Interestingly, when regarding the length transformation as real for the first time as a part of the original manifold, this is the root source of the $10E38M_\odot$ 'negative mass-energy' requirement, as is apparent in the manifold shaping and related energy tensor.

Terminologies and Descriptions of Terms Used

*I need to reiterate here that mass-energy, or any type of energy, is not required to produce a spacetime manifold that possesses the qualities described herein.* That is, a localized change in spatial geometry that extends out to great distances and results in a relocation at some distant point in space in a fraction of a second; as a manifold, requires no mass-energy to achieve. The only requirement, or 'fuel' for this alteration in spatial geometry of this sort is rapid detection and measurement, *only; two clock systems, observing each other.*

This may seem odd, however it was explained in several prior lengthy texts as the natural manifestation of the 'observer effect,' and what that means on a holographic, e.g., AdS Horizon Surface, where all points are superluminally isolated, but connected via that domain figuratively 'below' the Horizon Surface, the normal light distance between points. It is the observer, which can be any electromagnetic interaction, that empirically and unambiguously 'forces' Information to re-emerge from subspace [Ryu-Takayanagi Path $l$] to the Horizon Surface. If the rate of detection is greater than the normal light distances between points, Ryu-Takayanagi Path $l$, the Information re-emerges in a sense, prematurely, as a result of simple trigonometric limitations. In effect, the Zeno Effect is this alteration of the nature of the AdS Horizon Surface, vs that region figuratively below, sometimes referred to as the AdS subspace. This appears throughout the history of the double-slit experiment as seemingly anomalous timing of events and even notions of time reversals, superluminal phenomenon, and so on, but has a geometric solution on a Holographic Horizon vs the normal light distance between points.

Meaning, classically a seeming violation or not understood timing of events that appears or seems to violate causality is typically and conventionally explained via transcendental hypotheses, which is the root source of all notions of higher dimensionalities. The non-transcendent hypothesis is that the AdS Horizon Surface is constantly evolving via the vanishing Ryu-Takayanagi AdS Horizon Surface value epsilon [$\varepsilon$]. Typically regarded only on cosmological time scales, the actual vanishing rate of epsilon is apparent *on all scales*. Observation, which is any electromagnetic interaction, 'forces' systems from their eigenstates to eigenvalues. The more rapidly we 'force' systems from eigenstates to eigenvalues, *the less epsilon has evolved* in those shorter time periods. As a result, the local unitary value epsilon is of different conformal scale than 'normal' observation rate, the conformal scale of epsilon will be the function of detection rate. Because the unitary value epsilon defines the Planck length [$L_p$] and Planck time [$t_p$], we see what appear to be temporal anomalies. These do not occur in nature, only we can do that, observation.

I am not going to redescribe all of these things and rederive them here, just reference them at the end to the prior series of papers and texts and provide the scaffold of the spacetime manifold in several steps to the final solution. It is not complicated; the sequitur order of events merely has to be taken into account. There is a use of some self-similar relationships [fractals] that are novel, but not entirely challenging.

It must be understood that fractal relationships behave and must be treated very differently from equations of the conventional sort, which is a principle not universally understood in physics. For example, a classic Mandelbrot equation looks like:

$$Z' = Z^2 + C$$

57

Terminologies and Descriptions of Terms Used

There is no possible way to look at a value in a Mandelbrot and back calculate the origin, or even the immediate prior iteration. Fractals cannot be manipulated like equations.

The fractals, self-similar relationships, describe and define the relationships of conformal systems in a way that the various schools of thought have not foreseen. Conformal field Theory and conformal Gravitation are probably the most challenging and complicated mathematical systems in existence. However, if one applies a simple fractal, fifty pages of math just go away because they are redundant.

It is this conformal scaling that describes and defines all things in nature, fractals, and a few fundamental sequences. When the AdS Horizon Surface value epsilon, which is the defining quality and quantity for the unitary Planck intervals of length [Lp] and time [tp] is treated as fractal relationships, and we look at the act of 'observation rate' vs the AdS Horizon Surface value epsilon with respect to its vanishing scope and rate, we find that it is the difference in these conformal scales that looks very similar in nature to Schwarzschild and Lorentz transformations. I am not saying that Schwarzschild and Lorentz transformations are equated with this phenomenon, only that they share the same inherent property of, *scale differences,* e.g., dilation, contraction, and so on.

A change in the rate of progression of unitary time must be accompanied by an associate change in spatial geometry, else violate General Relativity in a way that has no prior description. This is the working principle of the 'engine.' By detection and measurement at some phenomenal rate, we alter, by convention, the rate of progression of unitary time, thus, the spatial geometry must also conform, else violate general relativity in some way that has no prior description. That is very straightforward. A 'Warp Drive' requires no logical sense nor amount of energy, Faster than Light 'propulsion' requires no energy whatsoever.

The self-similar relationships are solutions to Conformal Scale Invariance that do not invoke unobservable dimensionalities. *There can be no use of unobservable dimensionalities in a real and tangible 'propulsion' system where real motion is taking place.*

The actual physical mechanisms of the drive system were described in the prior paper on FTL Propulsion via the Zeno Effect. It is a phased array system that employs *non-destructive* detection and measurement of spin entangled particle pairs at a very high rate of speed in a volumetric, space-filling system of events by detecting emitted magneto-Bremsstrahlung radiation as simple RF. The more detailed 'engine' design is forthcoming.

Terminologies and Descriptions of Terms Used

## Terminologies:

*Qubit:* A single Temporal Element of one unitary Planck interval of time-length:

[Qubit, proper noun, not qubit or qubyte]: $(u_0 \to u_{0+1}) = \{L_p, t_p\}$

*N-Qubit:* Some integer quantity of *Qubits* that define a system. Where the *Qubit* represents a unitary Planck interval in one dimension, such as length only, the N-Qubit is a novel term that describes a system such as that described below, consisting of any number of *Qubits*.

The system's constraints are referred to as the 'worldsheet,' designated by $A_\Omega$.

It is a term used so as not to confuse with the conflicting descriptions regarding the 'qubit' as it appears in Information Dynamics across the literature. Here, the 'Qubit' is set aside, and the term N-Qubit is *the most primitive nature of Information* on a Planckian scale:

From the Bekenstein Limit:

$$N = S = \frac{A_\Omega}{4L_p^2}$$

Then setting $A_\Omega$ to 4-Qubit, e.g., $4L_p^2$:

$$N = \frac{A_\Omega}{4L_p^2} = \frac{4L_p^2}{4L_p^2} = 1$$

$$c \equiv \frac{1L_p}{1t_p} \mid \therefore$$

$$\text{к}N_Q = \text{к}\frac{L_p^2}{t_p^2} = c^2 \equiv \text{к}1 = A_\Omega$$

These are all derived in summary and at length in the Appendix on Defining the Qubit.

The term, $A_\Omega$ is the 'worldsheet,' a conventional name that merely defines the constraints of the system. Above, the constraints are Planckian in scale. The term koppa [к] is a police pseudo-operator that marks those variables that must remain fixed, else violate Preservation of Information Conservation.

The Maric Operator

## The Maric Operator

Note that here, **Phi [φ]** is the Maric Operator:

$$\varphi(\pm v) \equiv |v|$$

$$\varphi(\pm v) = |v|; \text{ then given } c = l/t \equiv 1;\ \varphi(\pm l) = |l|;\ \varphi(\pm t) = |t|;\ \text{therefore } \varphi(\pm 1) = |1|$$

The phi [Maric] operator has been described and defined in several prior papers in this series referenced at the end of this text. The phi operator was presented by Mileva Maric to Abrahm Joffe, who reviewed the 1905 'On the electrodynamics of moving bodies,' prior to the Einsteins submitting the paper for publication. Mileva Maric was the co-author of the famous 1905 paper, regarded as 'Special Relativity.' Her contribution is primarily this mapping of a value to its absolute value, as shown above.

The Maric Operator maps the absolute value onto a system that is in relative motion, of the sort $\varphi(v)\varphi(-v) = 1$, more specifically $\varphi(\pm v) = |1|$, where that motion is defined as being metered by the single quantized photon. This may seem a bit odd but is the strategy the Einsteins used in deriving 'Special Relativity,' as the mathematical scaffold for the constancy of the speed of light. Keep in mind that to this day there is no agreed upon mathematical description or hard set of mathematical derivations for the constancy of the speed of light, [nor has it been empirically proven] it is merely accepted as such and enforced.

In the 1905 strategy, system K [stationary] and system K' [in relative motion to K] are metered not by one another, but by the single quantized photon, $k$, the 'rod in motion, whose locally quantized meter stick exists at the upper [dilation] and lower ['contraction'] boundary constraints of the Limits at Infinity. As such, the systems K and K' are stated outwardly to 'be at rest relative to each other.'

See Appendix I: The Maric Operator for images of these portions of the 1905 'On the electrodynamics of moving bodies.'

Some derived rules and properties of the Maric Operator:

Here, the Maric Operator, which is defined in this text as applying to any system ≥2-unitary Planck intervals or 2-wave cycles of a native wave function [e.g., $4\pi$]:

- On scales greater than 2-Planck unitary intervals, the Maric Operator applies, and:

$$\varphi(\pm v) \equiv |v|$$

$$\varphi(\pm v) = |v|; \text{ then given } c = l/t \equiv 1;\ \varphi(\pm l) = |l|;\ \varphi(\pm t) = |t|;\ \text{therefore } \varphi(\pm 1) = |1|$$

*Describes the unipolar qualities of Gravitation and time.*

- On scales equal or less than 2-Planck unitary intervals, or 2-wave cycles of a native wave function [e.g., $4\pi$], the Maric operator *does not apply*, and:

The Maric Operator

$$\varphi(\pm v) \neq |v|$$

$\varphi(\pm v) \neq |v|$; then given $c = l/t \equiv 1$; $\varphi(\pm l) \neq |l|$; $\varphi(\pm t) \neq |t|$; therefore $\varphi(\pm 1) \neq |1|$

And the convention of sign [e.g., electromagnetic] applies to such systems. However, *does not apply* to the Zeno Effect [e.g., rapid detection and measurement]. This is an empirical observation, as demonstrated in the divergence from the Zeno to Anti-Zeno Effects.

It is because of the Maric Operator, which is defined in detail in Appendix I, most particularly of the form:

$$\varphi(\pm v) \equiv |v|$$

That the original Alcubierre derivation is non-sequitur. It is strongly advisable to read Appendix I so that one can see the snapshots of those portions, in context, of the original 1905 'On the electrodynamics of moving bodies,' and understand the nature of this Einstein-Maric approach. The entire meaning of this approach is that velocity, with respect to vector, is non-sequitur.

## LIGO and Preferential Frame of Reference

LIGO's detection of gravitational waves raises many interesting points. I have explained that this is an interferometer detecting its own change of state under a Schwarzschild transform and am satisfied by this argument. Here we keep in mind that LIGO cannot detect its own change of state as a result of its motion around the sun, e.g., classic Michelson-Morley experiment, but can and does detect its own change of state as a direct result of a change in spatial geometry.

Unambiguously, LIGO cannot detect a Lorentzian transformation under any circumstances.

This is performed via a type of pulsed lasering effect that compares the timing of intervals. However, again, this approach, if it were not a direct Schwarzschild transform but a Lorentzian transform, could not detect its own change of state as a direct result of its relative motion.

Basic rules and properties take the form:

1. When mass-energy is coupled to the local spacetime geometry, the *quantity* of N-Qubits may vary, provided mass-energy and Information Conservation are preserved.
2. In the case where spatial and temporal geometries are coupled to their local source of mass*energy, Preferential Frame of Reference applies for General Relativity under Schwarzschild transformations.

1. When mass-energy is not coupled to the local spacetime geometry, the *quantity* of N-Qubits must remain fixed, else violate preservation of Information conservation.
2. In cases where spatial and temporal geometries are *not* coupled to their local source of mass*energy, Preferential Frame of Reference *does not apply* for General Relativity under Schwarzschild transformations, and a system may *detect* its own change of state under a Schwarzschild transformation.
3. In cases where spatial and temporal geometries are *not* coupled to their local source of mass*energy, Preferential Frame of Reference *does not apply* for General Relativity under Schwarzschild transformations, and a system may *change* its own change of state under a Schwarzschild transformation.
4. As such, it is the *magnitude of* the unitary Planck intervals that changes, rather than the *quantity of* unitary Planck intervals.
5. There is no energy nor mass-energy equivalent sequitur to a description of the magnitude of the unitary Planck intervals changing.

This is the failing point of the Alcubierre and all subsequent attempts at localized spacetime manifolds. Regardless of the energy requirements, because they employ mass-energy in an attempt to alter spatial geometry, the spatial geometry is thus coupled to the mass-energy. As a result, no such system can alter its change of state under a Schwarzschild transform.

Some rules: as put forth in Quantum Information Dynamics Volume III

Some rules: as put forth in Quantum Information Dynamics Volume III

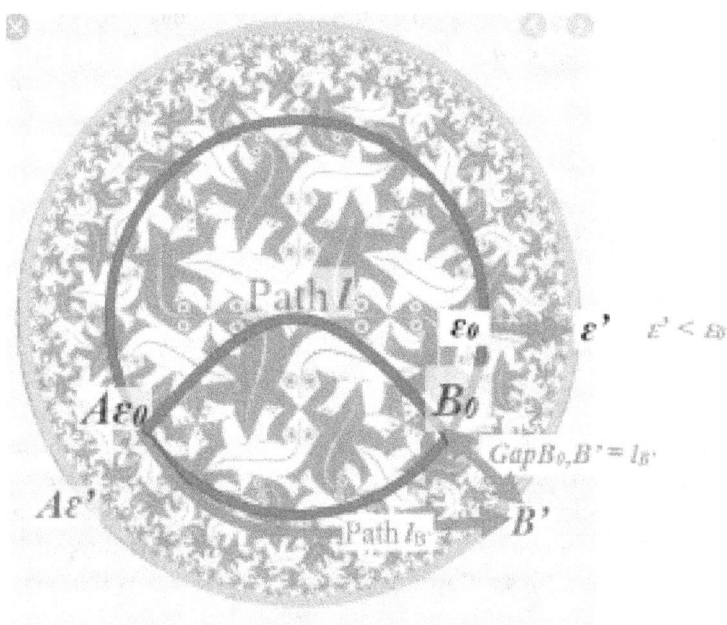

### The AdS System

Note that by conventional descriptions, Information is thought to traverse the red arc, the Ryu-Takayanagi Path $l$ [above in red]. However, this is impossible. Information leaving Alice at point $A\varepsilon_0$ toward Bob at $B_0$ must travel from the AdS Horizon Surface where the values epsilon [$\varepsilon$] are smallest, toward the inner region where the values epsilon are of greater conformal scale. *This is directing Information into the past, literally, not figuratively.* Information must take the green arc from Alice at $A\varepsilon_0$ to Bob, *who will not be at point $B_0$, but point $B'$ [prime]*.

The green arc at no point travels inward toward the values epsilon of greater conformal scale, thus never into the past. The green arc continually follows the evolution of the AdS Horizon Surface value epsilon [$\varepsilon$], labeled as Path $l_B'$, in green.

1. Ryu-Takayanagi Path $l$ is the normal light distance between points.
2. An eigen*value* can only exist *on* the AdS Horizon surface.
3. An eigen*state* can only exist [figuratively] 'below' the AdS Horizon Surface.
4. Any electromagnetic interaction, which is all observation, 'forces' [for lack of a better term] Information to emerge from figuratively below the AdS Horizon Surface by *constraining* them to be eigen*values*.
5. Entangled systems are in eigenstates, and as such can only travel [figuratively] 'below' the AdS Horizon Surface.
6. Entangled systems travel via the rigid red arc of convention, Ryu-Takayanagi Path $l$, [red arc], which is temporally frozen, static. Thus, transfer of Information is rather non-sequitur, as the entire system is temporally frozen.

Dr. William Joseph Bray

**Some rules: as put forth in Quantum Information Dynamics Volume III**

7. Bob's observation, [for non-entangled systems] which is any electromagnetic interaction, 'forces' the system into an eigenvalue, thus to the AdS Horizon Surface.
8. The rate at which Bob detects Information affects the rate that Information is forced into an eigenvalue, thus the time at which Information 're-emerges' at the AdS Horizon Surface. *This is the Zeno Effect. This will be an in depth focus of this text.*

### I. When Preferential Frame of Reference Applies

- Preferential Frame of Reference applies to all non-entangled systems.
- Non-entangled [usually macroscopic] systems can only transfer Information via the *fluid arc* [in green], *Path l'. [prime]*
- When Preferential Frame of Reference applies, Information takes the [Ryu-Takayanagi] Path *l' [l-prime]* which is the green arc. The Information leaves Alice at time $A\varepsilon_0$ [time-zero] and is observed by Bob at $B'$ [causal future from $B_0$]
- When Preferential Frame of Reference *applies*, which is true for all non-entangled and macroscopic systems, Information takes the green arc, from $A\varepsilon_0$ [time-zero] to $B'$ [time has progressed forward]. Epsilon [$\varepsilon$] has progressed [evolved] in its vanishing scope and rate defined by $c \equiv 1L_p/1t_p$. As a result of the vanishing scope and rate of epsilon, $\varepsilon' < \varepsilon_0$ [the new value epsilon-prime is less than epsilon at time zero] which leaves the 'Gap' indicated in the diagram [double headed green arrow], *which is always equal in magnitude to the normal light distance between points. The 'Gap' is the direct result of the present value epsilon being of lesser magnitude than the past magnitude of epsilon.*
  o *What is Temporally True in one system may not be true in all systems.*
  o The 'Gap' represents what *may seem like* atypical causality violations.

#### A. Thus, when Preferential Frame of Reference applies:

*What is Temporally True in one system is not true in all associated systems.*

This is the expected result of a double-slit experiment that does not turn out that way in a double-slit experiment, as for entangled systems. The 'scale' is only defined by Preferential Frame of Reference, which describes rather or not a system will traverse the conventional Ryu-Takayanagi Path *l* [PFR does not apply, e.g., entangled systems] or the postulate non-conventional Ryu-Takayanagi Path *l'* [prime], PFR applies, e.g., non-entangled systems.

### II. When Preferential Frame of Reference does not Apply: *For Entangled Systems Only*

- Preferential Frame of Reference *does not apply for entangled systems.*
- When Preferential Frame of Reference *does not apply*, Information traverses the [static] rigid red arc of convention, Ryu-Takayanagi Path *l [not prime]*. Information leaves Alice at time $A\varepsilon_0$ and is observed by Bob at time $B_0$ *[not B-prime]*. Both $A\varepsilon_0$ and $B_0$ represent the same instant in time, time-zero. This is actually the convention, albeit, not stated outwardly, and does not seem to appear, oddly, in any of the literature.

**Some rules: as put forth in Quantum Information Dynamics Volume III**

- When Preferential Frame of Reference *does not apply*, which is specific to entangled systems, Information takes the rigid red arc [of convention] of Ryu-Takayanagi Path *l*. In this case, *epsilon has not progressed* [evolved] because the information leaves at time-zero for $A\varepsilon_0$ and arrives at point $B_0$, which also represents time-zero.
  - Since the rigid red arc of convention represents the normal light distance between points, it is clear that the information has not travelled in a manner consistent with the normal progression of time. It has not, however, travelled via [Ryu-Takayanagi] Path *L*, which is superluminally forbidden.
  - What is Temporally True for one system is true for all systems.

*If the AdS Horizon Surface value epsilon [ε] is not evolving, time is not progressing.*

B. Thus, *in the case of entangled systems*, when Preferential Frame of Reference does not apply:

*What is true in one system is true in all associated systems because all systems are fixed at time-zero.*

1. Preferential Frame of Reference *does not apply for entangled systems*; thus, they traverse or exchange information via the conventional Ryu-Takayanagi Path *l*. *e*
   a. What is true in one system is true in all systems.
2. Preferential Frame of Reference *applies for all non-entangled systems*; thus, they do not traverse or exchange information via the conventional Ryu-Takayanagi Path *l* but traverse or exchange Information via the fluid Path *l'*. [prime]

**Then:**

- In the Delayed Choice Quantum Eraser, [and Delayed Choice Entanglement Swapping] because the system is entangled, Preferential Frame of Reference does not apply and what is true in one system must be true in all systems.
- Thus, in the DCQE, Information is traversing the conventional Ryu-Takayanagi Path *l*.
- Ryu-Takayanagi Path *l* is a non-dynamic, static system. Information starts with Alice at time-zero and reaches Bob, who is also at time-zero, via the conventional Ryu-Takayanagi Path *l*.
- Ryu-Takayanagi Path *l* is equated with the horizon value epsilon, which is only conventionally regarded as dynamic on cosmological time scales. However, this cannot be. The local effects of epsilon's dynamic nature is constantly observed, but not understood as a locally dynamic epsilon.

Some rules: as put forth in Quantum Information Dynamics Volume III

And the most essential Rules:

1. *It is only in the case where the AdS Horizon Surface value epsilon [ε] is changing that time proceeds forward.*
2. *If the AdS Horizon Surface value epsilon [ε] is not changing, then time remains frozen.*

The Delayed Choice Quantum Eraser, Delayed Choice Entanglement Swapping, and Lunar Landing Anomaly all share a common motive, Preferential Frame of Reference *does not apply* for The Delayed Choice Quantum Eraser, Delayed Choice Entanglement Swapping, and Preferential Frame of Reference *applies* for the Lunar Landing anomaly. Therefore, Information pseudo-traverses the conventional Ryu-Takayanagi Path *l*, which is a static, temporally frozen system. This is discussed at length throughout this text. However, the focus will be on Gravitation in this text.

III. **Thus, there are two distinct and separate conditions of which Preferential Frame of Reference does not apply:**

1. **The system is entangled.**
2. **The system is not coupled to some mass-energy source.**

## Zeno Effect Dynamics of the AdS Horizon Surface

### Zeno Effect Dynamics of the AdS Horizon Surface

What I am trying to get across here is related to my prior statement:

> Normal detection I am regarding in terms of not being a Zeno Effect. Under 'normal' conditions, Bob is *forcing* the Information to evolve from an eigenstate to an eigenvalue. There is no empirical evidence otherwise. An eigenvalue can only exist *on* the AdS Horizon Surface. Thus, what Bob is doing is throwing coinflips of the sort:

$$<detect|no\text{-}detect>$$

> The Zeno Effect will be described in terms of how often Bob *does not* **detect** an electromagnetic phenomenon.

We are scanning or interrogating the PN junction at 1-billion-times per second [1GHz]. The rate of Information is at 1KHz, meaning, we are observing a radioactive source, for example, that is emitting 1,000 events per second. Thus, 1:1,000,000 of the interrogations yields a positive result. We are actually capturing information at a much lesser rate than normal scans rates at 1000-times per second and vastly less information than our Anti-Zeno rates at 500 times per second [500Hz].

*These numbers are purely fictitious for argument's sake.*

At 1000 scans per second, we had no eigenstates, only eigenvalues. Like a radio signal, the Information will traverse Path $I'$, [prime, not the conventional Ryu-Takayanagi Path 1] and will be observed by our detector Bob, the PN junction, at point $B'$.

Meaning, for the 1GHz scan rate, 1:1,000,000 of our interrogations yield a positive result, compared to the Anti-Zeno rate, where half of the scans yielded a positive result. At the normal rate, every scan yielded a positive result.

This is the static nature of the PN junction, which holds true for electron multipliers and even photomultiplier tubes. The PN junction itself is an instantaneous event, literally. The instant an electromagnetic boson strikes the surface, and electron in the PN junction jumps from one orbital to another, which is by convention, *an instantaneous event*. Thus, it is not the PN junction that decides the scan rate. The scan rate is determined by how often we tell the software and hardware to interrogate the PN junction.

Keep in mind this is worded very similarly to the previous description of the below eigenvalues for the Anti-Zeno Effect. However, now we are *not*-detecting as the dominant characteristic of the eigenstate *being forced to* evolve to the eigenvalue. As a result, the evolution of the AdS Horizon value epsilon is greater under Zeno Dynamic conditions than epsilon is under 'normal' detection rates.

That is, less frequent evolution from eigenstates to eigenvalues forces Information to the horizon surface less frequently, and the evolution of epsilon therefore progresses *further*,

## Zeno Effect Dynamics of the AdS Horizon Surface

such that $\varepsilon_Z$ is much less than $\varepsilon'$, which is less than $\varepsilon_0$. Keep in mind that epsilon evolves regardless of what events may be occurring on the horizon, thus, less frequent *positive* observations as a result of increased detection rate results in epsilon having evolved to a greater degree.

- $\varepsilon'$ is less than $\varepsilon_0$, always.
- $\varepsilon_Z$ is $< \varepsilon' < \varepsilon_0$

I will use carbon-14 in these examples out of familiarity and realism in specific activity vs *possible and plausible* detection rates.

The entire principle may seem backwards. Again, by observing 1KHz events at a 'normal' rate, we get 1,000 out of 1,000 positive detections. By observing the same source, which is producing 1KHz events, at a rate of 1GHz, only one in one million interrogations of the PN junction will yield a positive result. The Zeno Effect is based purely on *not detecting events in a natural clock system*. Again, the consequence is that the AdS Horizon Surface value epsilon has evolved to a greater degree, meaning it is lesser in magnitude than observation at a 'normal' rate of observation or interrogation.

The Zeno Effect I am redefining as a *macroscopic phenomenon* where the detector, Bob, is coupled to the system being observed. This is by convention, the Zeno Effect is defined as the observing system, e.g., detector, coupled to the observed system, usually a natural clock, such as a radioactive source.

The Zeno Effect is then Bob, measuring the system described under normal detection rates as:

$$<detect|no\text{-}detect>$$

However, the 'detection' is an instantaneous thing where the electron is excited and makes the quantum jump to another orbital. In this case, by scanning my PN junction at some phenomenal rate, what Bob is doing is changing the ratio of *<detect|no-detect>* at 50/50 to some phenomenal value where the condition:

$$<\psi|no\text{-}detect>$$

The condition *no-detect* dominates. As a result, the evolution of the AdS Horizon value epsilon is greater under Zeno Dynamic conditions than epsilon is under 'normal' detection rates.

- $\varepsilon'$ is less than $\varepsilon_0$, always.
- $\varepsilon_Z$ is $< \varepsilon' < \varepsilon_0$

This brings to mind the simple question: can we produce a Zeno Effect on just one singular event? The answer is unknown at this time but seems highly unlikely. However, we do

## Zeno Effect Dynamics of the AdS Horizon Surface

such things in the variations of the 2-slit by sending, for example, one single photon at a time to the detector. These experiments have never been done.

*What then happens if you perform the Delayed Choice Quantum Eraser experiment under Zeno Dynamic conditions?*

That is also unknown but seems like a highly coveted task for a grad cadet. We discussed that in the last section. An entangled system cannot, as the postulate goes, produce a Zeno Effect, because the Information is traversing the conventional Ryu-Takayanagi Path $l$. Only non-entangled systems can traverse the pseudo-Ryu-Takayanagi Path $l'$.

Epsilon-prime is the result of the time of flight for the Information to traverse the normal light distance, which is Ryu-Takayanagi Path $l$. Epsilon-naught is the convention wherein we regard the path as rigid, but cannot be the case, because Bob is at any distance regardless of how small, from Alice.

Epsilon-Z [Zeno dynamic] is the result of the Zeno dynamic, $\langle \psi | no\text{-}detect \rangle$, with the condition no-detect dominating, thus the AdS Horizon value epsilon has evolved further than if the Information merely traversed the normal light distance of the rigid red arc of convention, Ryu-Takayanagi Path $l$.

## Normal Detection Rates Over Normal Light Distance

It is important to consider that the region figuratively below the horizon cannot change physical geometry as this would then demand physically changing the past. This is a key concept in understanding the progression of time, temporal mechanics, and how and why Information transferred between entangled systems differs from that of non-entangled systems.

Meaning, the geometry of the region figuratively below the horizon does not actually exist, but is merely a record, if you will, of past states of the horizon surface. Any physical change to the geometry figuratively below the horizon then would require physically changing the past. Although an interesting feature for science fiction nature does not behave this way.

We will use the Zeno Effect to demonstrate how the progression of time evolves on the AdS Horizon Surface. The Zeno Effect is perfect in this, in that it is Bob's monitoring of information sent to him by Alice that affects where on the horizon surface the information figuratively 're-emerges to the horizon surface. This will become clearer. The reason it occurs this way is because it is Bob, by any electromagnetic interaction of observation, who figuratively 'forces' the Information from its eigenstate to eigenvalue or values.

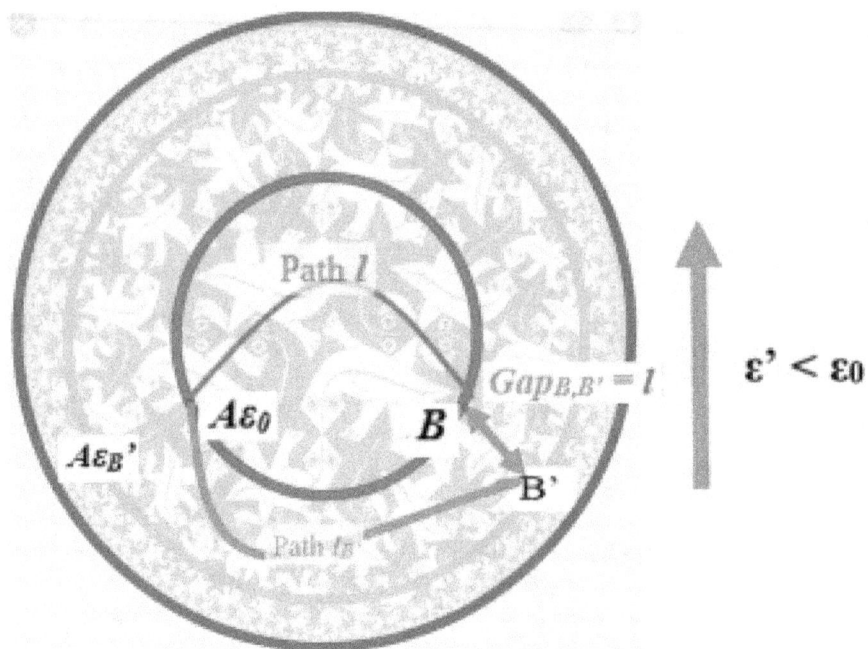

In the above image, the green arrow to the right indicates the state of epsilon as $\varepsilon'$ at some time in the future from event zero at point $\varepsilon_0$. The value $\varepsilon'$ is always less than $\varepsilon_0$. This is because, by convention, the value epsilon is vanishing. Again, normally regarded only on

Normal Detection Rates Over Normal Light Distance

cosmological time scales, the actual scope and rate of epsilon's vanishing magnitude is defined by $c \equiv 1L_p/1t_p$, where Planck length is designated $L_p$ and Planck time, $t_p$.

## Normal Detection Rates Over Normal Light Distance

**Normal Light Distance**

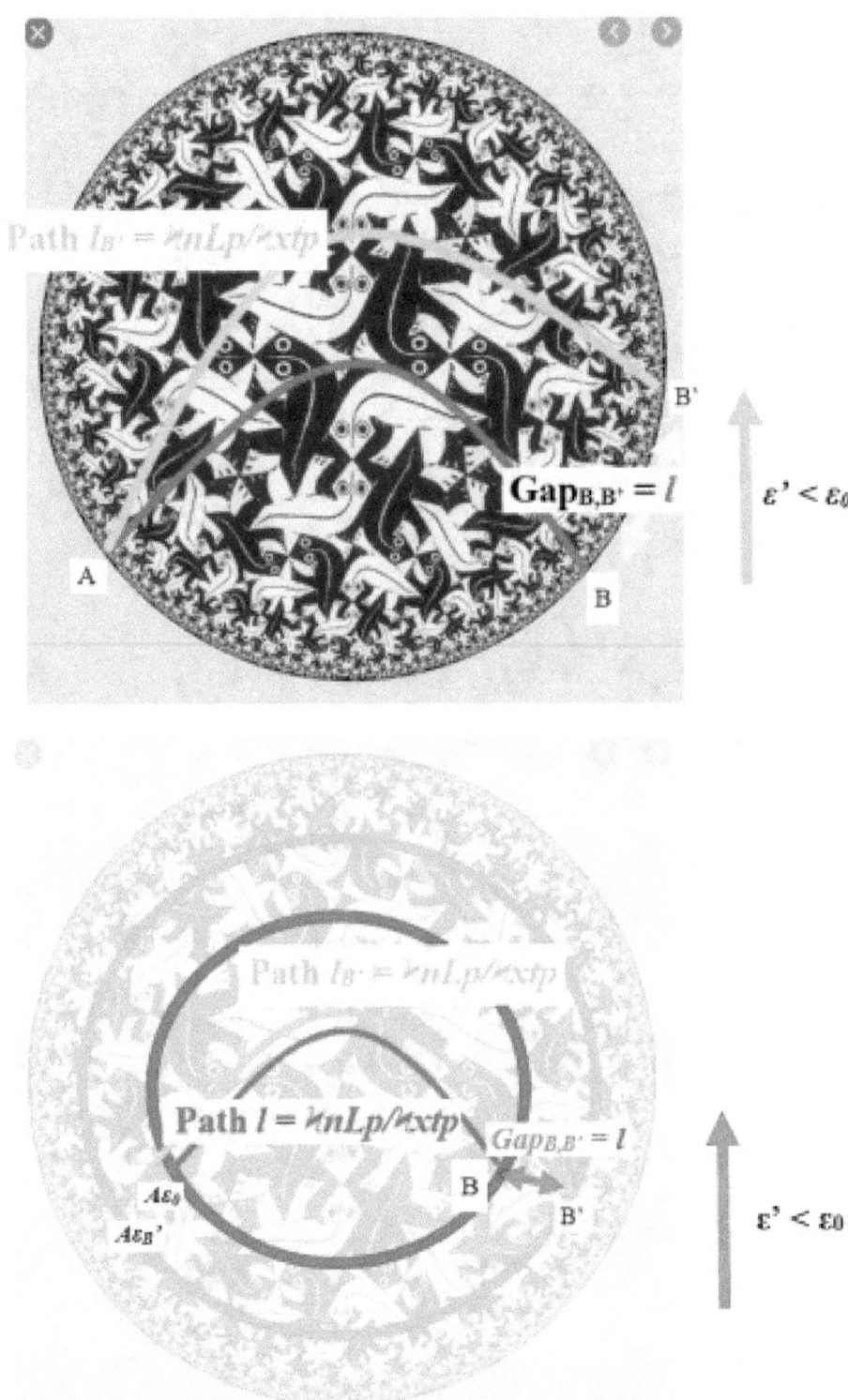

Normal Detection Rates Over Normal Light Distance

- At time zero Alice is at the point indicated as $A\varepsilon_0$ and Bob is at point B. The normal light distance between them is the red arc, Path $l$, is 100-light-seconds in length.
- Bob observes the Information Alice sends at a rate insufficient to produce a Zeno Effect.
- The Information travels from Alice at $A\varepsilon_0$ toward Bob and 100-seconds later Bob detects the first bit of the Information. At this time Bob is now at position B' because the Horizon Surface has evolved at a rate and scope of $c = 1Lp/1tp$. **Epsilon is $\varepsilon$' at point B' and is of lesser value than $\varepsilon_0$.**
- Thus, the arc takes the form of the yellow arc. The red arc must be regarded as fluid, taking the form of the yellow arc.
- This difference between Bob at point B and at B' results in a Gap [in green].
- The Gap [in green] is equal in magnitude to the red arc, Path $l$.
- Bob's value $\varepsilon_{Bob}$ is $\varepsilon_{Alice}$/Path $l$, [ $A\varepsilon_0$/Path$l$ ] the normal light distance between Alice and Bob because the Horizon value epsilon has evolved a factor of $c = 1Lp/1tp$ since Alice sent the Information at time zero. Thus, $\varepsilon_{Bob} < \varepsilon_{Alice}$
- The Gap$_{B,B'}$ is equal in magnitude to the normal light distance between Alice and Bob, Path $l$.
- His Locally Quantized Meter Stick as epsilon is thus $\varepsilon_{Alice}$/Path $l$, the normal light distance between Alice and Bob.
- He will require 10 of his unitary intervals [seconds] $\varepsilon_{Bob} = \varepsilon_{Alice}$/Path $l$ to read the 10-second packet of Information sent from Alice.
- His placement as the normal light distance **Path $l_{B'}$** is equal to **Path $l$**, as measured by his [Bob's] Locally Quantized Meter Stick, which is shorter than Alice Locally Quantized Meter Stick by a factor of $c = 1Lp/1tp$.
- The *quantity of unitary intervals* has not changed in this, only the magnitude of each unitary interval as a Conformally Scale Invariant value, described by the self-similar relationships of the Fractal Set.
- The progression of unitary time also dictates that epsilon has diminished *from* $G_B$, at time-zero, to time $G_{B'}$ because time has passed in terms of Time of Flight. *The velocity* of Information transfer has not changed, nor has the Time of Flight changed. Consequently, the *standard yellow arc is of the same value* [designated as Path $l_{B'}$] as the 'normal arc' Path $l_B$: *As Measured by Bob at positions B and B'.*

There are some causal anomalies, but not paradox, that the Gap$_{B,B'}$ produces that will be discussed in another section of this text.

## Zeno Dynamic AdS Horizon Surface Quality

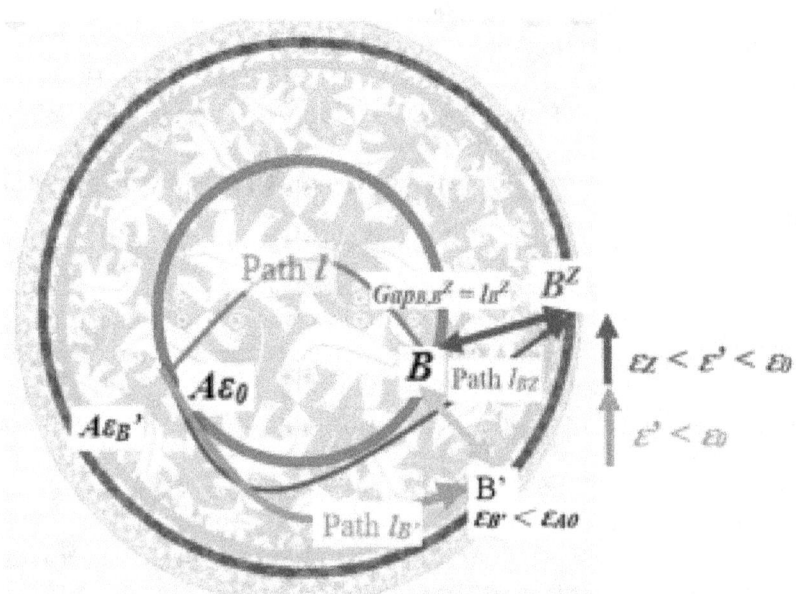

$$Path\ l_B^Z > Path\ l_{B'}$$
$$Path\ l_B^Z > [RT]\ Path\ l$$

- The Gap for $B$ to $B^Z$ [purple double-headed arrow] is *greater than* Path $l$ [not $l$-prime]. More time has passed according to Bob's locally quantized meterstick, which is fixed with his locally quantized clock.
- The *Path* for $l_B^Z$ *[purple arc]* is however, *still greater than* Path $l$ [not-prime, rigid red arc of Ryu-Takayanagi convention]. Bob's locally quantized meterstick at point $B^Z$ is smaller than it would be at $B'$, he thus measures a *greater distance coupled with longer time of flight* with his smaller meterstick.
- Furthermore, the *Path* for $l_B^Z$ is *greater than* Path $l_B$ which is the 'normal' observation rate, e.g., no Zeno Effect.
- $\varepsilon^Z < \varepsilon'$ The horizon value epsilon under Zeno conditions has evolved more than epsilon under 'normal' conditions. As equated to the unitary Planck intervals of length {Lp} and time {tp} the unitary intervals are *lesser in magnitude than* they would be under 'normal' conditions.
- $\varepsilon^Z < \varepsilon_0$ The horizon value epsilon under Zeno conditions has evolved more than epsilon as it was at time-zero. The unitary intervals are therefore lesser *in magnitude than* they were at time-zero.
- Bob is now at point $B^Z$. His distance from his original position is measured orthogonally but not orthonormally to his prior position, now on a later evolution of the AdS Horizon Surface.

Zeno Dynamic AdS Horizon Surface Quality

The higher rate of scanning and interrogating the PN junction misses less events. This however is not a mechanical thing with a mechanistic result. Bob is *forcing* Information to the AdS Horizon *less often* under these conditions by *forcing* the Information from an eigenstate to an eigenvalue, *less often*. Consequently, the horizon value epsilon evolves *more than* that of 'normal' detection and the horizon value epsilon is therefore of *lesser* magnitude. His Locally Quantized Meterstick is thus smaller than it would be under 'normal' conditions,' he thus measures time and distance as being *greater than* those unitary Planck intervals from the past.

Again, the notion that we are 'detecting' more or less rapidly is incorrect. The PN junction is merely a static thing that has no 'knob' on it to tell it to detect events more or less often. What we are doing is scanning or interrogating the PN junction more or less often.

- When we scan the PN junction more often [Zeno], we mostly yield negative results, or positive results *less often*. We are forcing Information from an eigenstate to an eigenvalue, and consequently to the AdS Horizon Surface, *less often*. Epsilon evolves therefore *more than* it would under 'normal' conditions and is consequently *smaller in magnitude* than epsilon under 'normal' conditions. Our smaller locally quantized meter stick measures Information from the past as being larger in both size {Lp} and time {tp}.
- When we scan the PN junction *less often*, we mostly yield positive results, or negative results *less often*. We are forcing Information from an eigenstate to an eigenvalue, and consequently to the AdS Horizon Surface, *more often*. Epsilon evolves therefore *less than* it would under 'normal' conditions and is consequently *larger in magnitude* than epsilon under 'normal' conditions. Our larger locally quantized meter stick measures Information from the past as being lesser in both size {Lp} and time {tp}.

We look again at the diagram: [following page]

## Zeno Dynamic AdS Horizon Surface Quality

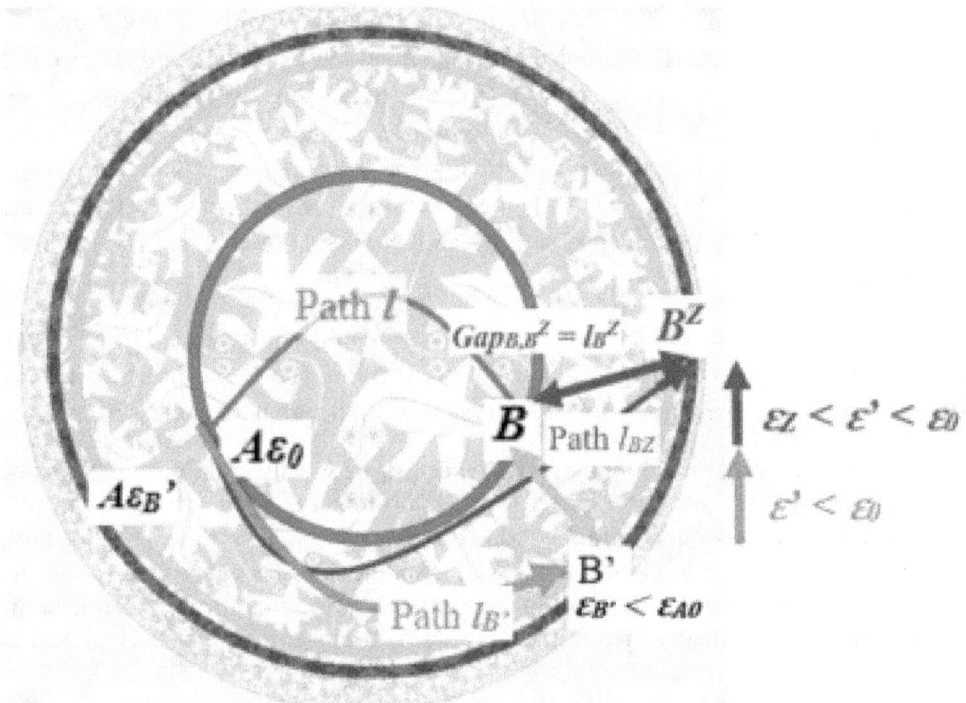

Path $l_B{}^Z$ > Path $l_{B'}$

Path $l_B{}^Z$ > [RT] Path $l$

Alice sends a packet of Information to Bob at time-zero, $A\varepsilon_0$. The Information travels via Path $l$ toward Bob at point $B$. During the Time of Flight, the value epsilon diminishes. The result of rapid observation is that the Horizon value epsilon is *more evolved* than epsilon-prime [$\varepsilon'$]. The key concept is that {B', $A\varepsilon_{B'}$}* is the *normal-time evolution* of the Horizon Surface as the normal light-distance between Alice and Bob [in green]. This is not however, the red arc of convention Ryu-Takayanagi Path $l$. [*$A_{\varepsilon B'}$ is the time coordinate for Alice at the same static value of the green arc for Bob at $B'$.]

Recall the postulate that only entangled Information of entangled systems can travel the red arc of convention, Ryu-Takayanagi Path $l$ [not-prime].

In terms of a Locally Quantized Meterstick, Bob's meter stick is more evolved in terms of epsilon, which means his meterstick is shorter than it would be at time $\varepsilon_0$. Similar to what we regard in Relativistics, his shorter local meterstick means he will use more unitary intervals on his meterstick to measure the package of Information sent from Alice. However, at point $B_Z$ [Zeno] his locally quantized meterstick is shorter than at $B'$. The value $\varepsilon^Z$ is of lesser magnitude than at $\varepsilon'$.

For example, the package of Information Alice sent is 10-seconds long and the Zeno Effect is a factor of 1.2. This means Bob will measure 12 seconds of Information. In order for this

Zeno Dynamic AdS Horizon Surface Quality

to be the case, we cannot imply that the Information has changed during its time of flight. That would require *physically changing the past*, the Information is in its original Conformal state. It must be that Bob's meterstick has decreased in magnitude from what he would normally measure at point B'. Epsilon being more evolved accounts for this change in Bob's Conformal value.

The Zeno Effect is the horizon value epsilon is more evolved at such time Bob takes the measurement than under normal light distance conditions. Thus:

- $\varepsilon'$ is less than $\varepsilon_0$, always.
- **$\varepsilon_z$ is $< \varepsilon' < \varepsilon_0$**

Epsilon-prime $\varepsilon'$ is always of less Conformal Scale than epsilon-naught $\varepsilon_0$. And note that in some cases I write these out because I am not confident the superscripts and subscripts are clearly visible. This is by convention with respect to the Ryu-Takayanagi description of the AdS Horizon Surface.

The evolution of the AdS system as described by Ryu-Takayanagi is based on this premise, that over time the value epsilon is diminishing, this is by convention. The only novel condition that is being described here is that the evolution of epsilon, normally regarded only over cosmological time scales occurs at a scope and rate defined by $c \equiv 1L_p/1t_p$.

For example, if Alice sends a signal to Bob that takes one year to reach him, we can agree that the cosmos has changed over that amount of time, regardless of the magnitude of that change. This is true all the way down to the smallest possible unitary distance, it is merely a matter of defining how much this change in magnitude is.

Bob's local meterstick cannot possibly be the same Conformal Value at the time he receives the package of Information, that would demand that epsilon is static, not dynamic. Here, we are simply reasoning out how much change in magnitude Bob's meterstick undergoes under what conditions.

Understand that the description of the 'Gap' is independent of the description of the Zeno Effect. Under standard measurement conditions the change in Conformal Scale is readily described by the conventions of redshift and so on. That is, there is insufficient distance between Earth and Moon to attribute any macroscopic effect on or between systems. The Zeno Effect is a description where Bob uniquely 'forces' Information to the Horizon surface by observation, which is any electromagnetic interaction.

- At time zero Alice is at the point indicated as $A\varepsilon_0$ and Bob is at point B. The normal light distance between them is the red arc, Path $l$, is 100-light-seconds in length.
- Bob observes the Information Alice sends at a rate sufficient to produce a Zeno Effect of factor 1.2.
- The Information travels from Alice at $A\varepsilon_0$ toward Bob and *120-seconds later* Bob detects the first bit of the Information. At this time Bob is now at position $B^Z$ because the Horizon Surface has evolved at a rate and scope of $c = 1L_p/1t_p$.

Zeno Dynamic AdS Horizon Surface Quality

- This 20-second increase in time *is associated with but not the cause* of the state of epsilon evolving more at point epsilon-Zeno [$\varepsilon_Z$] than it would be at point $\varepsilon_0$.

The most difficult thing to grasp is that the Zeno Effect, measurement at phenomenal rates, is *not detecting events, as the mechanism that drives the Zeno dynamics of the system, as they occur on the AdS Horizon Surface.*

The Horizon Value epsilon is more evolved at point $\varepsilon_Z$ than at $\varepsilon_0$.

- Epsilon is $\varepsilon^Z$ at point $B^Z$ and is of lesser value than $\varepsilon'$ which is of lesser value than $\varepsilon_0$.

$\varepsilon'$ is less than $\varepsilon_0$, always.

$\varepsilon_Z$ is $< \varepsilon' < \varepsilon_0$

  - Thus, the arc takes the form of the purple arc. The red arc must be regarded as fluid, taking the form of the purple arc.
  - This difference between Bob at point $B$ and at $B^Z$ results in a Gap [in purple].
  - The Gap [in purple] is *greater in magnitude* than the red arc, Path $l$.
  - The Gap [in purple] is *greater in magnitude* than the green arc, Path $l'$. *[prime]*

The smaller Conformal Magnitude of $\varepsilon_Z$ demands using more unitary intervals to take the measurement than at $\varepsilon_0$. For example, if the magnitude of my local meterstick 'shrunk' to 1/2x, I would need twice as many meters to measure some other system.

This might seem counter intuitive. The more rapidly Bob measures the Information the more rapidly one would assume the Information re-emerges, for lack of a better term, from the region figuratively 'below' the Horizon Surface. In this case, which is the intuitive, not the Zeno case, one would expect the horizon value epsilon to be less evolved, not more evolved than acquisition at a standard rate, as measured between Bob's unitary intervals of measure.

Intuitively we wouldn't expect any change in temporal rate of progression, e.g., no Zeno Effect is intuitive.

- Bob's value $\varepsilon_{Bob}{}^Z$ is $\varepsilon_{Alice\text{-}naught}$/Path $l_B{}^Z$, [ $A\varepsilon_0$/Path$l_B{}^Z$ ] Not the normal light distance between Alice and Bob because the Horizon value epsilon has evolved a factor of c = 1Lp/1tp since Alice sent the Information at time zero. **Thus,** $\varepsilon_B{}^Z < \varepsilon' < \varepsilon_0$
- The Gap$_{B,B}{}^Z$ is greater in magnitude than the normal light distance between Alice and Bob, Path $l$, and equal in magnitude to **Path $l_B{}^Z$**
- His Locally Quantized Meter Stick as epsilon is thus $\varepsilon_B{}^Z$ /Path $l_B{}^Z$, NOT the normal light distance between Alice and Bob, but greater in magnitude than the normal light distance between Alice and Bob at time zero. However, as indicated, Alice is also on a different position on the Horizon Surface than she was at time zero.

Zeno Dynamic AdS Horizon Surface Quality

- He will require 12 of his unitary intervals [seconds] $\varepsilon_B{}^Z$ to read the 10-second packet of Information sent from Alice.
- His placement as the normal light distance **Path $l_B{}^Z$** is equal to **Path $l_B{}^Z$**, as measured by his [Bob's] <u>Locally Quantized Meter Stick</u>, which is shorter than Alice Locally Quantized Meter Stick by a factor of c = 1Lp/1tp, for their respective positions on the Horizon Surface.
- The *quantity of unitary intervals* has not changed in this, only *the magnitude of each unitary interval* as a Conformally Scale Invariant value, described by the self-similar relationships of the Fractal Set.
    - A slower progression of unitary time also dictates that epsilon has diminished *more* than $\varepsilon_{B,B'}$ because more time has passed in terms of Time of Flight. *The velocity* of Information transfer has not changed, only the amount of Time of Flight has changed. Consequently, the *Zeno arc is of greater value* [designated as Path $l_B{}^Z$] than the 'normal arc' Path $l_{B'}$: *As Measured by Bob at positions B' and $B_Z$.*

The causal anomalies that can result from this Zeno $Gap_{B,B}{}^Z$ have never been characterized, nor have they been discussed in prior art. There will be a discussion of this in another section of this text.

The second demand is that there is a *real change* of length between Alice and Bob, the distance has increased, because Bob's local meterstick is of lesser Conformal Value than it would be under non-Zeno conditions.

## Zeno-Effect on the AdS Horizon Surface

## Zeno-Effect on the AdS Horizon Surface

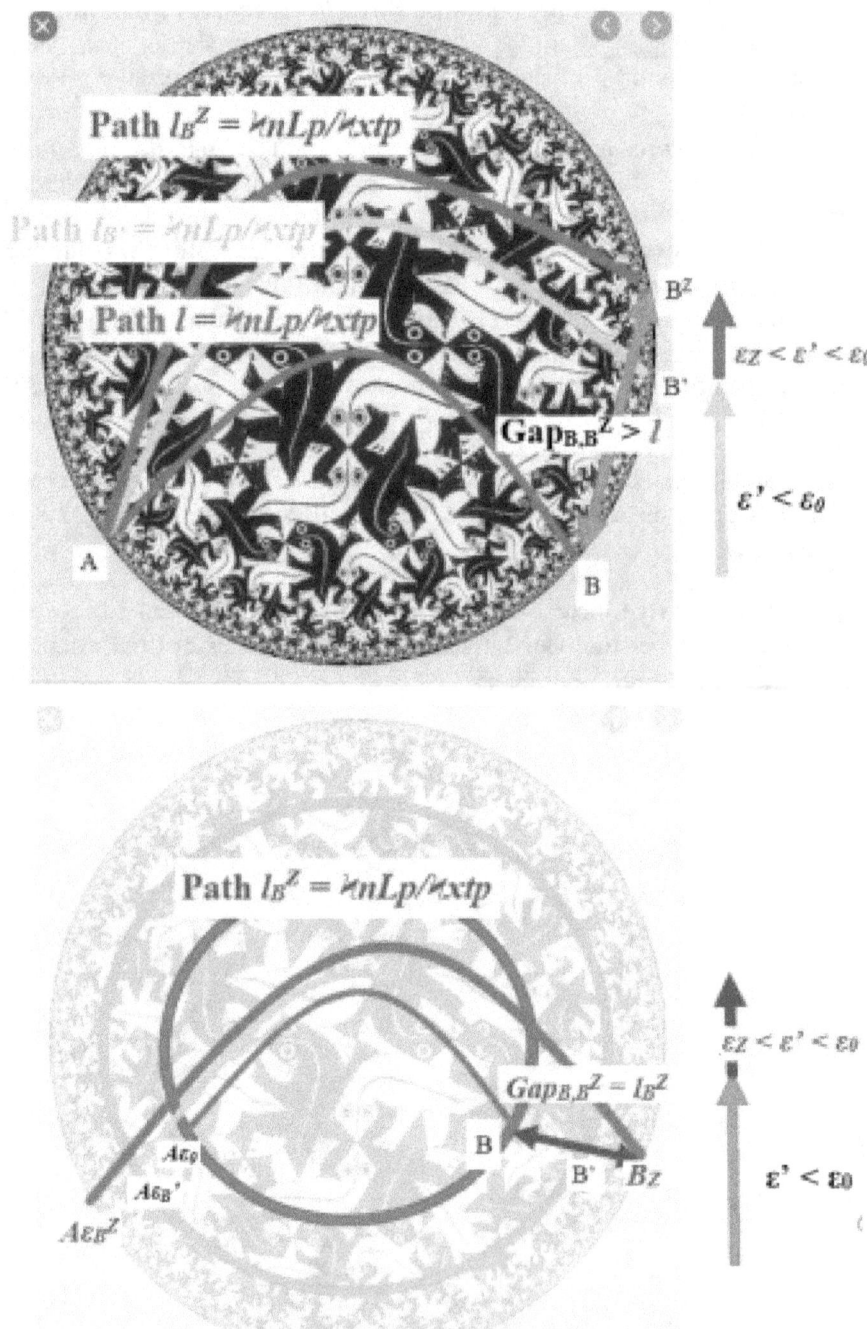

- At time zero Alice is at the point indicated as $A\varepsilon_0$ and Bob is at point B. The normal light distance between them is the red arc, Path $l$, is 100-light-seconds in length.
- Bob observes the Information Alice sends at a rate sufficient to produce a Zeno Effect of factor 1.2

## Zeno-Effect on the AdS Horizon Surface

- The Information travels from Alice at $A\varepsilon_0$ toward Bob and *120-seconds later* Bob detects the first bit of the Information. At this time Bob is now at position $B^Z$ because the Horizon Surface has evolved at a rate and scope of $c = 1Lp/1tp$. **Epsilon is $\varepsilon^Z$ at point $B^Z$ and is of lesser value than $\varepsilon'$ which is of lesser value than $\varepsilon_0$.**
- Thus, the arc takes the form of the purple arc. The red arc must be regarded as fluid, taking the form of the purple arc.
- This difference between Bob at point B and at $B^Z$ results in a Gap [in purple].
- The Gap [in purple] is *greater in magnitude* than the red arc, Path $l$.
- Bob's value $\varepsilon_{Bob}{}^Z$ is $\varepsilon_{Alice\text{-}naught}$ /Path $l_B{}^Z$, [ $A\varepsilon_0$/Path$l_B{}^Z$ ] NOT the normal light distance between Alice and Bob because the Horizon value epsilon has evolved a factor of $c = 1Lp/1tp$ since Alice sent the Information at time zero. Thus, $\varepsilon_B{}^Z < \varepsilon' < \varepsilon_0$
- The Gap$_{B,B}{}^Z$ is greater in magnitude than the normal light distance between Alice and Bob, Path $l$, and equal in magnitude to **Path $l_B{}^Z$**
- His Locally Quantized Meter Stick as epsilon is thus $\varepsilon_B{}^Z$ /Path $l_B{}^Z$, NOT the normal light distance between Alice and Bob, but greater in magnitude than the normal light distance between Alice and Bob at time zero. However, as indicated, Alice is also on a different position on the Horizon Surface than she was at time zero.
- He will require 12 of his unitary intervals [seconds] $\varepsilon_B{}^Z$ to read the 10-second packet of Information sent from Alice.
- His placement as the normal light distance **Path $l_B{}^Z$** is equal to **Path $l_B{}^Z$**, as measured by his [Bob's] Locally Quantized Meter Stick, which is shorter than Alice Locally Quantized Meter Stick by a factor of $c = 1Lp/1tp$, for their respective positions on the Horizon Surface.
- The *quantity of unitary intervals* has not changed in this, only the magnitude of each unitary interval as a Conformally Scale Invariant value, described by the self-similar relationships of the Fractal Set.
- A slower progression of unitary time also dictates that epsilon has diminished *more* than $\varepsilon_{B,B'}$ because more time has passed in terms of Time of Flight. *The velocity* of Information transfer has not changed, only the amount of Time of Flight has changed. Consequently, the *Zeno arc is of greater value* [designated as Path $l_B{}^Z$] than the 'normal arc' Path $l_{B'}$. *As Measured by Bob at positions $B'$ and $B_Z$.*

The causal anomalies that can result from this Zeno Gap$_{B,B}{}^Z$ have never been characterized, nor have they been discussed in prior art. There will be a discussion of this in another section of this text.

## Anti-Zeno Effect Dynamics of the AdS Horizon Surface

Again, it is important to consider that the region figuratively below the horizon cannot change physical geometry as this would then demand physically changing the past. Meaning, the geometry of the region figuratively below the horizon does not actually exist, but is merely a record, if you will, of past states of the horizon. Any physical change to the geometry figuratively below the horizon then would require physically changing the past.

What I am trying to get across here is related to my prior statement:

> Normal detection I am regarding in terms of not being a Zeno Effect. Under 'normal' conditions, Bob is *forcing* the Information to evolve from an eigenstate to an eigenvalue. There is no empirical evidence otherwise. An eigenvalue can only exist *on* the AdS Horizon Surface. Thus, what Bob is doing is throwing coinflips of the sort:

$$<detect|no\text{-}detect>$$

> Again, the Anti-Zeno Effect will be described in terms of *how often Bob does detect an electromagnetic phenomenon.*

The Zeno Effect I am defining as a macroscopic phenomenon where the detector, Bob, is coupled to the system being observed. The Zeno Effect is then Bob, measuring the system described under normal detection rates as:

$$<detect|no\text{-}detect>$$

However, the 'detection' is an instantaneous thing where the electron is excited and makes the quantum jump to another orbital. In this case, by *scanning* my PN junction at some phenomenal rate, what Bob is doing is changing the ratio of <detect|no-detect> at 50/50 to some phenomenal value where the condition:

$$<detect|\psi>$$

The condition *detect* dominates over that of *non-detect*. This may seem counter intuitive when we regard the Anti-Zeno Effect as a reduction of detection rate. However, this is not a detection rate issue, it is a scanning rate of a static PN junction. Again, the PN junction consists of orbitals that change instantaneously upon impacting of an electromagnetic boson, there is no sequitur change in PN junction rate. We are scanning or interrogating the PN junction, only, as the change in 'observation rate.'

*If you consider that perhaps 1-thousand events per second are striking the surface of the PN junction:* We have a carbon-14 source whose specific activity, distance from the PN junction and the PN's cross section to the carbon-14 source means exactly 1000 [beta] decays will strike the surface of the PN.

Anti-Zeno Effect Dynamics of the AdS Horizon Surface

**'Normal' Detection rate:**

A non-phenomenal scanning rate of the PN junction would be 1KHz, 1000-scans of the PN junction per second. The possibilities are:

$$<Detect|no\text{-}detect>$$

Setting efficiency aside momentarily, we consider that all of the scans, 1000 out of 1000, will yield a positive result.

There is no eigenstate:

$$<Detect|no\text{-}detect>$$

There are only eigenvalues.

*All of these numbers are fictitious, purely for argument's sake.*

**Zeno Effect:**

We are scanning or interrogating the PN junction 1-billion times per second. Only one in one million scans will yield a positive result. The vast majority of the scans [999,999 scans] yield *<no-detect>*. Thus, we are actually capturing far less information, the vast majority of the state is *no-detection*.

**Anti-Zeno Effect:**

We are scanning or interrogating the PN junction at 500-times per second. Half of the interrogations yield a positive result. This is obviously not an actual rate that would suffice to produce an Anti-Zeno Effect; however, this is merely a demo-argument. I need to keep things as straightforward as possible, thus we will act as if that 500Hz is a pulpable Anti-Zeno rate, for argument's sake.

We are actually capturing information at a *greater rate than normal scans* rates at 1000-times per second. However, there are no eigenstates in this, only eigenvalues. All Information will travel via the postulated Path *l'*, [*l*-prime] and all of the ancillary arguments hold true.

We are capturing vastly more information than our Zeno rates at 1-billion times per second. Meaning, half of our interrogations via the Anti-Zeno Effect yield a positive result, compared to the Zeno rate, where only one in a million scans yielded a positive result.

Effectively, we are actually missing real events. This could be regarded as true given the efficiency of a typical system. However, in this case it is not a random loss of Information but a systematic one.

This is the static nature of the PN junction, which holds true for electron multipliers and even photomultiplier tubes. The detectors themselves do not typically have on-off switching. Researchers are changing the rate at which they scan or interrogate the PN junction, rather they are aware of this or not.

Anti-Zeno Effect Dynamics of the AdS Horizon Surface

Keep in mind this is worded very similarly to the previous description of the above eigenvalue for the Zeno Effect. However, now we are detecting as the dominant characteristic of the eigenstate *being forced to* evolve to the eigenvalue. As a result, the evolution of the AdS Horizon value epsilon is lesser under Anti-Zeno Dynamic conditions than epsilon is under 'normal' detection rates.

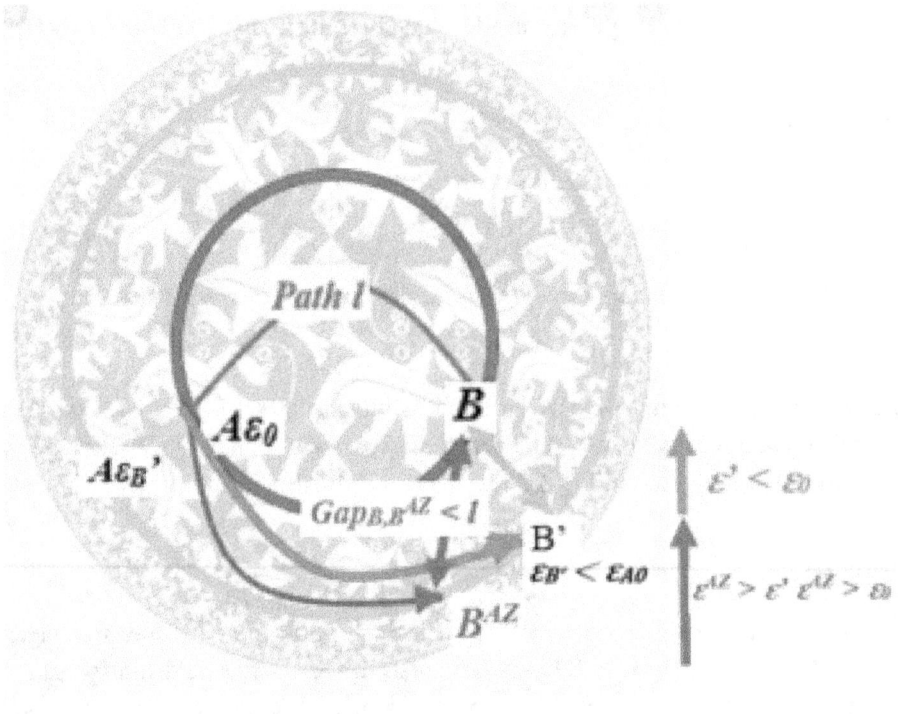

- $\varepsilon'$ is less than $\varepsilon_0$, always. That is standard.
- $\varepsilon_{AZ}$ is $> \varepsilon' < \varepsilon_0$ The Anti-Zeno horizon value epsilon has evolved *less than* the 'normal' horizon value epsilon.
- The Gap for $B$ to $B^{AZ}$ is *less than* Path $l$ [not $l$-prime]. Less time has passed according to Bob's locally quantized meterstick, which is fixed with his locally quantized clock.
- The *Path* for $l_B{}^{AZ}$ is however, *greater than* Path $l$ [not-prime]. Bob's locally quantized meterstick is larger than it would be at $B'$, he thus measures a *shorter distance coupled with shorter time of flight* with his larger meterstick.
- Furthermore, the *Path* for $l_B{}^{AZ}$ is *less than* Path $l_{B'}$. This is the key factor.
- $\varepsilon^{AZ} > \varepsilon'$ The horizon value epsilon under Anti-Zeno conditions has evolved less than epsilon under 'normal' conditions. As equated to the unitary Planck intervals of length {Lp} and time {tp} the unitary intervals are *larger in magnitude than* they would be under 'normal' conditions.

Anti-Zeno Effect Dynamics of the AdS Horizon Surface

- $\varepsilon^{AZ} > \varepsilon_0$) The horizon value epsilon under Anti-Zeno conditions has evolved more than epsilon as it was at time-zero. The unitary intervals are therefore *les in magnitude than* they were at time-zero.
- Bob is now at point $B^{AZ}$. His distance from his original position is measured orthogonally but not orthonormally to his prior position, now on a later evolution of the AdS Horizon Surface.

[Note again that I highlight the 'primes' because of the penchant to read these things on pads and even phones].

The lower rate of scanning the PN junction *misses events*. This however is not a mechanical thing with a mechanistic result. A mechanistic thing would include, for instance, the efficiency, which is a random loss of information, not a systematic one.

Bob is *forcing* Information to the AdS Horizon *more often* under these conditions by *forcing* the Information from an eigenstate to an eigenvalue, *more often*. Consequently, the horizon value epsilon evolves *less than* that of 'normal' detection and is therefore of greater magnitude. His Locally Quantized Meterstick is thus *greater than* it would be under 'normal' conditions,' he thus measures time and distance as being *lesser than* those unitary Planck intervals from the past.

Because Bob's Locally quantized Meterstick is *greater than* it would be under normal conditions, he measures events figuratively 'emerging' from figuratively 'below' the horizon as *smaller in magnitude*. That would be specifically, the unitary Planck intervals of length {Lp} and time {tp}.

I think this is a condition that is confounding to Zeno Dynamic experiments, wherein researchers cannot reproduce their own results, and no one has ever reproduced another researcher's results. This is also the confounding factor that prohibits researchers from discussing their experimental setups in enough detail to reproduce the experiments. Meaning, no one has quantitatively understood nor published any reconciliation with the expected number of decays per second under 'normal' observation conditions with those of Zeno and Anti-Zeno Dynamic conditions, vs. the Zeno or Anti-Zeno yielded number of decays per second.

You will note that at least one, if not both axes in a Zeno dynamic paper are normalized to some arbitrary value. This is non-quantitative. If we have carbon-14 as a source and our proximity and cross section to the detector should see 1000-decays per second, we should be able to report that, but no one has to date. Meaning, there is not a single peer reviewed paper on the Zeno Effect where the expected number of decays [given a radioactive source] or otherwise events per second is stated.

If our Zeno Dynamic yields perhaps 500-decays per second, no one has reported that, and our Anti-Zeno Dynamic greater than 1000-decays per second, also not reported quantitatively. In general, we see a plot with at least one axis normalized to some arbitrary units, and faster means more, with no possible way to back calculate if anything happened at all in the experiment or to what degree.

Anti-Zeno Effect Dynamics of the AdS Horizon Surface

Again, the notion that we are 'detecting' more or less rapidly is incorrect. The PN junction is merely a static thing that has no 'knob' on it to tell it to detect events more or less often. What we are doing is scanning or interrogating the PN junction more or less often.

- When we scan the PN junction *more often* [Zeno], we mostly yield negative results, or positive results *less often*. We are forcing Information from an eigenstate to an eigenvalue, and consequently to the AdS Horizon Surface, *less often*. Epsilon evolves therefore *more than* it would under 'normal' conditions and is consequently *smaller in magnitude* than epsilon under 'normal' conditions. Our smaller locally quantized meter stick measures Information from the past as being larger in both size {Lp} and time {tp}.
- When we scan the PN junction *less often*, we mostly yield positive results, or negative results *less often*. We are forcing Information from an eigenstate to an eigenvalue, and consequently to the AdS Horizon Surface, *more often*. Epsilon evolves therefore *less than* it would under 'normal' conditions and is consequently *larger in magnitude* than epsilon under 'normal' conditions. Our larger locally quantized meter stick measures Information from the past as being lesser in both size {Lp} and time {tp}.

This is every bit as painful as Special Relativity *but follows* the same exact set of rules in terms of what we perceive of another system that is moving at relativistic velocity.

The Zeno Effect does not change a thing that occurred in the past, namely, a radioactive emission of a beta particle that occurred moments ago. The Zeno Effect changes your Locally Quantized Meterstick, with the same exact result as it would under Lorentzian or Schwarzschild conditions. I am presenting the Zeno Effect as a phenomenon that has a real utility, namely, the artificial alteration of the progression of time on a benchtop scale is thus a macroscopic phenomenon that results in a real change in the geometry of space. The Zeno Effect, if properly investigated, would yield the artificial alteration of the geometry of spacetime, *void the use of any energy or mass-energy equivalent*. This was defined and derived at length in:

1. Artificial Alteration of Spatial Geometry via the Quantum Zeno Effect. December 2020 DOI: 10.13140/RG.2.2.13527.29601, DOI: 10.13140/RG.2.2.13527.29601/1 [updated file]
2. Faster Than Light Propulsion via the Zeno Effect. March 2021 DOI: 10.13140/RG.2.2.28575.48809

And again, this is not a mechanistic re-explanation of the Zeno Effect. It is the 'upside-down' explanation of yielded results, wherein we force Information from an eigenstate to an eigenvalue more or less often. And this equates to where and when Information thus emerges to the AdS Horizon surface.

Epsilon-prime is the result of the time of flight for the Information to traverse the normal light distance, which is Ryu-Takayanagi Path *l*. Epsilon-naught is the convention wherein we regard the path as rigid, but cannot be the case, because Bob is at any distance regardless of how small, from Alice.

## Anti-Zeno Effect Dynamics of the AdS Horizon Surface

Epsilon-AZ is the result of the Zeno dynamic, $<detect|\psi>$, with the condition *detect* dominating, thus the AdS Horizon value epsilon has evolved less than if the Information merely traversed the normal light distance of the rigid red arc of convention, Ryu-Takayanagi Path $l$.

What needs to be pointed out here is that the QAZE as depicted here [following page] shows that the Horizon Surface has evolved from time zero to time-$\{B^{AZ}, A\varepsilon^{AZ}\}$. However, as depicted, $\{B^{AZ}, A\varepsilon^{AZ}\}$ is *less evolved* than $\{B', A\varepsilon_{B'}\}$. The key concept is that $\{B', A\varepsilon_{B'}\}$ is the normal-time evolution of the Horizon Surface as the normal light-distance between Alice and Bob. A less evolved value epsilon is equated with a unitary Planck value of greater magnitude.

*All Paths are greater in magnitude than the conventional Ryu-Takayanagi Path $l$*, the conventional Ryu-Takayanagi Path that connects Alice *at time-zero* with Bob, who is also at time-zero. As stated, this will only apply to entangled systems.

$$\textbf{Paths } l_{B'}\ l_B{}^Z\ l_B{}^{AZ} > l$$

*Ryu-Takayanagi Path $l$ connects two static points, Alice and Bob, both at time zero, time has not progressed, is not dynamic, and is the rationale for only entangled systems traversing this path. Preferential Frame of Reference does not apply; what is true for Alice is true for Bob,* **true in all systems.**

*All non-entangled systems traverse* Paths $l_B \cdot l_B{}^Z\ l_B{}^{AZ}$, *time has progressed and is dynamic. All Paths $l_B \cdot l_B{}^Z\ l_B{}^{AZ}$ are greater than Path $l$. Preferential Frame of Reference applies. What is true for Alice* **may not be true** *for Bob [all systems].*

Anti-Zeno Effect Dynamics of the AdS Horizon Surface

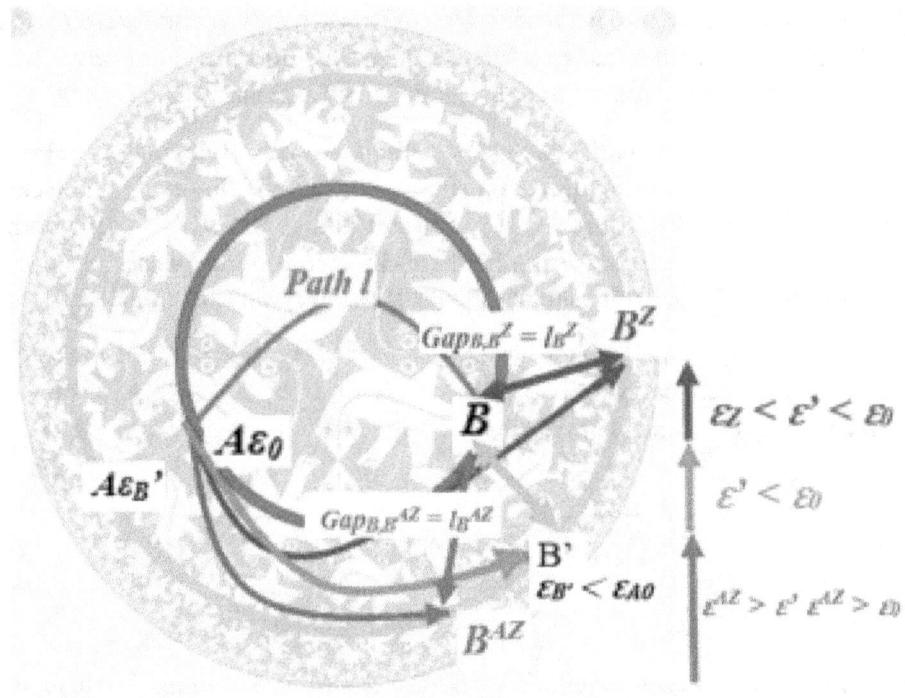

Path $l_B{}^{AZ}$ < Path $l_{B'}$   Path $l_B{}^Z$ > Path $l_{B'}$

Path $l_B{}^{AZ}$ > [RT] Path $l$   Path $l_B{}^Z$ > [RT] Path $l$

All values $epsilon_x{}^X$ are greater in magnitude than $A\varepsilon_0$: [which is at time-zero]

$$A\varepsilon_B \cdot A\varepsilon_B{}^Z \; A\varepsilon_B{}^{AZ} > A\varepsilon_0$$

All values $B_x{}^x$ are greater in magnitude than $B$ [which is also at time-zero]

$$B' \; B^Z \; B^{AZ} > B$$

All values $epsilon_x{}^x$ are greater in magnitude than $\varepsilon_0$:

$$\varepsilon' \; \varepsilon_Z \; \varepsilon_{AZ} > \varepsilon_0$$

This is because Path $l$, $A\varepsilon_0$, B, $\varepsilon_0$ all represent the conditions at time zero. The only values of lesser magnitude must therefore be before time-zero, which is a causal violation of the simplest sort. Recall that in this text there are no causal violations on any scale greater than 2-native wave functions. The very basic premise is that the Horizon is evolving at a scope and rate greater than convention has regarded it up to this point. However, as stated, the scope and rate are defined by $c \equiv 1Lp/1tp$. The value epsilon equated with $c^2$ puts things in a tighter perspective, these numeric examples make it clearer regarding the scope and rate. The self-similar, fractal relationships define the Conformal Scale Invariance of the system and is a valid description of how the Conformal Scale Invariance occurs at all.

## Anti-Zeno Effect Dynamics of the AdS Horizon Surface

All three scenarios, normal, Zeno, and anti-Zeno, are in terms 'as measured by Bob,' at his respective positions.

There is another issue that has to be clarified here. Given Path $l$ is the normal light distance between points, it might seem that the transfer of Information from Alice at time-zero to Bob at point $B^{AZ}$, taking only 8-seconds of time of flight would be a superluminal violation. However, this is not unlike 'length contraction,' as convention regards the distance between points as undergoing Lorentzian transformation. This [above description] is a Schwarzschild transformation; however, we can use a Lorentzian example.

It is Bob's Bob Value epsilon at point $B^{AZ}$, where the Horizon has evolved to at that moment, that defines the amount of Information in the package from Alice. It is the magnitude of the Gap that defines this 'Bob' Value. The Gap is always equal in magnitude to the apparent distance, in this case, Path $l_B{}^{AZ}$, which is of lesser magnitude than Path $l_B$, the normal light distance between points Alice and Bob *at time-ε'*.

*The normal light-distance between points Alice and Bob at time-zero is Conformally equal to the normal light distance between Alice and Bob at time-ε', according to the unitary Planck interval, which is $Lp'/tp' = \varepsilon'$, at time-ε'.*

$$\natural\varepsilon' \equiv \frac{Lp'^2}{tp'^2} = \natural A'_\Omega = \natural N'_Q = c^2 \equiv \natural_{path} L'$$

Where Path L' is defined as:

$$\varepsilon' = \frac{L'}{e^{\frac{3l}{e}}} = \frac{L'}{e^{\ln\left(\frac{L}{\varepsilon}\right)}}$$

$$\natural\varepsilon' \equiv \frac{L'}{e^{\frac{3l}{e}}} \equiv \natural_{path} L' = \frac{Lp'^2}{tp'^2} = \natural A'_\Omega = \natural N'_Q = c^2$$

$$\varphi(\pm L'_p) \rightleftharpoons |\sqrt{\frac{hG'}{2\pi c^3}}|; \quad \varphi(\pm tp') \rightleftharpoons |\sqrt{\frac{hG'}{2\pi c^5}}|$$

Anti-Zeno Effect Dynamics of the AdS Horizon Surface

$$\natural\varepsilon \rightleftharpoons \frac{Lp^2}{tp^2} \rightleftharpoons \natural A_\Omega \rightleftharpoons \natural N_Q \rightleftharpoons c^2 \rightleftharpoons \natural path L$$

$$\natural\varepsilon' \rightleftharpoons \frac{L'}{e^{\frac{3l}{e}}} \rightleftharpoons \natural path L' \rightleftharpoons \frac{Lp'^2}{tp'^2} \rightleftharpoons \natural A'_\Omega \rightleftharpoons \natural N'_Q \rightleftharpoons c^2$$

Note that the above are fractal relationships and cannot be manipulated like an algebraic expression. I also described this at length in the above noted papers. Briefly, there is no possible way to look at a value for $Z' \rightleftharpoons Z_a^2 + C$; looking at any value $Z'$ will never yield what iteration [not perturbation] of $Z_a$ under any circumstances.

If we look at a time dilation scenario, using a Lorentz dilation as an example, what Bob observes if Alice is moving away from him at [what works out to] ~0.553c, yielding a Lorentzian t-prime of 1.2. This is a scenario where Preferential Frame of Reference applies; it is a Lorentzian system. The altered temporal system is Alice, Bob is unyielding. In a Zeno system, both Alice and Bob are changing, it is the Information whose relative Conformal Scale remains constant.

However, Bob's Locally Quantized Meter Stick, as a result of the evolution of epsilon, varies in proportion according to the rate of observation [Zeno Effect] that he observes the Information. That is, it is Bob who *forces the Information* to 're-emerge' at the Horizon Surface by detection, which can be any electromagnetic interaction.

*It is Bob who decides when and where the Information 're-emerges' at the Horizon Surface. That is evident in the collective history of quantum mechanics.*

## Anti-Zeno Effect Dynamics of the AdS Horizon Surface

### Anti-Zeno Effect

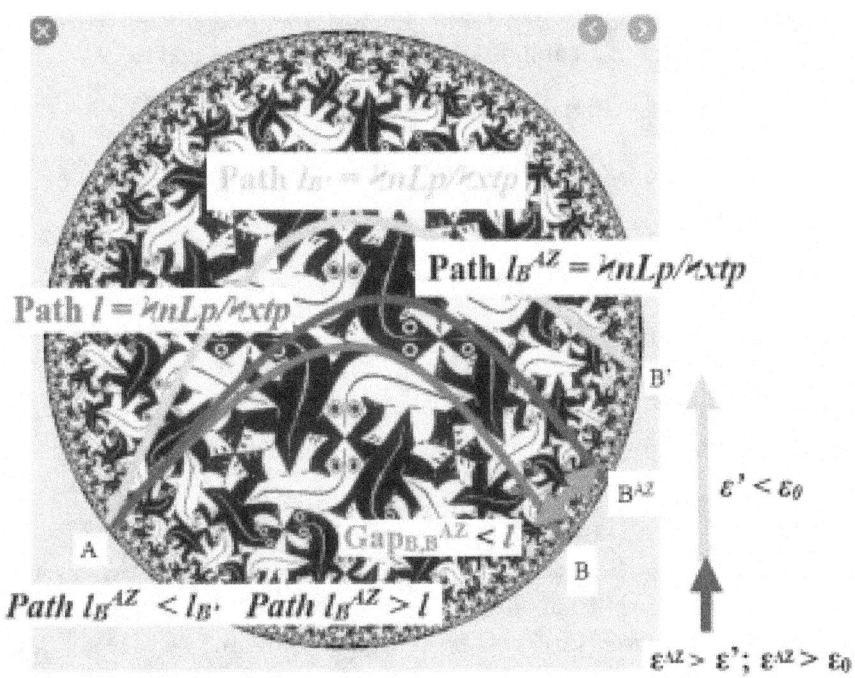

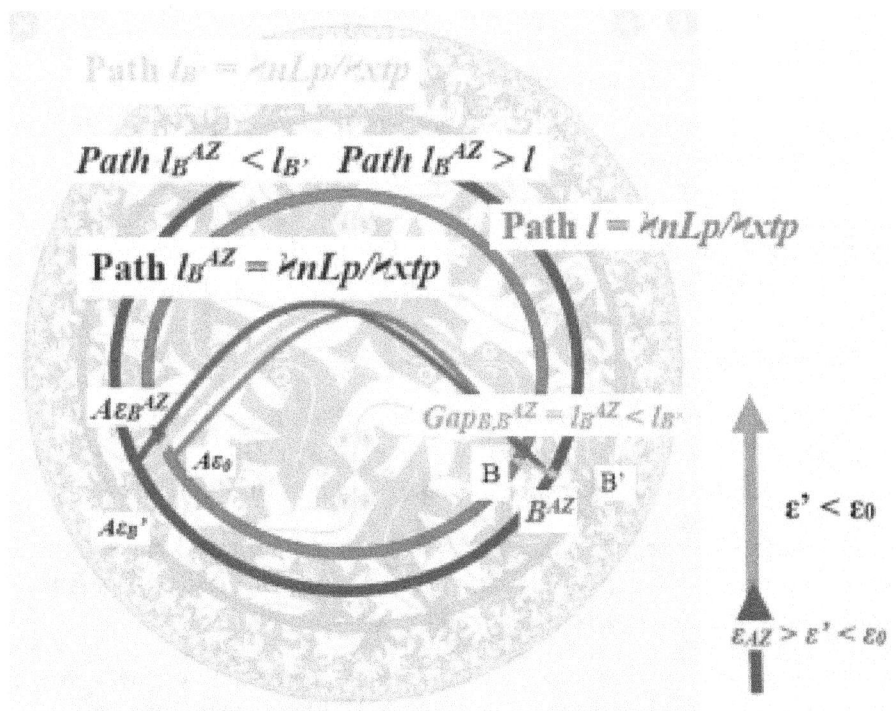

Anti-Zeno Effect Dynamics of the AdS Horizon Surface

What needs to be pointed out here is that the QAZE as depicted here shows that the Horizon Surface has evolved from time zero to time-$\{B^{AZ}, A\varepsilon^{AZ}\}$. However, as depicted, $\{B^{AZ}, A\varepsilon^{AZ}\}$ is *less evolved* than $\{B', A\varepsilon_{B'}\}$. The key concept is that $\{B', A\varepsilon_{B'}\}$ is the normal-time evolution of the Horizon Surface as the normal light-distance between Alice and Bob.

*All Paths are greater in magnitude than l.*

$$\text{Path } l_{B'} \cdot l_B{}^Z \, l_B{}^{AZ} > l$$

*All values epsilon$_x{}^X$ are greater in magnitude than A$\varepsilon$:*

$$A\varepsilon_{B'} \, A\varepsilon_B{}^Z \, A\varepsilon_B{}^{AZ} > A\varepsilon_0$$

*All values $B_x{}^x$ are greater in magnitude than B:*

$$B' \, B^Z \, B^{AZ} > B$$

*All values epsilon$_x{}^x$ are greater in magnitude than $\varepsilon_0$:*

$$\varepsilon' \, \varepsilon_Z \, \varepsilon_{AZ} > \varepsilon_0$$

This is because Path $l$, $A\varepsilon_0$, B, $\varepsilon_0$ all represent the conditions at time zero. The only values of lesser magnitude must therefore be before time-zero, which is a causal violation of the simplest sort. The very basic premise is that the Horizon is evolving at a scope and rate greater than convention has regarded it up to this point. However, as stated, the scope and rate are defined by $c \equiv 1Lp/1tp$. The value epsilon equated with $c^2$ puts things in a tighter perspective, these numeric examples make it more clear regarding the scope and rate. The self-similar, fractal relationships define the Conformal Scale Invariance of the system and is a valid description of how the Conformal Scale Invariance occurs at all. To date there is no agreed upon resolution, other than perhaps attempts to equate these to $H_0$, which is described in this text as 550 data points with a variance of 21% is a definitive null hypothesis.

At time-zero Alice sends [10-seconds of] Information to Bob, who is 100-light-seconds away, as measured by Path $l$, as the fixed, rigid red arc. During that 100-seconds of Time of Flight the Horizon Surface value epsilon evolves at a scope and rate of $c = 1Lp/1tp$ toward point $\{B', A\varepsilon_{B'}\}$. And at that point $\{B', A\varepsilon_{B'}\}$ the value epsilon-prime $[\varepsilon']$ is *smaller* in Conformal Scale than epsilon's starting value $\varepsilon_0$.

However, in the case of the Anti-Zeno Effect epsilon$^{AZ}$ has not evolved as much as epsilon-prime. Thus, the value $\varepsilon^{AZ}$ is *larger in Conformal Scale than $\varepsilon'$*. Epsilon$^{AZ}$, however, is of lesser value [Conformal Scale] than $\varepsilon_0$, because there is some evolution since time-zero.

The 'Gap' [in green] between $\{B, A\varepsilon_0\}$ and $\{B^{AZ}, A\varepsilon^{AZ}\}$ is less than the Gap between $\{B, A\varepsilon_0\}$ and $\{B', A\varepsilon'\}$. The causal anomalies [but not paradoxes] that arise from this take an interesting turn, albeit no one has characterized them to date.

- At time zero Alice is at the point indicated as $A\varepsilon_0$ and Bob is at point B. The normal light distance between them is the red arc, Path $l$, is 100-light-seconds in length.

Anti-Zeno Effect Dynamics of the AdS Horizon Surface

- Bob observes the Information Alice sends at a rate sufficient to produce an Anti-Zeno Effect of factor 0.8
- The Information travels from Alice at A$\varepsilon_0$ toward Bob and *80-seconds later* Bob detects the first bit of the Information. At this time Bob is now at position $B^{AZ}$ because the Horizon Surface has evolved at a rate and scope of c = 1Lp/1tp. Epsilon is $\varepsilon^{AZ}$ **at point $B^{AZ}$ and is of *greater* value than $\varepsilon$'** which is *of lesser value than* $\varepsilon_0$. However, $\varepsilon^{AZ}$ is of lesser value than $\varepsilon^0$. Thus $\varepsilon^{AZ} > \varepsilon' < \varepsilon_0$:

$$\varepsilon^{AZ} > \varepsilon'$$

$$\varepsilon' < \varepsilon_0$$

$$\varepsilon^{AZ} < \varepsilon_0$$

- Thus, the arc takes the form of the blue arc. The red arc must be regarded as fluid, taking the form of the blue arc.
- This difference between Bob at point B and at $B^{AZ}$ results in a Gap [in green].
- The Gap [in green] is *greater in magnitude* than the red arc, Path *l*.
- The Gap [in green] is *less in magnitude* than the yellow arc, Path *l*. The Gap {$B^{AZ}$, B} is less than the Gap between {B' B}.
- Bob's value $\varepsilon_{Bob}{}^{AZ}$ is $\varepsilon_{Alice-naught}$ /Path $l_B{}^{AZ}$, [ A$\varepsilon_0$/Path$l_B{}^{AZ}$ ] NOT the normal light distance between Alice and Bob because the Horizon value epsilon has evolved a factor of c = 1Lp/1tp since Alice sent the Information at time zero. Thus, $\varepsilon_B{}^{AZ} > \varepsilon' < \varepsilon_0$
- The Gap$_{B,B}{}^{AZ}$ is greater in magnitude than the normal light distance between Alice and Bob, Path *l*, and equal in magnitude to **Path $l_B{}^{AZ}$**
- His Locally Quantized Meter Stick as epsilon is thus $\varepsilon_B{}^{AZ}$ /Path $l_B{}^{AZ}$, NOT the normal light distance between Alice and Bob, but greater in magnitude than the normal light distance between Alice and Bob at time zero. However, as indicated, Alice is also on a different position on the Horizon Surface than she was at time zero.
- He will require 8 of his unitary intervals [seconds] $\varepsilon_B{}^{AZ}$ to read the 10-second packet of Information sent from Alice.
- His placement as the normal light distance **Path $l_B{}^{AZ}$** is equal to **Path $l_B{}^{AZ}$**, as measured by his [Bob's] Locally Quantized Meter Stick, which is shorter than Alice Locally Quantized Meter Stick by a factor of c = 1Lp/1tp, for their respective positions on the Horizon Surface.
- The *quantity of unitary intervals* has not changed in this, only the magnitude of each unitary interval as a Conformally Scale Invariant value, described by the self-similar relationships of the Fractal Set.
- A more rapid progression of unitary time also dictates that epsilon has diminished *less* than $\varepsilon_{B,B'}$ because more less time has passed in terms of Time of Flight. *The velocity* of Information transfer has not changed, only the amount of Time of Flight has changed. Consequently, the *Anti-Zeno arc is of lesser value* [designated as Path $l_B{}^{AZ}$] than the 'normal arc' Path $l_B$: *As Measured by Bob at positions $B_{AZ}$ and B'.*

All three scenarios, normal, Zeno, and anti-Zeno, are in terms 'as measured by Bob,' at his respective positions.

## Anti-Zeno Effect Dynamics of the AdS Horizon Surface

There is another issue that has to be clarified here. Given Path $l$ is the normal light distance between points, it might seem that the transfer of Information from Alice at time-zero to Bob at point $B^{AZ}$, taking only 8-seconds of time of flight would be a superluminal violation. However, this is not unlike 'length contraction,' as convention regards the distance between points as undergoing Lorentzian transformation. This is a Schwarzschild transformation; however, we can use a Lorentzian example.

It is Bob's Bob Value epsilon at point $B^{AZ}$, where the Horizon has evolved to at that moment, that defines the amount of Information in the package from Alice. It is the magnitude of the Gap that defines this Bob Value. The Gap is always equal in magnitude to the apparent distance, in this case, Path $l_B{}^{AZ}$, which is of lesser magnitude than Path $l_B$, the normal light distance between points Alice and Bob *at time-ε'*.

*The normal light-distance between points Alice and Bob at time-zero is Conformally equal to the normal light distance between Alice and Bob at time-ε', according to the unitary Planck interval, which is $Lp'/tp' = ε'$, at time-ε'.*

$$4ε' \equiv \frac{Lp'^2}{tp'^2} = 4A'_\Omega = 4N'_Q = c^2 \equiv {}_{4path}L'$$

Where Path L' is defined as:

$$ε' = \frac{L'}{e^{\frac{3l}{e}}} = \frac{L'}{e^{\ln\left(\frac{L}{ε}\right)}}$$

$$4ε' \equiv \frac{L'}{e^{\frac{3l}{e}}} \equiv {}_{4path}L' = \frac{Lp'^2}{tp'^2} = 4A'_\Omega = 4N'_Q = c^2$$

$$\varphi(\pm L'_p) \rightleftharpoons \left|\sqrt{\frac{\hbar G'}{2\pi c^3}}\right|;\ \varphi(\pm tp') \rightleftharpoons \left|\sqrt{\frac{\hbar G'}{2\pi c^5}}\right|$$

$$4ε \rightleftharpoons \frac{Lp^2}{tp^2} \rightleftharpoons 4A_\Omega \rightleftharpoons 4N_Q \rightleftharpoons c^2 \rightleftharpoons {}_{4path}L$$

$$4ε' \rightleftharpoons \frac{L'}{e^{\frac{3l}{e}}} \rightleftharpoons {}_{4path}L' \rightleftharpoons \frac{Lp'^2}{tp'^2} \rightleftharpoons 4A'_\Omega \rightleftharpoons 4N'_Q \rightleftharpoons c^2$$

If we look at a time dilation scenario, using a Lorentz dilation as an example, what Bob observes if Alice is moving away from him at [what works out to] ~0.553c, yielding a Lorentzian t-prime

### Anti-Zeno Effect Dynamics of the AdS Horizon Surface

of 1.2. This is a scenario where Preferential Frame of Reference applies; it is a Lorentzian system. The altered temporal system is Alice, Bob is unyielding. In a Zeno system, both Alice and Bob are changing, it is the Information whose relative Conformal Scale remains constant.

However, Bob's Locally Quantized Meter Stick, as a result of the evolution of epsilon, varies in proportion according to the rate of observation [Zeno Effect] that he observes the Information. That is, it is Bob who *forces the Information* to 're-emerge' at the Horizon Surface by detection, which can be any electromagnetic interaction.

*It is Bob who decides when and where the Information 're-emerges' at the Horizon Surface.*

## Describing Temporal Order of Events on the AdS Horizon Surface

## Describing Temporal Order of Events on the AdS Horizon Surface

Here, I make one more adjustment to the classic depiction:

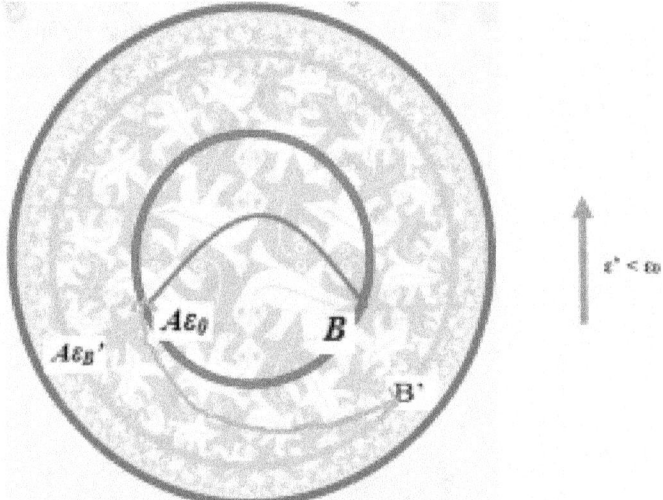

By taking the green arc, Path $l'$, we have taken the least quantity of unitary intervals {Lp, tp} and we have constantly moved toward the future, representing typical causality. Thus, we go back to the visual a few pages back and make the same adjustments:

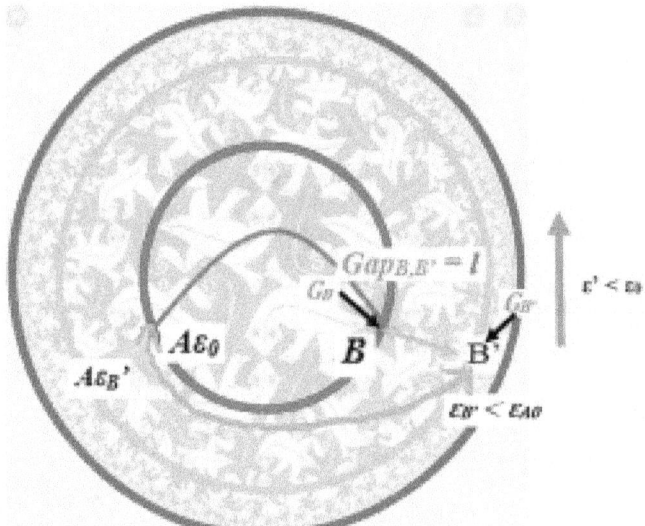

Now, everything makes sense. We are sending Information from Alice at $A\varepsilon_0$ to Bob who *will be at $B'$*. The green arc represents the path information will thus travel, and it is constantly 1) passing through only *decreasing* scales of unitary Planck intervals {Lp, tp} as values epsilon [ε] and 2) we are only sending Information consistently toward the future, which then represents typical causality.

### Describing Temporal Order of Events on the AdS Horizon Surface

Furthermore, it is a bit clearer how the 'Gap' [now designating as *Gp*] represents where Bob *was* as well as now representing where Bob *is*. The 'Gap,' *Gp* is thus equal in scale to the normal light distance between points, as moving causally toward the future, only. This 'Gap,' *Gp*, leads to a variety of seeming causality violations that are in fact directly related to the Lunar Landing Anomaly. However, because the Lunar Landing is regarded as 1) macroscopic and 2) explained off by your high school teacher, literally, as timing delays if explained or noticed at all. In variations of the 2-slit things are regarded more seriously, albeit ontologically.

If you study the diagram closely, you will realize that it is not possible for Bob to *not move*, figuratively, from position $B$ to $B'$. Again, regardless of scope and rate, epsilon is not a static thing, but dynamic, vanishing. The vanishing rate of epsilon to the best of my knowledge has never been dutifully derived or discussed in terms of a rate conjecture. In Quantum Information Dynamics vols I and II, I put forth some summary derivations. Essentially, it all goes back to the quantized definitions, $c \equiv L_p/t_p$, where $L_p$ is the unitary Planck length and $t_p$ is Planck time. We had:

$$\hbar\varepsilon' \equiv \frac{L'}{e^{\ln(\frac{L}{\varepsilon})}} = c^2 = \frac{\hbar L_p'^2}{\hbar t_p'^2} \equiv \hbar N_Q = \hbar A'_\Omega$$

$$\hbar\varepsilon' < \varepsilon_0$$

$$\hbar N_Q = \hbar \frac{L_p^2}{t_p^2} = c^2 \equiv \hbar 1$$

$$\varepsilon \equiv \frac{L_p^2}{t_p^2} = A_\Omega = N_Q = c^2 = E/m$$

The distance in time from Bob at point $B$ to point $B'$ is in fact equal in magnitude to the normal light distance of Path *l*. Points $B$ and $B'$ cannot be placed anywhere else in time. As Bob looks 'backward' from his new position at point $B'$ to where he was, there is in fact a temporal distance that is also equal in magnitude to where Alice was at time zero, $A_\emptyset$. This 'shell' that has figuratively moved away from the epicenter has a seemingly backward causality when we regard where Bob was at time zero and where he is now. He cannot be any other place than the normal light distance, as metered in time, from his prior position at point $B$, now at $B'$.

### Describing Temporal Order of Events on the AdS Horizon Surface

In the Lunar Landing Anomaly, this is how we established Preferential Frame of Reference. Preferential refers to frames of reference not being equal. This inequality is demonstrated when the train crash that occurs on the moon in Armstrong being cut off as he begins to speak, "Houston..[CAPCOM operator now cuts in; *'Ok everybody. T-1, stand by for T-1.'*]....ahh....", which does not occur on the recording as events unfolded on Earth [e.g., no train crash]. All that we hear from the Earth recording is, 'Houston...Ahh.....' with no explanation, nor would there be any if the moon based recording hadn't been made.

If all frames of reference were indeed equal, every series of events that occurred on the moon would occur on Earth, either both trains do or do not crash. What is true on Earth must be true on the moon. But it does not happen this way. Which brings us back to:

*When Preferential Frame of Reference applies, what is true in one system may not be true in another system. Information travels the green arc in the diagram a couple of pages back.*

And,

*For entangled systems, Preferential Frame of Reference does not apply, and what is true in one system is true in all systems, because entangled systems travel the conventional Ryu-Takayanagi Path l, the rigid red arc, from time zero [A&] to time zero [B, not B-prime].*

And just for the record I will reiterate that by 'rigid red arc of convention,' I am referring to:

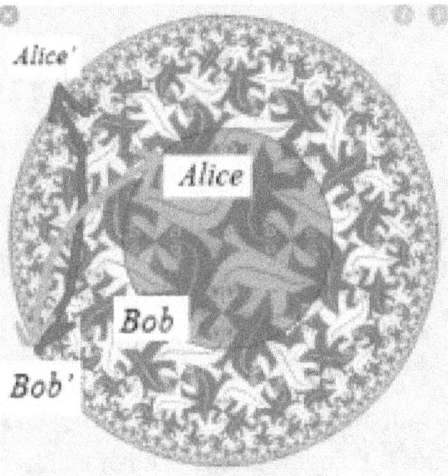

Where the red arc represents typical AdS descriptions of Ryu-Takayanagi Path *l* and the green arc is the postulate. Alice sends Information to Bob, who has moved from his position on the inner circle to the outer circle at *Bob'*. In that time, Alice has also moved from her position on the inner circle to her new position at *Alice'*. In order for Information to traverse the Path *l'* [*l-prime*, not *l*] without violating typical causality, the Information must traverse

### Describing Temporal Order of Events on the AdS Horizon Surface

the path in green, 1) taking the least quantity of unitary intervals [lizards] and 2) moving with typical causality, only moving into the future states of epsilon [$\varepsilon$].

It is a simple solution that I think is rather elegant. We have the dilemma of how Information traverses Ryu-Takayanagi Path *l*, which interestingly applies to entangled systems. In systems that are not entangled, they travel the green arc, whose derivation is later in this text. It is simply not possible for information to travel via Path *l* in non-entangled systems, because the starting and the endpoints both represent time-zero, the moment the Information is sent. If I have somehow not understood this or otherwise have not seen or found it in any of the literature, *that is agreed upon,* forgive me. I really have not seen anyone discuss the topic in this manner before, interestingly. The conventional path of information transfer in the AdS system doesn't seem to be an issue of discussion. It merely seems accepted in some way, albeit again, cannot logically be correct.

Furthermore, the green arc represents Information of the non-entangled sort traversing a path that is continuously moving toward the future, where the rigid red arc of convention has Information travelling first toward the past, and at some midpoint begins moving forward in time again, which is non-sequitur [illogical]. We can envision a system of this sort, but it cannot refer to Alice sending a signal from Philadelphia to Bob in Manhattan, her radio signal travelling toward the greater Conformal values [lizards] epsilon, into the past states of epsilon, meaning the past. Then we have the information at exactly the midpoint, turning temporal direction back toward the present. This in itself demands some sort of 'pre-destiny,' regarded as Predeterminism, Determinism, and so on. That is, it demands that the information knows its target [Bob, Manhattan] at the moment it is sent, in order to 'know' its midpoint in temporal time of flight. And again, the conventional description also demands that it reaches Bob instantaneously, at time-zero.

To make this clearer, we look again at the Bell Loophole-Free Test.

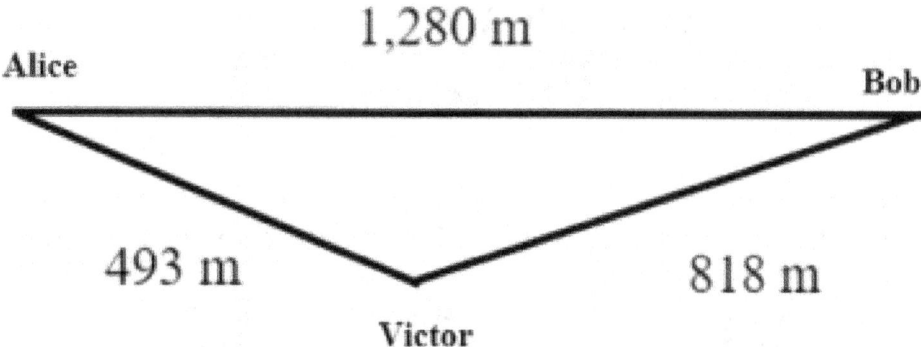

I know that the authors did not figure in the refractive index of 1.5 for the fiber optic cable, which limits the velocity of the photons to 0.6c, because of the first temporal statement:

> *The separation of the spins by 1,280 m defines a 4.27-μs time window during which the local events at A and B are space-like separated from each other (see the space–time diagram in Fig. 2b).*

## Describing Temporal Order of Events on the AdS Horizon Surface

- The authors are setting up a scenario where they are performing their counting in the above example at 4.27 microseconds after emission, however, at 0.6c in the fiber optical cable this should be a later value, 7.116 microseconds. Similar errors were made in all of the three legs of the system. Consequently, there is no back calculation to figure what they were counting as events.
- If photons are not traveling at c but at some lesser velocity due to, in this case, interaction with a medium, this voids Special Relativistic description altogether.

We have Alice, Bob, and Victor in three corners of a triangle. If Information travels via Ryu-Takayanagi Path $l$ [not prime], this demands that the Conformal Scale of the fiber optic cable changes:

- Alice to Bob
- Alice to Victor
- Bob to Victor

What I am trying to get across here is that at subluminal velocity of 0.6c in fiber optic cable the Conformal Scale changes in a manner where the values epsilon grows larger as Information travels from Alice to Bob, meaning it is moving into the literal past, violating causality, presumably in fiber optic cable. In addition, the idea that the Conformal Scale, which is equated with the unitary Planck intervals however you like [or not], *changes in the fiber optic cable* in a manner where the mid-length of the cable is of the largest Conformal values epsilon and at the ends, with Alice and Bob, are of the least magnitude, is illogical.

It is for this and a host of other reasons that this experiment cannot possibly represent the transfer of Information of entangled systems. By interacting with the medium, the photon

## Describing Temporal Order of Events on the AdS Horizon Surface

native wave functions are not in their native state. The idea that entangled systems of native wave functions in their eigenstates can interact with the medium, which is the case else they would not have a refractive index and remain in eigenstates is something that has to be tested following rigorous theory, that does not exist on the subject.

Meaning, the authors intent was to send Information in the form of photon native wave functions that must be in their eigenstates so as to fulfill the purpose of the Bell Loophole Free test. However, it is clear that native wave functions cannot be native wave functions as they cannot exist in the eigenstates in a medium win which they are interacting, namely the refractive index of about 1.5 which slows the photons to about 0.6c. In addition, it would seem apparent that the notion of being native wave functions in their eigenstates at 0.6c is anomalous, and any such assumption needs to be validated empirically before any such test result can be validated [Bell test]. It is a forgone conclusion that photon native wave functions in their eigenstates travel in vacuum and at $v = c$.

It seems to me they collected noise, and because binary noise is a nice coin-flip, got the expected result, but not because the test in any way represented what the authors intended.

There is no actual instantaneous transfer of Information. The postulate is that entangled systems do not observe Preferential Frame of Reference. 'What is true in one system is true in all systems' cannot be the case in fiber optic cable. This would suggest that what is true for the polymer molecules at one end of the cable is true for all polymer molecules in the entire cable, and in fact, in this case, all three cables. [Yes, I am aware that the cable is one molecule, that is not the point]. Just as a simple example, if we heat one end of the cable and see that it is not hot throughout the entire length of the cable that would validate the notion that what is true on one end of the cable may not be true for the other end of the cable.

This is what I was referring to earlier when I put forth the notion that an entangled system will transfer Information via the rigid red arc of convention, the conventional Ryu-Takayanagi Path $l$. An entangled system does not therefore obey the rules of Preferential Frame of Reference. A system that is not entangled, such as a macroscopic system, obeys Preferential Frame of Reference, and the path Information takes is the yellow, fluid arc, represented as Path $l'$. Path $l'$ is defined by the value epsilon at time-zero, when the Information is sent by Alice, to Bob not at $B$, but at $B'$. Thus, Information in a system that obeys Preferential Frame of Reference travels from Alice at time zero, $A$ to Bob at $B'$, also represented in the image as $\varepsilon$. The distance from Bob at point $B$ to Bob at $B'$ is that temporal 'Gap' that demands that what is true in one system [train-crash, CAPCOM cutting Armstrong off] may not be true in another system [no train-crash, Armstrong's otherwise inexplicable stutter].

As a mere matter of convenience, we look at some rearrangements of the sort: [the yellow highlights are to draw your attention to the terms being manipulated]

## Describing Temporal Order of Events on the AdS Horizon Surface

The derivations are more rigorous in Quantum Information Dynamics Vol I and II. Epsilon is equated to $c^2$ and as such, the unitary Planck intervals of length {Lp} and time {tp}:

$$\koppa\varepsilon' \equiv \frac{L'}{e^{\frac{3l}{e}}} = c^2 = \frac{Lp'^2}{tp'^2} = \koppa A'_\Omega$$

Where $A_\Omega$ is the worldsheet that defines the limits of the system. Koppa, again, is the police pseudo-operator that prevents changing the quantities of unitary intervals.

$$\koppa\varepsilon' \equiv \frac{L'}{e^{\frac{3l}{e}}} = c^2 = E/m$$

$$e^{\frac{3l}{e}} = e^{3\frac{e}{3}\ln\left(\frac{L}{\varepsilon}\right)/e} = e^{\ln\left(\frac{L}{\varepsilon}\right)}$$

And relates to mass-energy as:

$$\koppa\varepsilon' \equiv \frac{L'}{e^{\ln\left(\frac{L}{\varepsilon}\right)}} = c^2 = E/m$$

And energy per unit epsilon as:

$$m = \frac{Ee^{\frac{3l}{e}}}{L} = \frac{Ee^{3\frac{e}{3}\ln\left(\frac{L}{\varepsilon}\right)/e}}{L} = \frac{Ee^{\ln\left(\frac{L}{\varepsilon}\right)}}{L} = E/c^2 = E/\varepsilon$$

$$m = \frac{Ee^{\ln\left(\frac{L}{\varepsilon}\right)}}{L} = E/c^2 = E/\varepsilon$$

and the formative number of Qubits:

$$\koppa\varepsilon' \equiv \frac{L'}{e^{\ln\left(\frac{L}{\varepsilon}\right)}} = c^2 = \frac{\koppa Lp'^2}{\koppa tp'^2} \equiv \koppa N_Q = \koppa A'_\Omega$$

## Altering the Rate of Time Evolution on the AdS Horizon Surface

W. Bray DOI: 10.13140/RG.2.2.35703.75682

### Temporal Progression of the AdS System

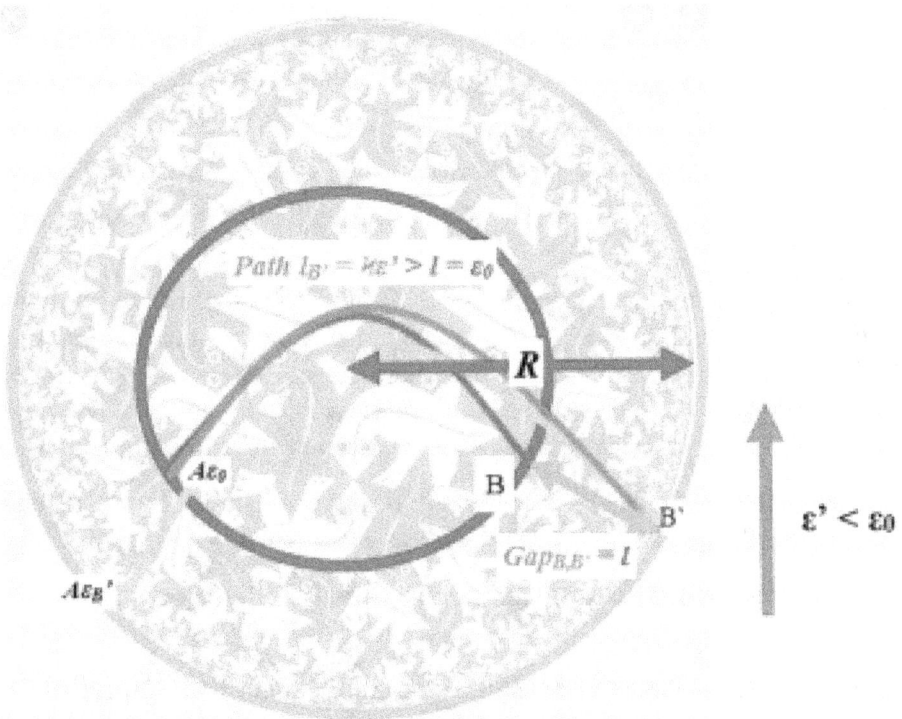

There is a general consensus among convention that regards an actual lateral velocity of growth of the three or four sphere in four or five dimensions, represented above by the large red arrow extending from the center of the visual metaphor to the edge. This has to be addressed, as it is an item of deep concern.

There is no *conventional* mathematical model, *as constrained to observable dimensionalities*, where the quantized horizon values epsilon can both decrease in size and retain Information Conservation. In general, convention regards the *quantity of* values epsilon as increasing. Given the value epsilon is by convention regarded as vanishing toward an infinitesimal [zero], this then demands that the violation of Information Conservation has an end result of infinite quantized values epsilon. That is, if we quantize the horizon value epsilon, regardless of how we chose to do it, and regard each value epsilon as [koppa, police operator] ϟε, an actual expansion of the three or four sphere would demand epsilon must increase in Conformal Scale, not decrease, according to the models of convention. The workaround is to invoke as many unobservable dimensionalities as possible, typically borrowed from String, m, n, and brane theories and so on, to make a proper accounting of things.

### Altering the Rate of Time Evolution on the AdS Horizon Surface

However, constrained to the 4-observable dimensions the entire AdS system does not work. This is a big problem. We have a model of the universe in its most primitive form requiring dimensionalities that never can be proven. The general attitude is exactly akin to any conviction, *since we looked indefinitely and found no evidence that these unobservable dimensionalities do not exist, we hereby dub them as 'existing.'*

In general, Information Conservation is constantly violated in the conventional descriptions of the AdS system. The idea that epsilon is quantized and that those Qubits are fixed in number, as they must be, is not addressed by any conventional approach. Instead, the general idea demands that as epsilon decreases in value toward an infinitesimal, the number of $n\varepsilon$ increases toward infinity, violating Preservation of Information Conservation. The idea that $kn\varepsilon$ is fixed does not appear in any prior art, to the best of my knowledge.

Simply, the number of Qubits that define and describe the cosmos is fixed and cannot change, nor can they be 'borrowed,' again, as there is no 'negative Information' to balance such accounting of Qubits via the HUP or any other means.

The conventional approach is to regard epsilon, as it progresses toward the ultraviolet, to increase in *quantity*, indefinitely. The artifact seems to originate in the 'velocity of recession' issue, regarded as $H_0$. The principle is a Lemaitre expanding cosmos hypothesis, e.g., Big Bang physics. The observation is indeed of an expanding cosmos. However, it must be clear to us by now that this physical expansion is not reconcilable with the AdS description in terms of an actual change in physical disposition.

For example, the 'interior' of the AdS system is still regarded as physically real, when it is nothing more than a cognitive record of prior states of the Horizon Surface.

### Relativistic-Like Conformally Scale Invariant Changes to Epsilon

### Observation Rate as the Determinant for Pseudo-Locality with respect to Time and Distance

Special Relativistic conditions are very difficult to explain, primarily because there are very few physicists who understand it correctly. General relativity is even more difficult because physicists invariably assign Preferential Frame of Reference, which does not at all apply to Schwarzschild type transformations in cases where such changes in spatial or temporal geometry are not coupled to a local source of mass*energy. To make matters even worse, Minkowski solutions are invariably applied to both, the sqrt(-1) having no real world relevance in observable dimensionalities, cannot possibly explain a *real permanent* disparagement between clock systems.

Frame of reference and Preferential Frame of Reference are not the same, not synonyms.

Bob speeds away from Alice: Preferential Frame of Reference means that there is a preferred reference between Alice and Bob, such that Bob is regarded as the one in motion, not Alice. However, the entire explanation, if one had read the original 1905 Einstein-Maric 'On the electrodynamics of moving bodies,' is that the selection of who is in motion as Bob speeds away, Alice or Bob, does not exist. Bob, although

## Altering the Rate of Time Evolution on the AdS Horizon Surface

he is speeding away, deliberately set in motion, can be and is regarded as at rest, observing Alice as speeding away.

The correct frame of reference is the cosmos, as proper motion. Preferential Frame of Reference is Bob, if he is in *proper motion* through normal spacetime. If, for example, Bob and Alice are heading toward or away from one another at some velocity v/c, then there is no relative transformation between them. *However*, both will undergo a Lorentz transformation with respect to some fixed point that is not in proper motion.

For example, both Alice and Bob depart Earth in opposite directions at 0.866c, their t-prime values are 2X, *with respect to Earth, but not each other*. If they follow the same exact but opposite paths outbound and then back to Earth, say, 1-lightyear, their clocks will agree with each other. However, their clocks will be permanently disparaged from those clocks on Earth by a factor of two. If we set them off at 90-degree angles to one another from Earth, the same is true. It is only when Alice and Bob have different proper motion that their clocks differ. All, for the most part stationary clocks in the universe will agree with those on Earth, but not Alice and Bob's clocks.

It is only when Alice is regarded as stationary in the cosmological frame of reference with little or no proper motion [some v/c < Bob's *proper velocity*] that Alice and Bob's clocks become permanently disparaged.

This was exemplified in the Dead Twin Paradox. If Bob is the only sentient being ibn the cosmos on his way from Earth to Wolf 359 [8-lightyears distance] at velocity 0.866c, he will measure, second for second on his clock a smooth transition toward a 4-year endpoint. Bob otherwise has no single [but 32 permutations] answer for what Bob sees on his own clock, second for second along the journey. The length transformations also take the same form but are reciprocated from convention.

Problematically, most every explanation I have seen and read goes into relative acceleration, e.g., some turnaround point, that has nothing to do with anything. It is just grossly incorrect. In the original Einstein-Maric 1905 derivation they set all systems, K [stationary] and K' [speeding] as measured against *k,* 'the rod in motion.' The rod in motion is derived as being at constant velocity c [constancy of the speed of light] and is thus the single quantized photon. All systems are at rest because the single quantized photon [which Einstein published just five months prior] is derived as existing at the Limits at Infinity and thus the only immutable meter stick in the cosmos. There is no portion of the 1905 where K [stationary] and K' [speeding] determine one another's states, all measurements are against *k*.

Proper motion, then, goes into the photon as being the only valid meter stick, whose scaffold is fundamentally spacetime.

What is being described here is that distance, and no other factor, is a Preferential Frame of Reference. This is true to the extent that various causal anomalies result, as demonstrated in the history of the 2-slit, as well as a host of other phenomenon that will be discussed in another section of this text.

## Preservation of Information Causality

Time is progressing from point $B$ at time $\varepsilon$ to time $\varepsilon'$ at point $B'$. Note that the vector of this is not across or below the AdS Horizon Surface, but *orthogonally* from some past state of the Horizon to some later state. Simply, if Information could traverse into the region regarded as 'below' the AdS Horizon Surface, it would thus be traveling acausally into the past. Convention has it this way, but this cannot be the case without violating Preservation of Causality:

*Preservation of Information Causality:*

*Causality cannot be violated on any scale.*

*Preservation of Information Conservation:*

*Information cannot be created, destroyed, nor 'borrowed.'*

Reverse Causality is a violation of Information Conservation on any and all scales greater than 2-native wave functions. The final sums adding up on paper is not the point:

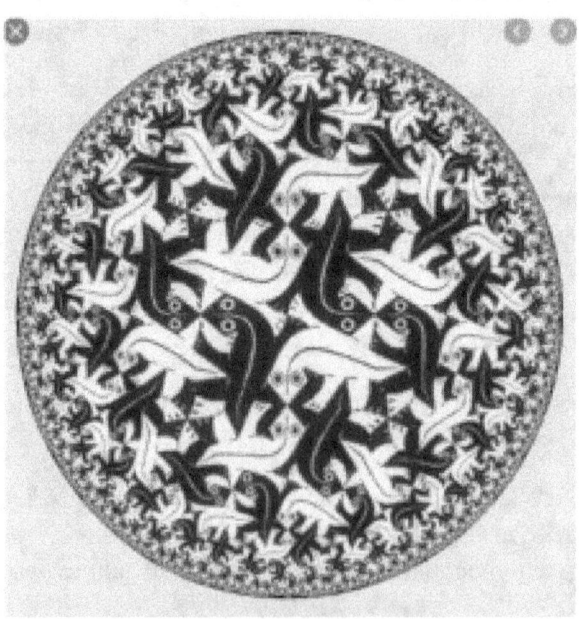

Convention has this odd way of thinking on the lines that, there are more 'lizards' [unitary Planck intervals, values epsilon, what have you] on the horizon edge than in the middle. This is a gross violation of Preservation of Information Conservation in the worst possible

Preservation of Information Causality

way. The *quantity of* unitary [Planck, what have you] intervals cannot change. Thus, it is the magnitude of a fixed number of unitary intervals that is changing over time. Conventional thinking then, would simply subtract the *quantity of* unitary intervals going back to the first moments of the Big Bang and on paper be satisfied with the result that conservation has been obeyed.

There is and always was a fixed *quantity of* unitary intervals, which change in magnitude over time. Reversing causality in the Escher 'field of lizards' above requires going, not figuratively but literally, down into the center of the AdS System. The region literally 'below' the AdS Horizon Surface *no longer exists, it is only a record of past states of the horizon.*

*Reverse causality on this scale would require destruction of the current state to yield some prior state, which is Information, and be 'destruction of Information.'*

Thus:

***Preservation of Information Causality Conservation:***

*Causality cannot be violated on scales greater than 2-native wave functions.*

Again:

The Maric Operator, which maps the absolute value onto a system: [see the link for detailed descriptions]

$$\varphi(\pm v) = |v|$$

**then given $c = l/t \equiv 1$**

$$\varphi(\pm l) = |l|$$

$$\varphi(\pm t) = |t|$$

**therefore $\varphi(\pm 1) = |1|$**

And the rules of the **Maric Operator** apply such that:

1. If $A_\Omega$ is $> 2\{L_p, t_p\}$ the Maric Operator takes the form $\varphi(\pm 1) = |1|$.
2. If $A_\Omega$ is $\leq 2\{L_p, t_p\}$ the Maric Operator takes the form $\varphi(\pm 1) \neq |1|$.

In case 1, this describes the unidirectional quality of Gravitation and time.

In case 2, the conventional rules of the Heisenberg Uncertainty Principle apply, and the convention of the *emergent effect* of charge applies. That is, on scales where the worldsheet

Preservation of Information Causality

$A_\Omega$, which defines the constraints and scope of the system, is less than or equal to two native wave functions, the convention of charge applies.

The difference is that case 1 describes the unipolar quality of Gravitation and time and case 2, purely a function of limiting the scale to two native wave functions, charge applies.

This orthogonal projection is key in understanding that reference to the 'Gap,' designated $Gp$. It is this Gap in time that describes why what is true in one system may not be true in all systems, which is a description of non-entangled systems. Convention regards vectors across and below the AdS Horizon Surface but does not consider this orthogonal projection. The orthogonal projection, as we will see, is the demand of the progression of normal time, regardless of scope and rate.

At time zero [$A\varepsilon_0$] Alice sends a signal. As the Information traverses the distance from Alice at point $A\varepsilon_0$ at time $\varepsilon_0$ time-zero toward Bob the value epsilon is diminishing in real-time. Again, the vanishing rate of epsilon is typically regarded on cosmological timescales. The red arc designates the convention of the passage of Information figuratively below the Horizon Surface. However, as epsilon diminishes, Bob can no longer be at point $B$, because 1) time has progressed and 2) the Horizon Value $\varepsilon$ has thus diminished in scale. If Bob still is at point $B$, it is because the system {Alice, Bob} is entangled.

- At time zero Alice is at the point indicated as $A\varepsilon_0$ and Bob is at point $B$. The normal light distance between them is the red arc, Path $l$, is 100-light-seconds in length.
- Bob observes the Information Alice sends at a rate insufficient to produce a Zeno Effect, e.g., normal observation.
- The Information travels from Alice at $A\varepsilon_0$ toward Bob and 100-seconds later Bob detects the first bit of the Information. At this time Bob is now at position $B'$ because the Horizon Surface has evolved at a rate and scope of $c = 1Lp/1tp$.
- Epsilon is now $\varepsilon'$ at point $B'$ and is of lesser value than $\varepsilon_0$. [Note the upward pointing green arrow]
    - Thus, the arc takes the form of the green arc. The red arc must be regarded as fluid, taking the form of the green arc.
    - Furthermore, the green arc represents typical causality, in that the Information is not travelling inward toward the values epsilon of higher Conformal Scale, literally, convention has the Information travelling into the past. Literally, convention has information travelling toward the past, 'below' the horizon and at exactly the midpoint begins travelling back forward in time; however, not to Bob's new location on the new horizon, which has expanded, but only to time-zero.
    - This difference between Bob at point $B$ and at $B'$ results in a Gap [the small green double-headed arrow]. The Gap, $Gp$, is always orthogonal [but not orthonormal] to Bob's past location on the horizon surface.
    - The Gap [in green] is equal in magnitude to the red arc [of convention], Path $l$, *always*.
- Bob's value $\varepsilon_{Bob}$ is $\varepsilon_{Alice}$/Path $l$, [ $A\varepsilon_0$/Path $l$ ] the normal light distance between Alice and Bob because the Horizon value epsilon has evolved a factor of $c = 1Lp/1tp$ since Alice sent the Information at time zero. Thus, $\varepsilon Bob' < \varepsilon Alice\text{-}0$ at point $B'$. Meaning,

Preservation of Information Causality

  epsilon at Bob's new position on the horizon has progressed exactly as much in time as the Information has taken to travel to him, the normal light distance between points, [conventional] Path $l$. Thus, epsilon at his new location $B'$ is smaller thant it was at point $B$, as well as epsilon for Alice at time-zero, $A\varepsilon 0$.
- The $Gap B, B'$ is equal in magnitude to the normal light distance between Alice and Bob, conventional Ryu-Takayanagi **Path** $l$. The $Gap B, B'$ is orthogonal but not orthonormal to the AdS Horizon Surface.
- He will require 10 of his unitary intervals [seconds] $\varepsilon_{Bob} = \varepsilon_{Alice}$ /Path $l$ to read the 10-second packet of Information sent from Alice.
- His placement as the normal light distance **Path** $l_{B'}$ is equal to **Path** $l$, as measured by his [Bob's] Locally Quantized Meter Stick, which is shorter than Alice's Locally Quantized Meter Stick by a factor of $c = 1Lp/1tp$ when Alice was at $A_B$.
- The *quantity of unitary intervals* has not changed in this, only the *magnitude* of each unitary interval as a Conformally Scale Invariant value, described by the self-similar relationships of the Fractal Set.
    - The progression of unitary time also dictates that epsilon has diminished *from $G_B$,* at time-zero, to time $G_{B'}$ because time has passed in terms of Time of Flight, and the vector is orthogonal to the Ads Horizon Surface.
    - *The velocity* of Information transfer has not changed, nor has the Time of Flight changed. Consequently, the *standard green arc is of the same value* [designated as Path $l_{B'}$] as the 'normal arc' Path $l_B$: *As Measured by Bob at positions B and B'.*

The key factor is that Bob's Locally Quantized Meterstick is [$\varepsilon$] smaller at time $B'$ [Bob-prime]. The *key factors* in the Locally Quantized Meterstick are:

- Only integer values of one's Locally Quantized Meterstick are possible. Meaning, one cannot measure some non-integer value of, for example, Planck lengths of some other system as this violates quantization.
- Each system has a unique set of locally quantized conditions and will therefore measure a *different quantized integer value of any other system; however, the quantity cannot differ, else violate Information Conservation.*
    a. If my Locally Quantized Meter Stick is 10X shorter in Conformal Scale than yours, I will measure your meter as being 10 of my meters. This is not a differing in quantity as all quantities are accounted for.
- It is thus the magnitude of the unitary Planck intervals of length {Lp} and time {tp} that change.
- Quanta [Information] can be said as axiom, are not created and/or destroyed regardless of treatment of this set of conditions, even if regarded as purely artifacts of observation, *only*. That is, if such is a mere artifact of observation, *then the demand is that I am observing the creation/annihilation of quanta [Information].*
- *Any treatment of measurement of one's own state under any Lorentz and/or Schwarzschild transformation as artifact thus negates any sequitur determination of the state of any and every other system, all transformations are real, not artifact.*

## Preservation of Information Causality

There are some causal anomalies, but not paradox, that the $Gap_{B,B'}$ produces that will be discussed in other sections of this text. The Lunar Landing Anomaly is the going example in this text.

In the simplest sense, wherein we can regard 'observation,' rather, 'detection' as any electromagnetic interaction, we look at the most fundamental description of a conventional detector as a PN junction. A PN junction will register a detection event by, ultimately, an electron changing its orbital momentum, e.g., the valence electron orbital. Disregarding a population of events, we focus on the *single quantized photon*. In this case, there will be a single change in orbital momentum, which by convention is an instantaneous event.

We look at a standard detection scenario that will be a 'coinflip' of the sort:

$$<detect|no\text{-}detect>$$

Again, disregarding a population of events and focusing steadfastly on the single quantized photon, a standard detection scenario is this coinflip of <detect|no-detect> eigenstate at a conventional scanning rate.

Normal detection I am regarding in terms of not being a Zeno Effect. Under 'normal' conditions, Bob is *forcing* the Information to evolve from an eigenstate to an eigenvalue. There is no empirical evidence otherwise. An eigenvalue can only exist *on* the AdS Horizon Surface. An eigen*state* can only exist 'below' the AdS Horizon Surface. Thus, what Bob is doing is throwing coinflips of the sort:

$$<detect|no\text{-}detect>$$

The Zeno Effect will be described in terms of *how often Bob **does not** detect an electromagnetic phenomenon*. **Not detecting** the eigenstate allows the AdS Horizon Surface to evolve further. As we will see, the result of this is that the Information figuratively 're-emerges' at a point where epsilon has diminished in magnitude more, thus the green arc extends out further.

Some rules:

- An eigen*value* can only exist *on* the AdS Horizon surface.
- An eigen*state* can only exist [figuratively] 'below' the AdS Horizon Surface.
- Any electromagnetic interaction, which is all observation, 'forces' [for lack of a better term] Information to emerge from figuratively below the AdS Horizon Surface by *constraining* them to be eigen*values*.
- Entangled systems are in eigenstates, and as such can only travel [figuratively] 'below' the AdS Horizon Surface.
- Entangled systems travel via the rigid red arc of convention, Ryu-Takayanagi Path *l*.
- Non-entangled [usually macroscopic] systems can only transfer Information via the *fluid arc*, **Path *l'***.
- Bob's observation, which is any electromagnetic interaction, 'forces' the system into an eigenvalue, thus to the AdS Horizon Surface.

Preservation of Information Causality

- The rate at which Bob detects Information affects the rate that Information is forced into an eigenvalue, thus the time at which Information 're-emerges' at the AdS Horizon Surface:

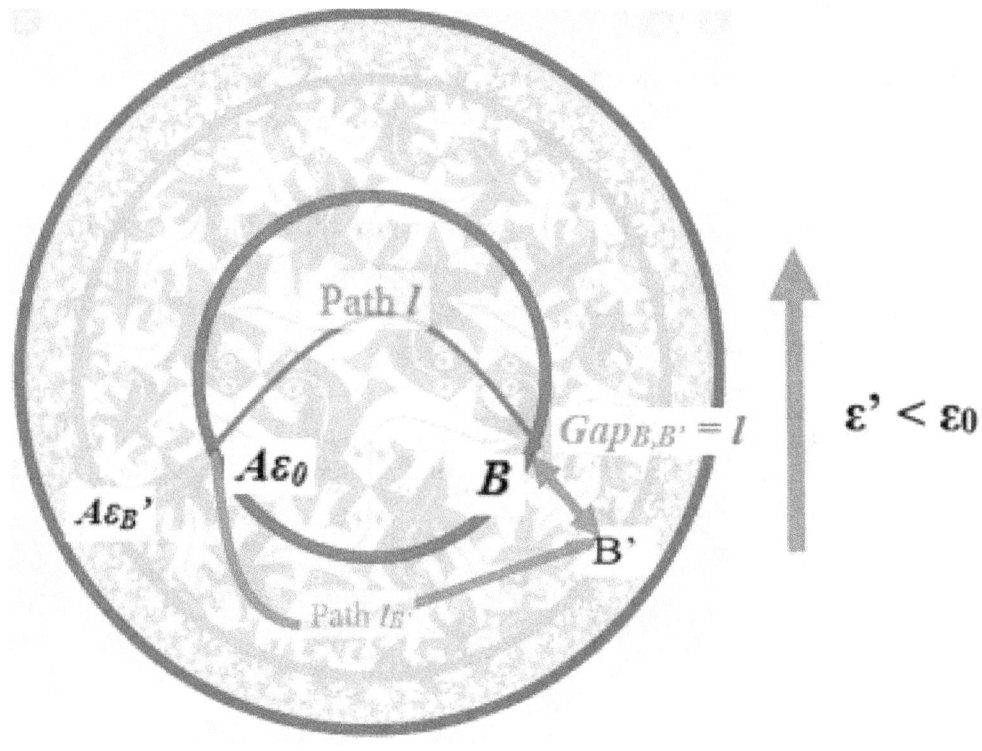

Path $l_{B'}$ > Path $l$

We will go over each step one at a time, 'normal' observation, Zeno and Anti-Zeno observation rates. The color codes help with keeping track of a busy looking image.

Please review the appendix, Defining the Qubit, which equates the horizon value epsilon to the unitary Planck intervals. The very key factor is:

$$\varepsilon = c^2 = \frac{E}{m} = \frac{Lp^2}{tp^2} = A_\Omega = N_Q$$

In the image above:

The world sheet $A_\Omega$ defines the constraints of the system, provided the *quantity of* Information in terms of Qubits is not changed, else violate Preservation of Information Conservation. The derivations of the orthogonal projection of the AdS Horizon Surface is in another appendix: Introduction to AdS Horizon Surface Eigenfunction and Orthogonality.

111

Preservation of Information Causality

- $\varepsilon' < \varepsilon_0$ The Horizon has evolved.
- $A_{\varepsilon_B'} < A_{\varepsilon_0}$ Alice has moved orthogonally to the AdS Horizon Surface.
- $B' < B$ Bob has moved orthogonally to the AdS Horizon Surface.
- $\varepsilon_{B'} < \varepsilon_{A0}$ Bob's new position 'above' the past horizon state means that his value epsilon is less than Alice's at the time she sent the Information at time-zero.
- $G_{B,B'} = Path\ l$: The value $G_{B,B'}$ is orthogonal to the AdS Horizon Surface. The value $G_{B,B'}$ is always equal to conventional Path $l$.
- $Path\ l_{B'} > Path\ l$    The conventional Ryu-Takayanagi Path $l$ is from Alice at time-zero to Bob who was also at time-zero.

The orthogonal derivations of the horizon state are in Appendix H, Orthonormality and Orthogonality. The section that follows are the derivations for Eigen-Normality and Conformal Scaling.

A Note on the Bray Spacetime Manifold

## A Note on the Bray Spacetime Manifold

The Faster Than Light Propulsion system is based on the premise that local spacetime is not coupled to some source of mass-energy. Thus, it can both observe and create its own Schwarzschild transform system. *What is true in one system [inside the Bray spacetime manifold] is true in all systems [outside of the Bray spacetime manifold].*

Thus, there are two distinct and separate conditions of which Preferential Frame of Reference does not apply:

1. The system is entangled.
2. The system is not coupled to some mass-energy source.

The Bray spacetime manifold is inherently different from Alcubierre's supposition in 1994. The Bray spacetime manifold is based on rapid detection and measurement only, and has no sequitur relationship with energy, thus no negative energy. The energy or mass-energy requirement to create [I use the term, 'paint'] a Bray spacetime manifold *is zero*. There is no logical description of energy in the spacetime manifold.

The Alcubierre manifold was based on the Star Trek-like premise of continuous motion, e.g., a logical description of velocity. That, in terms of the AdS Horizon Surface description, is *illogical*, as this then demands the Alcubierre manifold to traverse Ryu-Takayanagi Path *L*:

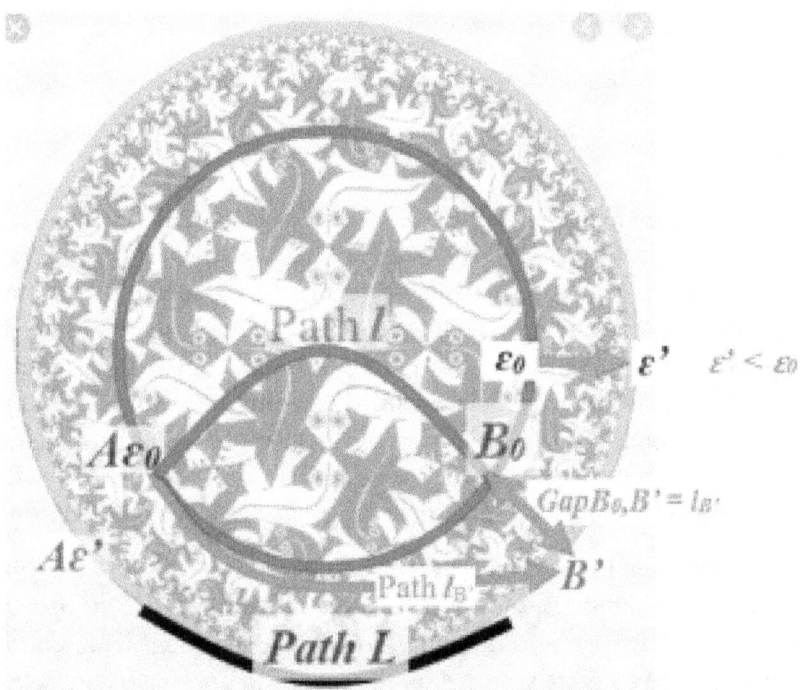

## A Note on the Bray Spacetime Manifold

The conventional Ryu-Takayanagi Path $L$ is by convention, superluminally forbidden. There is no sequitur [logical] description of position nor concepts of motion *on the AdS Horizon Surface*. Fundamentally, the AdS Horizon Surface value epsilon is continuously vanishing toward the ultraviolet. Motion can have no rational description in such a system that is continuously changing conformal scale whilst being traversed. The 'negative energy density' of the Alcubierre manifold is a very roundabout result of this very thing. The logical Platonic 'shape' of a thing under these conditions is ultimately shrinking away to nothingness as it travels.

In a more logical sense, even if we digress to thinking of epsilon as evolving only on cosmological time scales, consider the following example:

> I make an 'Alcubierre' type manifold so good it can travel 1-billion times the speed of light. Over the course of 1-year of travel at such a velocity, the cosmological value epsilon has changed by nearly 10% of the age of the cosmos, because that is the physical distance I presumably have traveled *'on'* the AdS Horizon Surface. The demand then is that my 'ship' is now some whacky mathematical relationship akin to 10% *smaller than when I departed*. There is no rational conventional $\Delta S$ solution for this phenomenon. How can I maintain an Alcubierre type manifold that must change conformal scale with each meter I have travelled?

The 'negative energy density' of Alcubierre's manifold is ultimately described by the portion of the image above at 3:00

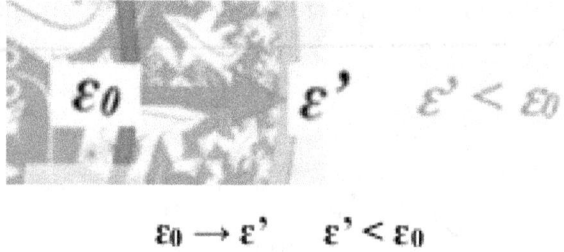

$$\varepsilon_0 \to \varepsilon' \quad \varepsilon' < \varepsilon_0$$

Epsilon-prime is always the smallest possible value, epsilon-naught is always *some past state of epsilon*. The past states of epsilon are always greater in magnitude than the present AdS Horizon Surface value of epsilon. 'Contracting' length in the manner Alcubierre described is a logical impossibility, as this would demand *extending the manifold into the future state of the horizon*. In a sense, 'time travel' is thus required to construct the leading edge of the Alcubierre manifold, which incidentally, was upside down and backward. Travelling into the future at any rate greater than the evolution of epsilon is thoroughly impossible and regardless of the mountains of science fiction 'peer reviewed' papers on the subject, is simply absurd. To travel exactly two Planck intervals ahead of the cosmos actually does require the exact mass-energy of the cosmos to do.

Conventionally, epsilon is only regarded [by convention] as vanishing on cosmological time scales. However, this is incorrect. Epsilon was derived as vanishing, *at the speed of light*. The 'constancy of the speed of light' was based in those works on this premise. Keep in mind there is no sound derivation for the constancy of the speed of light in conventional circles, it is merely enforced. The hypothetical relationships of the various hypotheses, such as Minkowski spacetime, do not validate themselves by way of the data that invoked the hypotheses.

**A Note on the Bray Spacetime Manifold**

The actual derivation for the constancy of the speed of light was in the 1905 Einstein-Marie paper, 'On the electrodynamics of moving bodies.' However, in modern times no one alive has demonstrated having actually read and understood this paper firsthand. I will include an Appendix which shows their original 1905 derivation here in this text.

The Conformal downsizing of epsilon is the result of a self-similar [fractal] relationship, *with itself*. That is far too lengthy to go into again in this text and will be a subject of focus in a future volume of Quantum Information Dynamics.

The Bray spacetime manifold, however, *is not coupled to the AdS system*, its local 'worldsheet,' $A_Ω$ is not coupled to the cosmological worldsheet, $A_Ω$. The $A_Ω$ is the conventional Ryu-Takayanagi AdS Horizon Surface. However, I have derived this to be any system of a scale greater than or equal to:

$$ℷN_Q = A_Ω = ℷ\frac{Lp^2}{tp^2} = c^2 = E/m \equiv ℷ1$$

The Bray spacetime manifold changes the *conformal scale* of the *local worldsheet* $A_Ω$. Note that here it is equated with $c^2$ and thusly $E/m$. Changing the conformal scale of $A_Ω$ is a fractal process, not an energetic one. Thus, there is no koppa [ϟ] police pseudo-operator in front of $A_Ω$.

I have this problem with convention that is difficult to get across. Convention looks at a Lorentz and/or Schwarzschild transformation and in this lazy fashion does not consider this issue:

> I have a meterstick. I send it off at a velocity or drop it into a Gravity Well such that its length changes [plus or minus, does not matter] by a factor of 2. The distance travelled by the meterstick and/or the distance from the center of mass are real and as such, all transformations are real, and cannot be artifact of observation. [Redshift is an example of how 'real' the Lorentz/Schwarzschild transformations are].

> Did I lose or gain Planck intervals? Doing so would be a gross violation of Preservation of Information Conservation and even conventional ΔS entropies:

$$Lp = \sqrt{\frac{hG}{2\pi c^3}}; \quad tp = \sqrt{\frac{hG}{2\pi c^5}}$$

The value $h$ is obvious, the value $G$ is in SI units of $m^3/Kgs^2$, thus, mass-energy. The Planck intervals are defined by energy. Thus, no quantity of Planck intervals can change without violating every conservation law.

## A Note on the Bray Spacetime Manifold

It is the magnitude of the intervals that changes, there is no other answer. I will describe why and how later in this text, as it is an essential process in the operation of the 'engine' design.

In the case of the Bray spacetime manifold, the local worldsheet is the spacetime manifold. This *artificial worldsheet* will be described herein as being constructed by producing a large population of beta particle events.

The beta particle events are sourced from a 'Particle Fountain,' the central nacelle, which is essentially a very hot source of beta, such as, I typically use $^{14}C$ for its familiarity, albeit, for this engine design is far from the nominal choice. The Specific Activity of $^{14}C$ works out to only 166.5G decays per second per gram. However, it is sufficient for argument's sake. Thus, for my fictitious central nacelle, I have 1Kg $^{14}C$ [166.5T dps] as the natural clock system, the beta source.

Specific Activity is given by:

$$\frac{dps}{g} = \frac{4.17E23}{t_{1/2}sec \times MW}$$

Keep in mind there are much 'hotter' isotopes, however, the higher the Specific Activity the shorter the half-life. Rubidium-75 for example is very hot at about 3E20 dps per gram. It however has a half-life of 19 seconds, which is unmanageable.

The particles are steered by means of the Near Field Emissions of the primary nacelle. A secondary Near Field from the second nacelle 'wiggles' the charged beta particles as they traverse the Near-Field I path. This causes them to emit magneto-Bremsstrahlung radiation. The magneto-Bremsstrahlung radiation is of a frequency specific to the Power of the Near Field(s) at that radial distance from the nacelles. The detection of the magneto-Bremsstrahlung is non-destructive, allowing for continuous interrogation of individual particles, rather than destructive detection such as is a conventional limitation to conventional Zeno Effects. Conventional Zeno experiments have never been able to produce *quantifiable results for this reason*. No Zeno experiment has ever published a quantifiable result in both axes, but invariably normalizes to some arbitrary, and frankly, useless scale.

The detectors are also housed in nacelles I through III. The detectors are a network of Koch antennas, which enables them to receive the broad spectrum necessary to identify the exact frequency of magneto-Bremsstrahlung radiation being emitted by each particle.

In this manner, a 'sphere' of natural clock systems is created. The sphere of natural clock systems [beta particles from decay events] is repeatedly interrogated at very high rates. This causes the temporal progression in the space filling volume of the sphere to change at will, as determined by the rate of observation of each portion of the manifold. As the rate of temporal progression of each portion of the manifold changes, the spatial geometry must thusly conform, else violate General Relativity in some manner that has no prior art.

**A Note on the Bray Spacetime Manifold**

The Bray spacetime manifold works by extending the unitary Planck intervals out to great distance. It is thus a type of *'Jump Drive,' not a description of continuous motion.* Obviously, computing power and speed is then the limiting factor to the scale of the FTL Jump.

The effect of rapid observation on the AdS Horizon Surface must be understood to comprehend the working principles of the Faster than Light Jump Drive. Again, there is no energy associated with this.

The Zeno Effect is a case where the local spacetime geometry is not coupled to any mass-energy or mass-energy equivalent. In such case, as was derived in prior [Bray] papers, it is the *magnitude of* the fundamental unitary Planck unitary interval that changes. Although this postulate is demanding, the self-similar relationships are equated with the AdS Conservation of Conformal Scale relationships. That is, the most fundamental premise of AdS/CFT is that the unitary Planck interval does in fact change with time, regarded as epsilon [ε]. The AdS/CFT description has the horizon value epsilon vanishing away toward the ultraviolet. This cosmological 'expansion' is not a change in the quantity of Planck unitary intervals, as this would be a gross violation of Information Conservation laws. Rather, the unitary Planck interval in the AdS/CFT model are diminishing in magnitude, in real time, at a scope and rate defined by $c \equiv 1L_p/1t_p$, where $L_p$ is the Planck length and $t_p$ the Planck time. This relationship *is not coupled to any local nor global mass-energy or mass-energy equivalent.* Rather, it is spacetime itself that is changing.

For a more detailed description of the AdS/CFT local quantization, see Appendix: Local Quantization and The Conditions of the Locally Quantized Meterstick.

When spacetime is changing, this demands then that, given the *quantity of* N-Qubits must remain fixed, else violate Preservation of Information Conservation, the demand then is that the Conformal Scale of the unitary Planck intervals {$L_p$, $t_p$} change.

The Zeno Effect, as derived in prior [Bray] papers, is the non-ontological realization of the 'observer effect.' In this case, as described in the prior text [2], the Zeno Effect is *forcing* Information to [figuratively] 're-emerge' at the AdS Horizon Surface. There is no mass-energy sequitur to this description, the unitary Planck length ($L_p$) and time ($t_p$) change Conformal Scale accordingly by way of a fractal relationship shown later in this text and derived in several prior [Bray] papers and texts. Alterations in the number of N-Qubits that define a system is dependent on the mass-energy conversion to and from mass-energy to some non-fixed values of N-Qubits that define spacetime. Alteration of the unitary Planck intervals of time and length have no such association with mass-energy. This is by convention in the AdS/CFT model.

Thus, the Zeno Effect, rather than a classic Schwarzschild transformation that results from local spacetime being coupled to local mass-energy, alter the Conformal Scaling of the unitary Planck intervals {$L_p$, $t_p$} by artificially altering the functional Ryu-Takayanagi path lengths $L$ and $l$:

### A Note on the Bray Spacetime Manifold

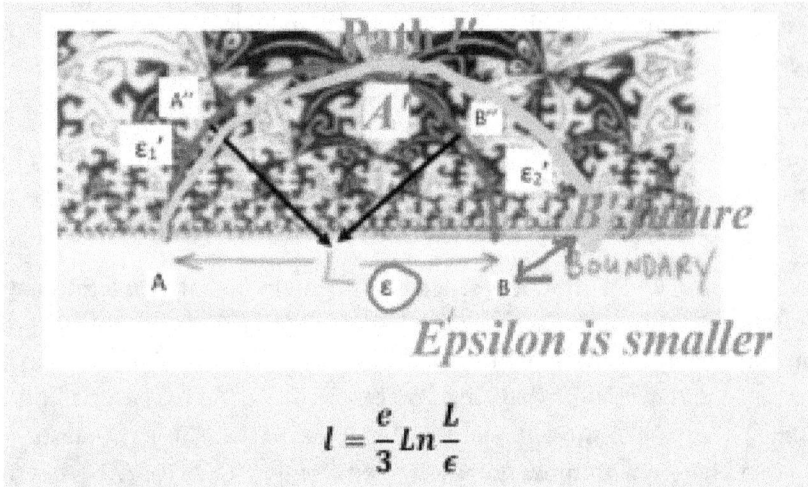

$$l = \frac{e}{3} Ln \frac{L}{\epsilon}$$

The value epsilon was derived as the unitary Planck intervals thusly:

Starting from the Bekenstein-Hawking Limit:

$$N = \frac{A_\Omega}{4Lp^2}$$

The obvious issue here is that the Qubit remains undefined in the hard sense of $N_Q$. Thus, we can simply manipulate in a rather elementary way. When the worldsheet $A_\Omega$ that defines a system is equal to the distribution, $4Lp^2/4Lp^2$ it thus defines the single quantized Qubit:

$$N = \frac{A_\Omega}{4Lp^2} = \frac{4Lp^2}{4Lp^2} = 1$$

$$c \equiv \frac{1L_p}{1t_p} \Big| \therefore$$

$$N = \frac{A_\Omega}{4tp^2} = \frac{4tp^2}{4tp^2} = 1$$

$$N = \frac{4Lp^2}{4Lp^2} = \frac{Lp^2}{Lp^2} = \frac{tp^2}{tp^2} = 1$$

## A Note on the Bray Spacetime Manifold

$$c \equiv \frac{1L_p}{1t_p}$$

$$ϟN_Q = ϟ\frac{Lp^2}{tp^2} = c^2 \equiv ϟ1$$

Koppa [ϟ] is a police operator that forbids the penchant to violate Preservation of Information Conservation. This includes 'borrowing,' as there is no 'negative Information' sequitur to such violation. The term ϟ$Lp^2/tp^2$ is intended to mean ϟN($Lp^2/tp^2$), where ϟN is the same integer value as ϟ$N_Q$.

The term $A_Ω$ refers to the 'worldsheet' of the system. The system can be the entirety of the AdS Horizon Surface or any portion of it, down to 1$(Lp/tp)^2$ *area* of that portion of Horizon. The term $c^2$ allows for this to be equated with every conventional relationship in physics, as well as quantization, these equated relationships appear throughout this text. The term ϟ1 refers specifically to <u>1-N-Qubit of Information</u>, which of course is then rendered as n(ϟ1) to describe larger systems. This definition will be equated with a host of fundamental equations and relationships.

$$l = \frac{e}{3}Ln\frac{L}{\epsilon}$$

Mass-momentum is not yet sequitur on the Planck scale, namely, 2-Planck intervals, thus:

$$σxσv \geq \frac{h}{4π}$$

Given, Path L is superluminal v ⯈ c:

$$[\text{Path}]\ L ⯈ \varepsilon > 0$$

$$\text{HUP} ⯈ A_Ω$$

Path L cannot be greater than ε because this is superluminally forbidden. That is, referring to some unitary value that represents the local meter stick cannot exceed some superluminal expression: a meter cannot be described ct < *l*. It also must be greater than zero, for obvious reasons. If we abide quantization, epsilon cannot be less than {Lp, tp}:

$$\varepsilon ⯇ \{\text{Lp, tp}\} \equiv (u_0 \pm u_{0+1})$$

Else, violate the lower constraint of the unitary Planck interval.

### A Note on the Bray Spacetime Manifold

$$\varepsilon \not\succ \{Lp, tp\} \equiv (u_0 \pm u_{0+1})$$

Else, represents a superluminal violation of $c \equiv 1Lp/1tp$.

$$\therefore$$

$$L \equiv I\varepsilon$$

$$\frac{Lp^2}{tp^2} = c^2 = A_\Omega$$

$$\varepsilon \equiv \frac{Lp^2}{tp^2} = A_\Omega = N_Q = c^2$$

*The key term:* $c \equiv Lp/tp = 1 \therefore Lp = tp$

Is true when constrained to the single Planck interval, thus the term

$$\int HUP = \int 2Lp = Lp^2$$

$$c \equiv Lp/tp = 1 \therefore Lp = tp$$

Momentum on a Planck scale is non-sequitur because mass has not yet been defined on a 2-unitary interval scale, thus:

$$\sigma x \sigma v = 2Lp \times 2tp = 4Lp^2$$

$$\sigma x \sigma v = 2Lp \times 2tp = 4Lp^2 \geq \frac{h}{2Lp}$$

Total Temporal Gravitic and Faster Than Light Propulsion Vol I: Theory and Principles

A Note on the Bray Spacetime Manifold

$$HUP \equiv \varepsilon \equiv \frac{Lp^2}{tp^2} = A_\Omega = \hbar N_Q = c^2 \equiv {}_{path}L$$

$$\hbar\varepsilon' \equiv \frac{Lp'^2}{tp'^2} = \hbar A'_\Omega = \hbar N'_Q = c^2 \equiv {}_{\hbar path}L'$$

Where the limit Path L' is defined as:

$$\varepsilon' = \frac{L'}{e^{\frac{3l}{e}}}$$

$$\hbar\varepsilon' \equiv \frac{L'}{e^{\frac{3l}{e}}} \equiv {}_{\hbar path}L' = \frac{Lp'^2}{tp'^2} = \hbar A'_\Omega = \hbar N'_Q = c^2$$

And

$$\varepsilon \equiv Lp^2/tp^2 \equiv c^2 \equiv \hbar N_Q \equiv A_\Omega \equiv HUP \equiv 1 = 2\,(Lp/tp) \equiv {}_{\hbar\hbar Path}L$$

$$Lp^2/tp^2 \equiv c^2$$

$$c^2 = E/m$$

$$\therefore$$

$$\varepsilon \equiv Lp^2/tp^2 \equiv c^2 = E/m \equiv \hbar N_Q \equiv A_\Omega \equiv HUP \equiv 1 = 2\,(Lp/tp) \equiv {}_{\hbar\hbar Path}L$$

$$Lp^2/tp^2 \equiv c^2 = E/m$$

It's just substitution. However, highlighting the key terms:

$$\varepsilon \equiv Lp^2/tp^2 \equiv c^2 = E/m \equiv \hbar N_Q \equiv A_\Omega$$

## A Note on the Bray Spacetime Manifold

Placing our police pseudo-operators:

$$HUP \equiv \xi\varepsilon \equiv \frac{Lp^2}{tp^2} = \xi A_\Omega = \xi N_Q = c^2 \equiv \xi path L$$

$$\xi\varepsilon' \equiv \frac{Lp'^2}{tp'^2} = \xi A'_\Omega = \xi N'_Q = c^2 = E/m \equiv \xi path L'$$

Where Path L' is defined as:

$$\varepsilon' = \frac{L'}{e^{\frac{3l}{e}}}$$

Can also be written

$$\varepsilon' = \frac{L'}{e^{\frac{3l\prime}{e}}}$$

$$e^{\frac{3l}{e}} = e^{\frac{3\frac{e}{3}\ln\left(\frac{L}{\varepsilon}\right)}{e}} = e^{\ln\left(\frac{L}{\varepsilon}\right)}$$

thus

$$\varepsilon' = \frac{L'}{e^{\ln\left(\frac{L}{\varepsilon}\right)}}$$

$$\xi\varepsilon' \equiv \frac{L'}{e^{\frac{3l}{e}}} \equiv \xi path L' = \frac{Lp'^2}{tp'^2} = \xi A'_\Omega = \xi N'_Q = c^2$$

## A Note on the Bray Spacetime Manifold

This was derived at length in the prior text and will not be a subject of focus in this text. The math that describes these relationships was derived in depth in the prior text, as well as several prior papers and texts, and will only be summarized here.

In summary of this, the Zeno Effect artificially alters the Horizon Surface values $\{Lp, tp\}$ rather than change the *quantity of* them. Mass-energy is non-sequitur to this description. Only detection and measurement are required to alter the unitary Planck intervals $\{Lp, tp\}$. Furthermore, given that the time dilation-like behaviors observed in the lab can nearly bring quantum systems to a complete halt, the factor at which this change in unitary Planck intervals occurs is many orders of magnitude greater than the application of mass-energy or equivalent can practically achieve.

Both Type II [subluminal] and Type I engines have no regard for mass, meaning, the amount of mass they are 'moving,' for lack of a better term. Where chemical rockets, for example, are extremely mass limited, is in fact the issue with chemical rockets, hauling millions of tons to the lunar Surface for the purpose of, for example, mining Helium-3 from the regolith, is impossible. However, The Type II [subluminal] engine has no such regard for mass and moving millions of tons of infrastructure and strip-mining equipment to the lunar Surface for the purpose of harvesting the regolith's rich helium-3 is very possible, and in fact, a legitimate goal.

The Type II subluminal engine alters the geometry of spacetime many orders of magnitude greater than any practical amount of mass [e.g., a black hole horizon] without the association of mass-energy, as mass-energy is a non-sequitur to the Zeno Effect. Thus, strip-mining the regolith from the Moon's Surface is completely feasible using this form of propulsion.

The Type II gravitic 'engine' is a Surface effect, constraining alterations in spatial geometry to a highly localized *Surface area*. The Type I superluminal research 'engine' is a space-filling, volume of quantum events that is loosely localized to the test volume of spacetime.

In general, the 'engine' design is a type of misnomer in that there is no actual motion taking place, with respect to the Type I superluminal engine. It is a superpositioning system on a large scale that requires zero energy to achieve. The modified, corrected, and quantized Alcubierre metric is used as a scaffold, where the unitary intervals $\{Lp, tp\}$ are dilated to extreme magnitudes, superpositioning the system between two points at great distance. It is more of a 'jump drive' type of approach, rather than regarded as fluid motion.

I think the term 'jump drive' is perhaps a [borrowed sci-fi] term that is more logical to the details of operation of the system. Rather than thinking in terms of 'speed,' e.g., 'warp speed,' the level of computer technology available will define the magnitude of the 'jump.' The proof of principle design is simply to demonstrate that a 'jump' does in fact occur under the special conditions of the applied Zeno Effect affect the spatial geometry in such a way that the system superpositions from one locality to another in perhaps a few Planck intervals of time.

Keeping in mind that $c \equiv 1Lp/1tp$ [where Lp refers to the Planck length and tp refers to the Planck interval of time], 'jumping' between localities over any distance greater than 1-Planck length in 1-Planck interval of time demonstrates faster than light 'Re-locality.'

## A Note on the Bray Spacetime Manifold

In this portion of the text that describes the most recent tentative design, I shall refer to the Phased Array in the conventional sense, and this will apply to both far and near field applications of the phased array. I will use the term *Inverse Phased Array* to refer to the detection system, which is an array of two or more detectors that pinpoints the locality of a particle, wave, or field effect via the sequitur use of the term, inverse phased array, like any large array. However, the actual scale of the system will vary from proof of principle to production of a prototype. I will rely heavily on Radiation Cyclotron Frequency and magneto-bremsstrahlung radiation as non-destructive emission and detection schemes as per reference [1]. The Project 8 team in reference-1 managed to detect single electrons as beta particles trapped in a desktop scale cyclotron [e.g., 'trap'] by detecting the Radiation Cyclotron Frequency emitted from tritium decays. This is both critical in being a natural clock system as well as the ability to detect single beta particles, e.g., single electron detection.

An Inverse Phased Array [IPA] system will replace the, or rather be upscaled to, the Radiation Cyclotron Frequency [RCF] type detection system Project Team 8 employed in their scheme for detection of single beta decays in the tritium natural clock system. The IPA will be used to pinpoint and characterize individual beta particles. Each particle will have a *history* associated with it. Each particle will be interrogated repeatedly via a non-destructive detection scheme of observing the magneto-bremsstrahlung radiation emitted by the particles as they traverse a directed path outward from the source, referred to as the Particle Fountain.

There are two Phased Array systems that operate in tandem in directing the particles' paths. Phased Array I [PAI] is the 'pointing' array and Phased Array II [PAII] is the 'wiggler.' PAI directs the particles outward from the Particle Fountain in a spherical patter and PAII wiggles or undulates the particles as they move outward. It is this wiggling that produces magneto-bremsstrahlung radiation that is detected via a trinary array of detectors, two in the nacelles and one in the main body. The PAI and PAII take advantage of the Near Field Effects associated with Electric and Magnetic fields to alter the disposition of the beta particles.

The key feature is that this type of RCF/MB [Radiation Cyclotron Frequency, Magneto-Bremsstrahlung] detection scheme is non-destructive, meaning that an individual particle can be measured repeatedly, unlike conventional natural clock systems that as a result of destructive measurement, can only be measured once. This is a distinct advantage given that in order to create a stable, robust, and reproducible Zeno Effect, a high detection/measurement rate is necessary. Thus, the *history* of each particle is cumulative as the detection scheme moves forward in detecting subsequent natural clock decays and events.

The types of emissions employed will of course all be hypothetical, and I intend to put forth at least two variants. My primary focus will be on beta decays, as these are easy to deal with in a Radio Cyclotron Frequency detection scheme, and electron-positron pair production for essentially the same reason.

Briefly, The Zeno Effect Related to the Bray-Alcubierre Metrics

### Briefly, The Zeno Effect Related to the Bray-Alcubierre Metrics is this:

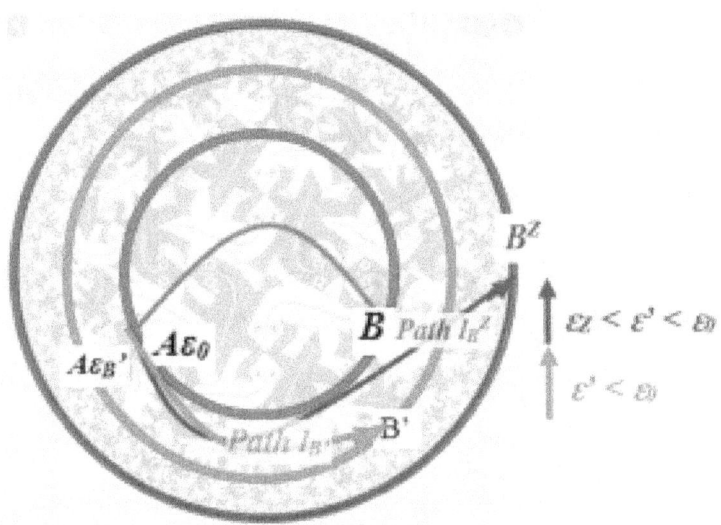

The Zeno Effect, as described in detail in QID Vol III, is *not detecting an eigenstate*. I used the example in that text that Bob sets up a $^{14}C$ source of 1-microgram, which thus yields 165KHz decays per second. The distance to the detector and cross section facing the detector are about 1cm each, which then yields 1KHz events entering the detector. If Bob interrogates the detector at 1KHz, then disregarding efficiency, each interrogation will yield an eigenvalue. However, if he interrogates the detector at a rate of 1MHz, only one in 1-thousand interrogations will yield an eigenvalue. At 1GHz interrogation, one in 1-million interrogations will yield an eigenvalue, and so on. Bob is setting up a phenomenon of the sort where his observation, which is any electromagnetic interaction, is of the sort:

$$\langle detect|\psi|no\text{-}detect\rangle$$

The condition 'detect,' for lack of a better word, 'forces' the eigenstate to an eigenvalue. At that point all bets are off, no further action will yield any phenomenon for that particle. If the condition

$$\psi(no\text{-}detect)$$

Is allowed to persist, the system will remain in eigenstate. Thus, in the diagram above, we follow the purple arc out, where epsilon $[\varepsilon]$ has evolved *further*. Because epsilon has evolved further, the magnitude of epsilon is smaller. I equated epsilon with the Planck intervals in several prior texts as:

From the Bekenstein-Limit:

$$N = S = \frac{A_\Omega}{4 L p^2}$$

Briefly, The Zeno Effect Related to the Bray-Alcubierre Metrics

Was derived:

$$ϟN_Q = ϟ\frac{Lp^2}{tp^2} = c^2 \equiv ϟ1$$

Koppa [ϟ] is the police pseudo-operator that marks elements that cannot be changed in quantity, else, violate Preservation of Information Conservation.

Then from the Ryu-Takayanagi postulate:

$$l = \frac{e}{3} Ln \frac{L}{\epsilon}$$

Given, Path $L$ is superluminal $v \not> c$:

$$L \geq \epsilon > 0$$

- $\epsilon \not\leq \{Lp, tp\} \equiv (u_0 \pm u_{0+1})$, else, violate the Planck [lower] limit.
- $\epsilon \not> \{Lp, tp\} \equiv (u_0 \pm u_{0+1})$, else, would demand that epsilon is superluminal.

$$\therefore$$

$$L \geq l\epsilon \; ; \; L \equiv n\epsilon$$

$$\epsilon \equiv (u_0 \pm u_{0+1})$$

$$\epsilon \equiv \{Lp, tp\}$$

And:

$$\epsilon = c^2 = \frac{E}{m} = \frac{Lp^2}{tp^2} = A_\Omega = N_Q$$

The key elements:

$$ϟ\epsilon = \frac{ϟLp^2}{ϟtp^2} = A_\Omega = ϟN_Q$$

The highlighted term $A_\Omega$ is the local 'worldsheet,' which defines the constraints of the system. The koppa [ϟ] in front of epsilon indicates that it is an element that cannot be changed in quantity, else, violate Preservation of Information Conservation, as are the unitary Planck intervals, and the quantity of Qubits.

Briefly, The Zeno Effect Related to the Bray-Alcubierre Metrics

The goal is to remap the sphere into the Bray Spacetime Manifold:

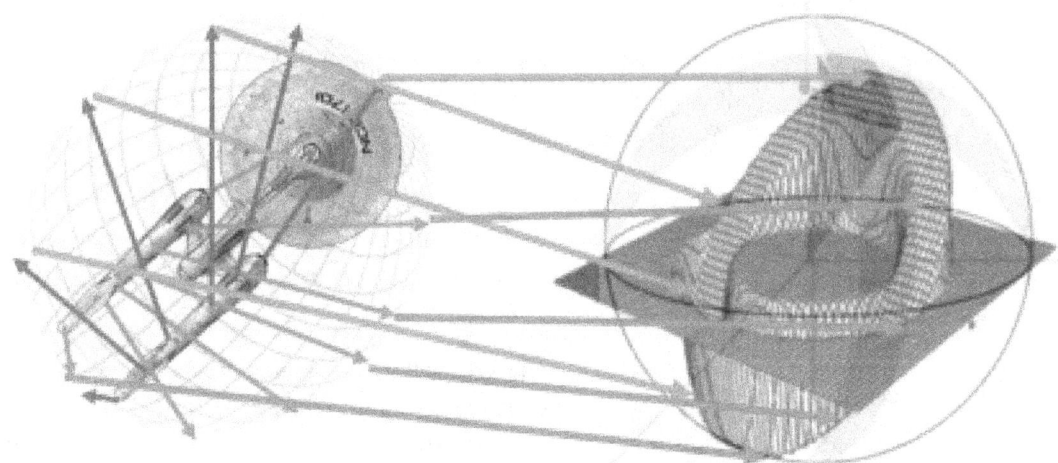

I removed a few of the arrows to make it a bit easier to look at and perhaps make sense of. Each point on the sphere is being 'pulled in' by altering the path length, which is a function of the *new rate of progression of time* for each beta particle path.

Our starship starts off in a more or less sphere that is occupied by a large population of quantum events, such as beta particles, each of which is the result of some natural clock system, we just use tritium for simplicity's sake. The measurement rates of each particle, which because it is a non-destructive detection and measurement process that looks at the, in this example, magneto-Bremsstrahlung radiation [MBR], are carried out *repeatedly* for each particle emitted. This is an important distinction because a much more robust Zeno Effect can be created by measuring the same event, rather than separate events. There is no prior art in the literature where systems are measured repeatedly, instead individual events, such as beta decays, are measured once. This has in fact been a major failing point in Zeno dynamic experiments and in part a reason why no Zeno result has ever been quantifiable.

That is, in any published paper regarding, for example, a beta emitting natural clock system, some radioactive source is measured at a high rate of speed. However, each decay from say, rubidium, is a separate event of an eigenstate that is measured, thus collapses to an eigenvalue. Once that beta particle is measured and subsequently collapses to an eigenvalue, measuring it again is pointless, *the way the experiments have been carried out to date*. Meaning, researchers are relying on the eigenstate to eigenvalue clock of a quantum system, thus calling it the *Quantum* Zeno Effect.

That practice has a Zeno Effect on the radioactive mass, the source. However, what researchers do not understand is that this is a perturbative effect. If you 'slow' the rate of radioactive decay, the quantity of beta particles decreases. In turn, you are measuring these at some phenomenal rate and causing the beta events to 'collapse' for lack of a better word from eigenstates to eigenvalues. This, in turn has a second order effect on the radioactive source, and so on. If you affect the radioactive source, there must be a perturbative effect on the emissions, which in turn are measured and 'forced' to eigenvalues, in turn affecting the source, and so on.

Briefly, The Zeno Effect Related to the Bray-Alcubierre Metrics

A Zeno Effect should always be a linear response. If there is a second order response, either the source is undergoing a perturbative effect, or it is just artifacts of the detector and computational systems.

The Zeno Effect is defined as the detector/detection system as coupled to the natural clock system; else no Quantum Zeno Effect would occur. Given the scale of such systems are benchtop, not quantum scale detectors, the rationale that the Zeno Effect is a quantum scale phenomenon is erroneous. The Zeno Effect, as discussed in the prior paper [2] was redefined as a macroscopic phenomenon that is observed on cosmological scales *as the natural clock system of the universe*. Nonetheless, the Zeno Effect, rather than the 'Quantum' Zeno Effect, is not a phenomenon limited to quantum scaled events.

Meaning unambiguously, the Zeno Effect is the cosmological clock. This was described and derived in depth in [2] and will not be a major subject of focus in this text. In brief, the local worldsheet $A_\Omega$ can be of any constraint, out to the cosmological horizon. It is by observation, which I define as any electromagnetic interaction, that figuratively 'forces' Information from the region *figuratively below* the AdS Horizon Surface to the surface by constraining them from eigenstates to eigenvalues. That part is important in the engine design and is the reason we want to create a Zeno Effect for *each particle* rather than some micro-system, such as a radioactive source. More detail on that as we move forward.

Lorentz and Schwarzschild transformations are examples where alterations in time and length occur on a macroscopic scale. The evolution of the AdS/CFT Horizon value epsilon is an example of this occurring on a cosmological scale. Obviously, if one limits their experiments to quantum scaled systems, the associated effects are thus limited to quantum scales. No one to date has presented this approach of performing the experiment on macroscopic scales, oddly. Even in light, again, of the fact that the Zeno effect is characterized as the natural clock system coupled to the detection scheme, which is then a benchtop scale.

The advantage to measuring a system non-destructively, thus being able to measure the same system repeatedly without collapsing the eigenstate to some eigenvalue, is an issue and a practice that allows us to upscale the phenomenon to any scale, provided we can engineer such scales and methods. Destructive detection methods that affect the radioactive source can thus only be limited in scale to the radioactive source to the detector path length, typically on scales of perhaps a centimeter or so. Nonetheless, this is not a 'quantum scale.'

It is not the *science* that is difficult, it is getting 21st century technology to do the science.

And I have to state clearly here that this *is not new technology, this is a struggle with existing technology to do new Science*. The ability to alter the progression of time and subsequent geometry of space *without mass-energy is fundamentally new science, as fundamental as fire*.

The phased array system 'paints' [generic term] the Alcubierre spacetime manifold by measuring a large population of space filling events, in these examples tritium for argument's sake, by adjusting the observation/detection rate at each point on the Surface *and internal volume* of the sphere that is occupied by particles moving at high velocity away from the engine source. For

Briefly, The Zeno Effect Related to the Bray-Alcubierre Metrics

example, we look at the apex of the sphere at the front of the 'ship,' at a given accelerated rate, according to the Bray Metric Algorithm that appears throughout this text, and *pulls* the front of the sphere inward by contracting space, or pushes outward by dilating space:

It should be noted here that the upward spire in Alcubierre's original metric is not a 'negative energy' phenomenon, as commonly conceived. The upward spire in fact *occurs in nature*:

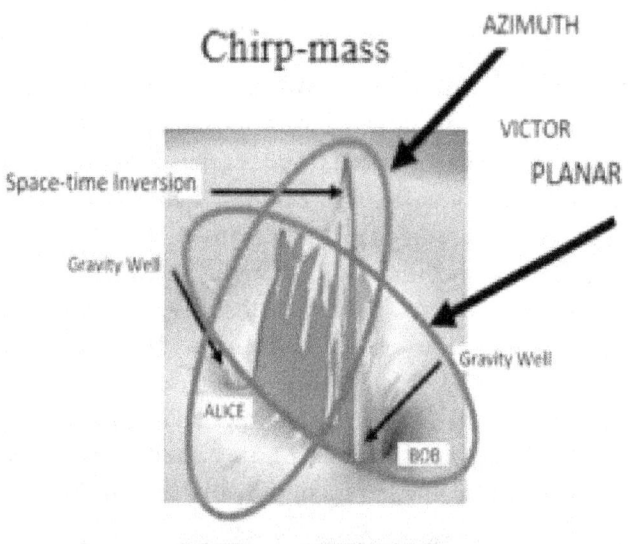

LIGO event GW170817

Briefly, The Zeno Effect Related to the Bray-Alcubierre Metrics

The physical relationship with the Bray Metric and the 'Splitting Point' described in detail in [2], where the Zeno and Anti-Zeno Effects *diverge:*

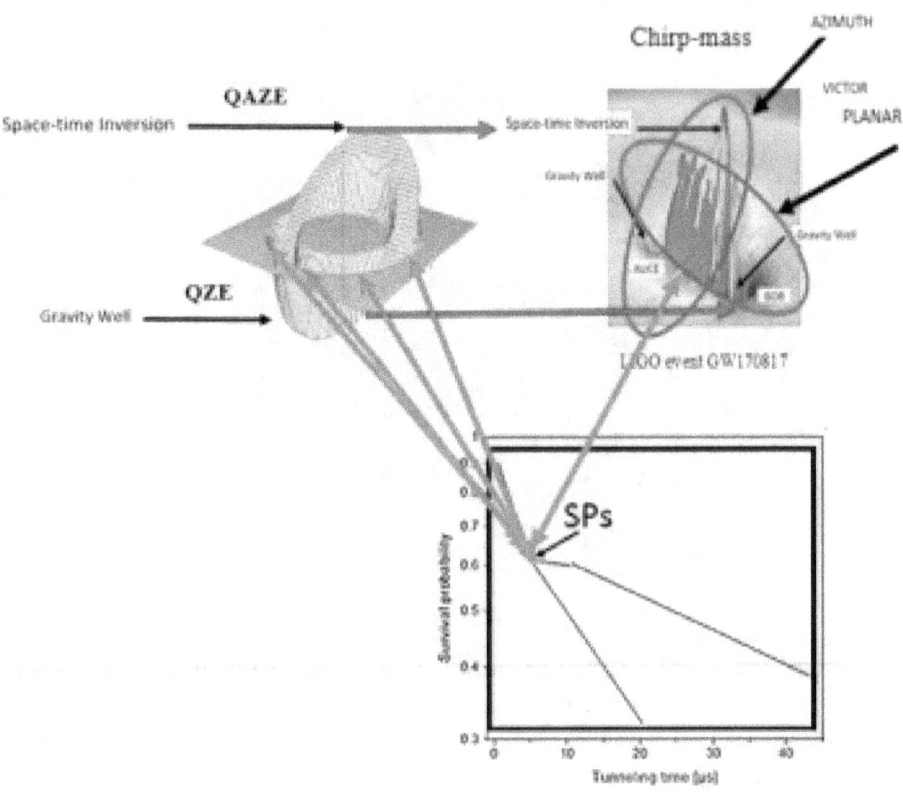

The Bray Metric is described and derived in detail in later sections of this text.

Typically, the Alcubierre spacetime manifold is drawn upside down and backward, because of the aberration of having the Schwarzschild metric regarded as 'contracting' as time dilates, which is a misnomer of General Relativity that insisted Preferential Frame of Reference applied to Schwarzschild transformations in all cases. Preferential Frame of Reference only applies to systems where the spatial geometry is coupled to the local mass, such as the Tensor gradient around a star or any massive body. LIGO's detection of Gravity Waves works by detecting its own change of state under a Schwarzschild transformation, because the wave [spatial geometry] is not coupled to its source of mass, hundreds of millions of light years away, hundreds of millions of years ago. This was what Wheeler predicted in 1955, and the entire rationale for building LIGO. And again, LIGO cannot detect its own change of state under the Lorentzian transformation as a result of our orbit around the sun, e.g., it is a classic Michelson-Morley interferometer. This was discussed at great length in the prior text on Artificial Alteration of Spatial Geometry via the Zeno Effect [2].

It is important to keep in mind that all charged particles in motion of any type, not just cyclotron induced, create Cyclotron Radiation as a result of being in motion in a magnetic field. The power of each electron in motion is given by:

Briefly, The Zeno Effect Related to the Bray-Alcubierre Metrics

$$\frac{-dE}{dt} = \frac{\sigma_t B^2 v^2}{c u_0}$$

For convenience's sake:

$$\sigma_t = \frac{8\pi}{3}\left(\frac{q^2}{4\pi\varepsilon_0 mc^2}\right)^2$$

Using $\varepsilon_0 \sim 8.85E\text{-}10\ C^2/Nm^2$ and $u_0 \sim 4\pi E\text{-}7$ in SI units of $Kgm/s^2 A^2 = Kgm/C^2$ [C being coulombs not $c$]. It all works out to $1.588E\text{-}14 Js$, which is about $2.4E19 h$, which is in the fW range. Both induction and SQUIDs are suitable.

Where $\sigma_t$ is the Thompson cross section [for a single electron $\sim 6.65E\text{-}29 m^2$], which is perpendicular to the electron's direction of motion, resulting in a polarized hv emission. B is the magnetic field strength; v is the velocity perpendicular to the magnetic field and $u_0$ is the permeability constant. The engine, because of its characteristics, must be operated in a vacuum. The benchtop proof of principle is not too difficult to test in this manner, however a prototype can only be characterized in the vacuum of [outer] space.

Detection of single beta particle in the fW range is in [1].

The amount of energy produced by a single electron, e.g., beta particle, is sufficient to produce enough signal to be detected by an array of detectors, as indicated above in the calculation for a single electron. This is critical, because this type of detection scheme is non-destructive, allowing for each beta particle to be measured repeatedly at a rate limited only by computation characteristics. This is necessary in order to produce a sufficient and quantifiable Zeno and Anti-Zeno effects. The type of EM emission will be simple electromagnetic dipole radiation.

I am not going to get into the phasing equations for a detection array because this will be purely dependent on the physical dimensions, materials, and arrangement of such arrays. Phased array or simply detector arrays are fairly standard.

In the case of the engine design, also keep in mind that the particles will not be moving in a cyclic fashion, but a curved line from the point of the emitters. The power emitted with respect to angle of the electron's motion is given by:

$$P(\gamma, \theta) = \frac{1}{4\pi\epsilon_0}\frac{2}{3}\frac{e^4}{m_e^2 c}B^2(\gamma^2 - 1)\sin^2\theta$$

Where $\theta$ is the pitch angle of the motion of the electron to the magnetic field. In these examples I use $^{14}C$ as a standard, which emits at about 156 KeV, producing a power of roughly 10 fW at a pitch angle of 90-degrees to the detector, assuming a 1T field strength. Because the field strength will

Briefly, The Zeno Effect Related to the Bray-Alcubierre Metrics

vary considerably for a full-scale prototype, significantly less in a benchtop proof of principle, the actual power at the detector has to be back calculated and figured into the algorithms.

Coupling to Mass-energy

## Coupling to Mass-energy

Basic postulates take the form:

- The first demand is that the *length transformation is thus real, else could not be detected via an interferometer, or any tangible method.* [e.g., LIGO detection of gravity waves].
- The second demand is that coupling to *local* mass*energy is not a requirement of spatial geometry. [e.g., gravity waves]
- The third demand then, is that mass*energy is not required to artificially alter spatial geometry.
- The fourth demand is that LIGO's detection of gravity waves is a direct detection of Schwarzschild transformations. This is demonstrated by the fact that LIGO cannot detect its own motion around the sun via a Lorentz transformation.
- There is no logical case where a Lorentz transformation is not coupled to its source of motion.


- When mass-energy is coupled to the local spacetime geometry, the *quantity* of N-Qubits may vary, provided mass-energy and Information Conservation are preserved in these exchanges and transformations.
- In the case where spatial and temporal geometries are coupled to their local source of mass*energy, Preferential Frame of Reference applies for General Relativity under Schwarzschild transformations. Thus, all of the classic rules and observations apply.

- When mass-energy is *not coupled* to the local spacetime geometry, the *quantity* of N-Qubits must remain fixed, else violate preservation of Information conservation.
- In cases where spatial and temporal geometries are *not* coupled to their local source of mass*energy, [e.g., gravity wave] Preferential Frame of Reference *does not apply* for General Relativity under Schwarzschild transformations. Thus, the rules of convention do not apply.

The Zeno Effect is the case where the local spacetime geometry is not coupled to any mass-energy or mass-energy equivalent. In such case, as was derived in prior [Bray] papers, it is the *magnitude of* the fundamental unitary Planck unitary interval that changes. Furthermore, these changes do not require mass*energy or equivalent. This is demonstrated by the artificial alteration of temporal progression, regardless of scale, by rapid detection, only. Scale is misunderstood in the Zeno Effect, which is defined as the detector [observing system] coupled to the observed system, which is typically on a meter scale, not a quantum scale.

Given there is no scale associated with General Relativity, [only remains unresolved], alteration of temporal rate without an associated change in spatial geometry would be a violation of General Relativity in such a way that has no prior description.

The scale of the Zeno Effect is not Planckian, as suggested by unagreed upon and diverse hypotheses, but macroscopic. This demand is that the Zeno Effect is defined, by convention, as the observed system [herein $r$] coupled to the observing system, e.g., detector [herein $R$]. In every case, the observed system

## Coupling to Mass-energy

$r$ is coupled to the detection system $R$ on a conventional benchtop, perhaps meter scale. The Zeno Effect is thus not a Planckian phenomenon, but macroscopic.

Although this postulate is demanding, the self-similar relationships are equated with the AdS Conservation of Conformal Scale relationships. That is, the most fundamental premise of AdS/CFT is that the unitary Planck interval does in fact change with time, regarded as epsilon [$\varepsilon$]. The AdS/CFT description has the horizon value epsilon vanishing away toward the ultraviolet. This is not a change in the quantity of Planck unitary intervals, as this would be a gross violation of Information Conservation laws. Rather, the unitary Planck interval in the AdS/CFT model are diminishing in magnitude, in real time, at a scope and rate defined by $c \equiv 1L_p/1t_p$, where $L_p$ is the Planck length and $t_p$ the Planck time. This relationship *is not coupled to any local nor global mass-energy or mass-energy equivalent*. Rather, it is spacetime itself that is changing.

When spacetime is changing, this demands then that, given the *quantity of* N-Qubits must remain fixed, else violate Preservation of Information Conservation, the demand then is that the Conformal Scale [e.g., magnitude] of the unitary Planck intervals $\{L_p, t_p\}$ change. In addition, there is no energy associated with this change, even hypothetical.

Furthermore, there is no description of the Zeno Effect that is predictive, nor effective by any means. There is not even a paper on the experimental aspects of the Zeno Effect sufficient *to reproduce the experiment*. There is no description that is constrained to observable dimensionalities. There is no math of any sort that describes the Zeno Effect whatsoever.

The derivations of the Zeno Effect with predictive qualities will appear in later volumes of this series of texts. In general, the entire Zeno system can be regarded as a highly localized phenomenon, e.g., spacetime manifold. As such, the corrected and quantized manifold would be described as temporal and associated spatial geometries by:

Quantizing then over the localized domain:

- The term $c^4$ becomes $(L_p^4/t_p^4)$
- The term $G$ becomes $G^*$ (relativistic), then in relation to $R$, designated as $G^{**}$.
- **The velocity $v$ becomes quantized as $nL_p/xt_p$.** Note that these are integer values, $nL_p$ and $xt_p$ as describing velocity $< c$.
- The term $v_x^2$ can take the form $((nL_p/xt_p)/(1L_p/1t_p))^2$
- The term $(y^2 + z^2)$ becomes $(L_{py}^2 + L_{pz}^2)$
- The determinant metric $g^2$ will be treated like $T^*$ (the Trace Matrix*) and simply referred to as $g^{*2}$
- The term $r_s$ (radius of the center of the phenomenon) is represented by $r_s(t) = [(x-x_s(t))^2 + y^2 + z^2]^{1/2}$ essentially replaced by quantized values such that:

$$[(L_px - L_px_s(t'))^2 + L_{py}^2 + L_{pz}^2]^{1/2}$$

Coupling to Mass-energy

**For the GR component(s) I** *assign:*

$$L_p' = L_{p0} / \sqrt{1 - \frac{2G'M}{n'L_p(\frac{1Lp}{1tp})^2}}$$

$$t_p' = t_{p0} / \sqrt{1 - \frac{2G'M}{n'L_p(\frac{1Lp}{1tp})^2}}$$

Note that these transformations are from the perspective *inside the spacetime manifold, looking out at flat spacetime.*

**FRACTAL SET ASSIGNMENT of GR Components:**

R|(G''|{t'', Lp''})R

$$L_p'' \rightleftharpoons L_{p0} / \sqrt{1 - \frac{2G''M}{n'L_p(\frac{1Lp}{1tp})^2}}$$

$$t_p'' \rightleftharpoons t_{p0} / \sqrt{1 - \frac{2G''M}{n'L_p(\frac{1Lp}{1tp})^2}}$$

The inner term (1Lp/1tp) does not frack, as it is by definition, unity. Also, Lp is not inverted, *as I assign*, this is a requirement for the hard definition, c = 1Lp/1tp. Also note that no primes are assigned as this value is by definition unalterable. This is where the aberration of the negative sign and the manifold being upside-down and backward originates. In any case, the ASTM *requires it to be so*. Regardless of convention, this is the assignment.

135

Coupling to Mass-energy

Then:

$$R \, \alpha \, t''$$

$$f_{R:r}(\pm k) \mapsto \frac{\left(\frac{Lp^4}{tp^4}\right)}{8\pi G'} \frac{\left(\frac{\frac{nLp}{xtp}}{\frac{1Lp}{1tp}}\right)^2 (Lp^2{}_y + Lp^2{}_z)}{4g'^2[(Lp_x - Lp_{xs}(tp'))^2 + (Lp^2{}_y + Lp^2{}_z)]} \left(\frac{df}{dr}\right)^2$$

Here, $R$ again is the observing rate and $t''$ [double-prime] is the resulting Zeno dynamic transformation. The term $f(R:r)$ is the function of resulting spatial coordinate change, as per the spacetime manifold. This spacetime manifold requires no energy or mass-energy equivalent for the artificial alteration of temporal and spatial geometries.

The value $k$ is the resultant mapping function where $+k$ is the QZE (downward wells, longer causal geometric path) and $-k$ (upward spire, shorter causal geometric path) is the QAZE: for $R \, \alpha \, t''$. That is, $\pm k$ explicitly refers to the mapping rate of observation, $R$, where I previously noted what I referred to as the 'Splitting Point' in the Raizen et al graph from 'normal' observed progression of time, to a QZE *or* a QAZE:

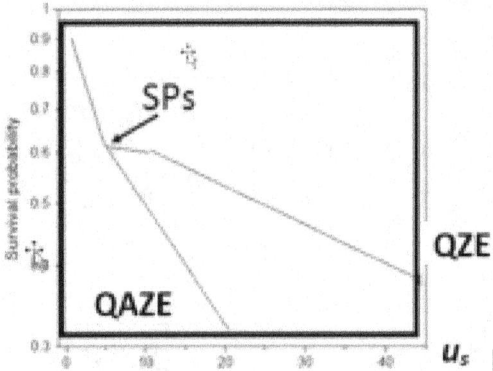

[Fischer M.C., Gutiérrez-Medina B. and Raizen M.G., "Observation of the Quantum Zeno and Anti-Zeno Effects in an Unstable System", Phys. Rev. Lett. 87, 040402 (2001)]

The above is the first mapping of the QZE vs the QAZE. The point I marked $Sps$ is this 'splitting point,' where the QZE and QAZE deviate via the Zeno dynamic function of observation rate vs resulting temporal rate, $t''$.

And for the record I must state that convention regards epsilon in a rather strange manner. Epsilon [Ryu-Takayanagi] is regarded conventionally as vanishing toward the ultraviolet whilst spacetime is also expanding. However, this is a gross violation of Preservation of Information Conservation in that this requires the *quantity of* values epsilon to increase without bound. This cannot be the case; else the cosmos is somehow violating information Conservation in the oddest way. The horizon value epsilon is

Coupling to Mass-energy

decreasing in Conformal Scale; however, the postulate is that this is a self-similar relationship, not one of figurative 'expansion,' of the cosmos in any conventional sense.

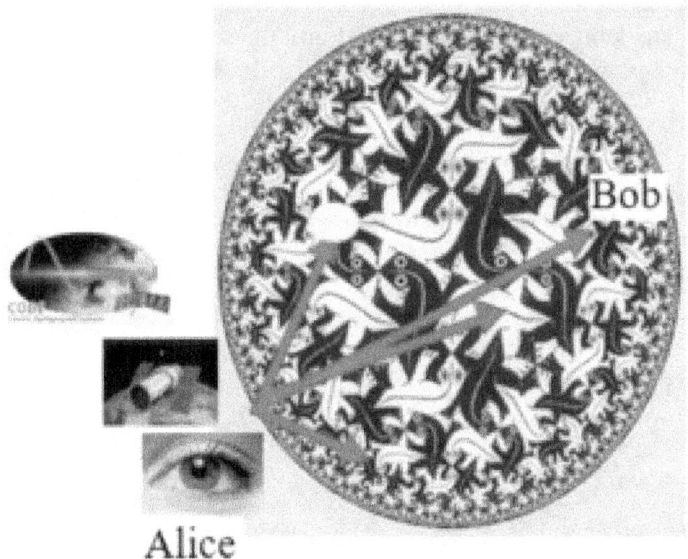

In the visual metaphor above, the 'lizards' represent values epsilon over cosmological time frames. Toward the center by convention the *magnitude of* the unitary Planck intervals are larger and as we progress toward the horizon the values epsilon diminish in size. There is no convention regarding what epsilon actually is in any quantized sense, which is why I quantized epsilon and equated epsilon with the other global values, such as {Lp, tp} again:

$$\natural \varepsilon' \equiv c^2 = \frac{E}{m} = \frac{Lp'^2}{tp'^2} = \natural A'_\Omega = \natural N'_Q = \frac{L'}{e^{\ln\left(\frac{L}{\varepsilon}\right)}} = \natural_{path} L'$$

$$\therefore$$

$$\natural \varepsilon' = \frac{E}{m} \equiv T^{00}$$

The problem is that as equated to the unitary Planck intervals {Lp, tp}, this requires epsilon to be increasing *in quantity* without bound as the Horizon Surface expands, *by convention*. I want to address this issue later, possibly with more definitive derivations for another text as they can be rather lengthy. In short, the quantity of unitary intervals cannot change, else violate Preservation of Information Conservation. Convention has not provided any *agreed upon* means by which the Conformal Scale of epsilon, or of the unitary Planck intervals that describe our [per Escher visual metaphor] 'lizards' can change scale without evoking unobservable dimensionalities. The test here will be to describe Conformal Scaling as constrained to 4-observable dimensions, only.

This is true of the Planck intervals of length and time {Lp, tp} in the grand case of General Relativity as AdS/CFT Correspondence, and as we shall see, thus also extends to Lorentz transform.

Coupling to Mass-energy

- *No real change can occur in an unobservable dimensionality.*
- The "negative-1" Law of Thermodynamics: Information cannot be created or destroyed, *nor borrowed*. [Susskind-Hawking, 'Information Paradox']
- The unitary Planck intervals of length {$L_p$} and time {$t_p$} represent Information and the *quantity of them* between any two points in spacetime is fixed.

*The physical quantities of Information do not change.*

Thus, a different approach to understanding the AdS system has to be considered.

## Detection Principles in Radiation Systems and Associated Electronic Designs

### Choice of Zeno Systems to Produce A Stable Zeno Effect

The obvious stage, given all of the conspicuous absence of such information in the 50-year history of the literature, is to select good Zeno Effect systems. Given there is no such information available, this goes back to square 1, determining a stable Zeno Effect by trying a number of proposed systems. What systems to propose? The systems should not have a high natural progression rate. For example, in the genre of atomic and nuclear excitations and decay, cesium-133 clock rate of about 9GHz is likely too high to even warrant investigation, as this might require detection rates in the high terahertz ranges, as we will discuss, is not actually technology of known description. Cesium-133 is at least not a good starting place for that reason.

As will be discussed and shown, the practical upper limit for detection and measurement at a high and variable rate will be systems in the 10-100KHz ranges. This can be adjusted by doping lesser quantities of, for example, C-14, but not practical in our example of cesium. In the case of cesium, the extinction coefficient for any one atom of cesium-133 is simply too fast for 'rapid detection' and measurement. In the simplest sense, would require a detector specific for that wavelength photon well into the THz range. There is no commercially available transfer line, metal nor fiber optic, that can transfer the information at that rate. The technology does not exist.

Sadly, we may look at a few examples of such claims of high detection rate in the Zeno literature that turn out to be such issues of detection and data transfer rates, dead-time, and pile-up. Because these issues take on second order curve characteristics, they look very much like a Zeno Effect, but are electronic issues.

Having various flavors of transistors and gates that operate in the hundreds of gigahertz range is exceptional news. However, we will look a bit later at the data [information] transfer rates in wire and fiber optic cable as being limited to the single gigahertz range. For example, I may have developed a miracle gate that operates in the 200-GHz range, however, the micro-wiring that connects that gate to the other gates on a silicon chip is severely limited to the single GHz range. That chip, regardless of the miraculous speeds of the individual gates will be limited to its slowest component, which is in this case the micro-wiring that connects millions of such high-speed gates together. If the millions of gates were connected by micro-fiber optic, we would still be limited to the 10-GHz range, because that is the max information transfer rate in fiber optic.

Fiber optic is limited by refractive index, most of common types are roughly ~1.5, limiting the photon velocity to 0.6c. For 1-meter, that means the one-way trip takes 3.3-nanoseconds at $v = c$ and 5.5-nanoseconds in fiber optic.

A micro-chip is wired with typically 20-nanometer interconnects. A billion such circuits ads up to 20-meters of microscopic wiring. However, a typical chip has about 100-Km of wiring in it. Fiber optic has about a tenfold advantage over wire. This extensive interconnecting of billions of individual circuits is a much greater limiting factor than how fast a single gate can operate.

## Detection Principles in Radiation Systems and Associated Electronic Designs

When all of this is summed together, we have to limit our detections for the Zeno Effect down in the single digit mega-Hertz range, if not kilohertz.

Unfortunately, upon reviewing hundreds of experimental setups over the years there are a few things that are apparent:

- Experimenters are not aware or knowledgeable of the limitations of the individual and thus collective hard and software they are using.
- The data transfer rates [wire and fiber optic] are significantly below any reasonable expectation of, for example, a radioactive source:
- The radioactive source with respect to the number of Curies [typically milli or micro] are not stated, are likely unknown, as this would require scintillation counting the source
- The overall distance from source to detector is never stated, also likely unknown
- The specifics of detector design are not stated
- The processor speeds are never stated
- The algorithm response time, speed, is never stated, likely unknown
- There is altogether insufficient information to reproduce the experiment

The Law of Thumb is that, unless proprietary or confidential, an experiment should include sufficient information that someone skilled in the art can reproduce the experimental setup. This does not appear in any case with the Zeno Effect in the literature.

Consequently, the greatest 'paradox' is tens of thousands of papers on the subject without two that represent reproducibility, the very backbone of what is called *science*.

And most unfortunately, of those papers I have been able to reverse engineer based on several assumptions, a significant portion of the Zeno literature of reported experimental findings are data and computational dead-time and pile-up; which looks exactly like a Zeno Effect. This will be explained in detail.

Since the only detectors of known detection rate are limited to the single gigahertz range, then we need to keep the system limited to a natural temporal progression in perhaps the kilohertz range, maybe in the low megahertz. For example, if our natural system is at 1-MHz and detector capable *of a known 1 GHz*, then observation at 1000-times the natural progression rate does not sound very impressive. If we do not know the fine details of the detector's characteristics, as those discussed earlier, there is no point in moving forward, we are only generating yet more unknowns.

Detector and Data Transfer Characteristics

## Detector and Data Transfer Characteristics

Thus, the first step is identifying detector characteristics. As discussed, we have signal-rise time, exponential signal decay time, signal-to-noise return to zero time, and so on: which leaves our THz detector effectively operating in the 10-GHz range.

Sensitivity in every possible genre of mode of detection must be known, *on a known scalar quantity*. This is a statistical issue. For instance, in the history of single photon and single electron detection, these properties of the detector are never stated, again because of the unlikelihood that they are known. How many single events must be detected before statistically one signal gets pumped up-stream to the circuitry? There are researchers who have reported such values, but these are limited to fields of electrical and RF engineering literature. The fact that none of these appear in the references for Zeno papers may indicate that the researchers are unaware of these characteristics, introducing, again, more unknowns. In this case, the inability to reproduce some prior experimental setup becomes pervasive across the Zeno literature.

In general, the most concise Zeno papers are coming out of Quantum Computing. In general, this would be the requirement that results *must be real and observable, and reproducible*, as this is an industrial approach to technological ends.

In the simplest sense, the Heisenberg Uncertainty Principle *demands*, in this case that is clearly a single photon event, that the best possible theoretical detection can be ½ of the photons that strike the detector surface. That is a rather vague claim, but also elementary. Then, as we expand out into the macroscopic dimensions of such a circuit, which is invariably some form of PN junction, the probability of a single photon striking the surface *that results in* crossing the PN junction is at best again cut in half. Then we expand out into the domain of the PN junction, which is not a single atom but a mass of doped silicon, and the matrix of the statistic drops off exponentially from ½ all the way down to whatever the actual number of atoms is. Thus, there is no way to confirm single photon detection. You cannot build an emitter that hypothetically emits 1-photon at a time and validate it with a detector that is also hypothetically able to detect one photon at a time: the two hypotheses validating one another: Remains as two separate hypotheses.

- The exacting number of events a detector must have in its reservoir before up-pumping a signal to be processed must be known. This must be known for each wavelength or energy level being detected.
- The signal rise time, exponential decay rate, signal to noise zero time, must be known.
- These values above at each frequency of photon excitation must be known, if the detection is of an EM event.
- Adjustment of detection rate must result in a known rate of detection.

In general, the Zeno literature is characterized, when a radioactive source is involved, in not stating, possibly not knowing, the decay rate of the isotope *that reaches the detector*. For example, carbon-14 at $1\mu Ci$ is specific. This information is never included in the Zeno literature. Again, there are instances in Quantum Computing where exacting figures are reported, but not sufficient

## Detector and Data Transfer Characteristics

to reproduce the experiment. The inability or unwillingness to report enough information to reproduce the experiment is a major obstacle in the study of Zeno Dynamics.

If the observed clock $r$ is to be a radioactive source, the first investigations should employ a radioactive isotope with a reasonably slow decay rate as measured in decays per second, e.g., Becquerels. 1 Bq = 1dps. The reason, again, is that the choice and design of detector will likely be limited to a practical 10-GHz response time, rather than the technical specifications of the manufacturer's claim of being in the THz range. The THz claim is based on loose assumptions regarding the statistical impossibility that is clearly evident in the graphs.

Since 10-GHz is at the utmost upper end of fiber optical information transfer, it cannot be the useful detection rate. Distance also figures into fiber optic data transfer rates. As a general rule:

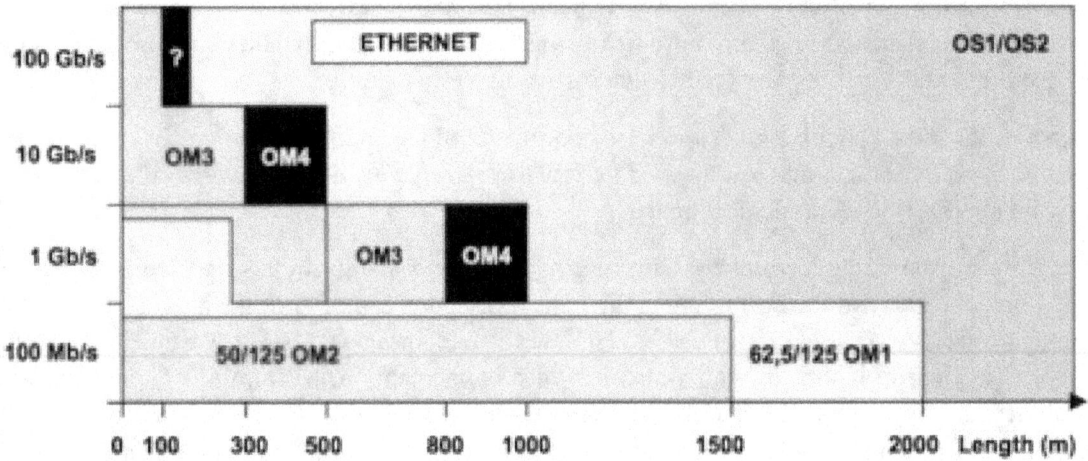

It is highly unlikely the conventional researcher in Zeno Dynamics is aware of or takes these distance issues into account. Where these distances are likely a much greater scale than one would surmise on a benchtop scale, they of course extrapolate downward and more importantly, simply do not appear as a limiting factor in the history of the Zeno literature.

In the simplest sense, conventional fiber optic cable has a refractive index of about 1.45, which limits the velocity of the photons to about 0.6c. Over 1000-meters this reduces transfer rate to about 15-MHz, unless exotic data compression is used. We will look later at the 'Loophole-free Bell Test,' where we find that this factor disqualifies the experiment. The problem being that the researchers treated the system, apparently unwittingly, as if the data transfer was at the speed of light, not accounting for the fiber optic refractive index.

The reported rates of faster speeds are of exotic systems that will not appear in academic laboratories. Data compression, low latency, and a host of variables that researchers in Zeno Dynamics do not know make such systems impractical, perhaps with the exception of people in the field of Quantum Computing.

When we extrapolate down from processing time and algorithmic design, we end up with a more practical upper limit of perhaps 500MHz, unless exotic data compression and other tactics are

### Detector and Data Transfer Characteristics

employed. This would be the practical data transfer rate on a laboratory scale using fiber optic cable.

$$1\text{-Curie is } 3.7\text{E}10 \text{ dps}, 37\text{GBq}, 37\text{GHz}.$$

$$1\text{mCi is } 3.7\text{E}7\text{Hz}, \text{ or } 37\text{MHz},$$

$$1u\text{Ci} = 3.7\text{E}4\text{Hz}, \text{ or } 37\text{Khz}.$$

However, the individual quantum scale events, atom-for-atom come into play in the Quantum Zeno Effect. So, we first identify a long half-life that translates to a slower temporal progression. Just a list of numbers and relationships for clarity's sake:

- $1\text{Ci} = 3.7\text{E}10 \text{ dps} = 37\text{Ghz}$
  - $1\text{mCi} = 3.7\text{E}7 \text{ dps} = 37\text{Mhz}$
  - $1u\text{Ci} = 3.7\text{E}4\text{dps} = 37\text{Khz}$
  - $100\text{-nCi} = 3.7\text{E}3 \text{ dps} = 3.7\text{Khz}$
  - These values are independent of the radioactive isotope or type.
- $1\text{Bq} = 1 \text{ dps}$
- $N = N_0 e^{kt}$ ; k is decay rate.
- $k = -\ln(2)/t_{1/2} \sim -0.693/t_{1/2}$

The negative sign simply indicates the sample is decreasing over time and can essentially be ignored. We'll use a 1-gram carbon natural sample, which is 1ppt C-14. The decay constant for C-14 will be, first converting from years to seconds, and rounding carbon's mas off to 12:

- $k = 0.693/5730\text{y} = 0.693/260752207208\text{s} = 3.835\text{E}-12/\text{s}$ ['per second].
- $1\text{g}/12 \times 1\text{E}-12 \times 6.022\text{E}23 \text{ g/M} = 5.018\text{E}10$ atoms C-14 in one-gram natural carbon. We want to know how many decays per second from that:
- $3.835\text{E}-12/\text{s} \times 5.018\text{E}10 = 0.192 \text{ dps}$.
- Thus, one trillionth of a gram of C-14 can expect 0.192dps. In Hertz it's just the reciprocal 5.196Hz.

So, we have the option to do it by mass as above or by Curies, which more readily converts to frequency.

- $78\text{Rb} = 345\text{s} [t_{1/2}]$. $k = 0.693/345\text{s} = 0.002/\text{s}$.
- If we had 1-trillionth of a gram 78-Rb would be; $1\text{E}-12/78 \times 6.022\text{E}23 = 7.72\text{E}9$ atoms 78-Rb.
- The dps would be $0.002/\text{s} \times 7.72\text{E}9 = 15.44\text{E}6/\text{s}$, or 15.44 MHz.

## Detector and Data Transfer Characteristics

The experimental expectation values of the frequency of events is at least 1,000 times higher than any rate of computational transfer of information and processing speed.

If we go back to the uCi route of the discussion, we need 100-nCi 78-R, which is going to be in the thousands of atoms. This dilution factor is akin to homeopathy, thousands of serial dilutions to get to the product. This is completely do-able, however, the cost of doing such with a radioactive isotope is not in the researchers' grasp, certainly not in an academic setting. I can only conclude that if some 78-Rb experiment actually took place, the amount of isotope in Curies was not known, which would explain why it was not reported. The dead-time and pile-up issues, the inability to calculate expectation values for the isotope, choice of detectors, hardware, firmware, software, transfer lines, and so on, that they did not report quantitative results. This is problematic across the Zeno literature.

The beauty of C-14 is also as a beta emitter, getting hard specifics on detector design should not be a problem. It will be an EM detector. We target 100-nCi is 3.7Khz event rate, which is ideally suited for our detector and electronics limitations. We can easily sweep into the high MHz range confidently seeking a Zeno Effect.

The discussion was of a micro-circuit consisting of some natural clock source and an associated detector appropriate for that source. In the case of radioactivity, for a beta emitter this is not at all different from any other EM phenomenon. In addition, with C-14, a population of 10-million atoms is not a difficult thing to dope into a circuit.

I want to dope 10-million microcircuits into a single chip with a surface area of perhaps one or two square millimeters, is also not at all difficult. The hypothesis is that by employing 10-million Quantum Zeno Effect systems we can create a small but macroscopic change in spatial geometry on a very constrained scale extending perhaps ten or more microns from the surface of the circuit. The idea is to test that hypothesis by passing laser over the surface to see if the 'beam' is deflected, and/or potentially placing it in or incident to a laser 'beam' in an interferometer, which is my method of choice. An interferometer will tell us by phase shift if there is any change in spatial geometry.

Simply, this is too easy to do, not to do it. And if the hypothesis is correct, the payoff of *artificial manipulation of spacetime geometry with zero energy* is incalculably huge.

Details of Zeno Design in the Case of Radioactive Detection

## **Details of Zeno Design in the Case of Radioactive Detection**

The design of the first stable Zeno Effect first incorporates the detector and its specifics, suitable for the task, which in this case will be beta emission detection at the energy specific to C-14. The beta emission is 156.5 KeV. It will not however emit the beta 'particles' in a straight line for the detector. Thus, the detector surface cross section and distance have to be taken into account.

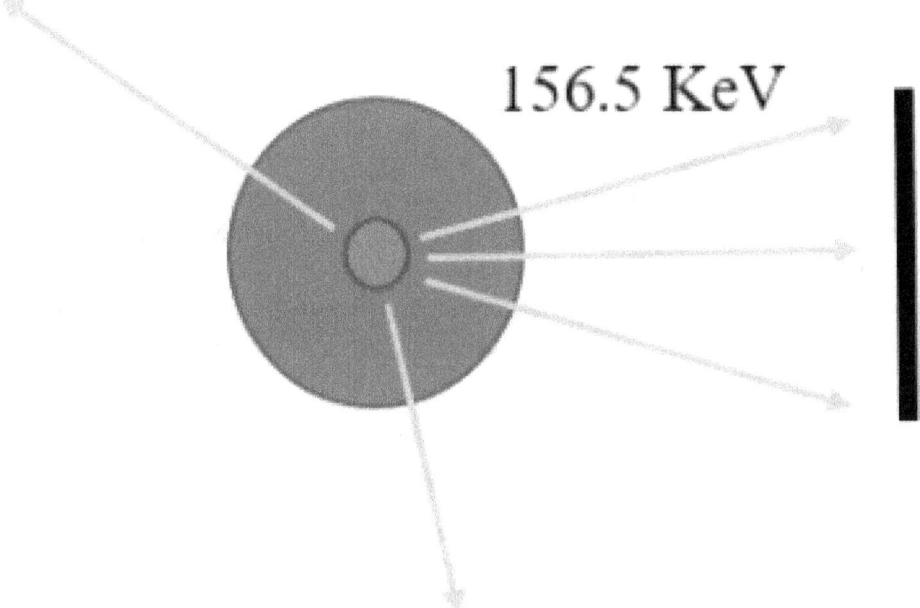

Note that the vectors of the beta emission are not *circular*, but spherical, {x, y, z}. $A = 4\pi r^2$ defines the number of decays per second at 156.5 KeV drops to, then the cross section of the detector surface. I'm just going to invent some numbers here for simplicity.

The detector surface is 10-mm from the radioactive source with a surface area of 10-mm². The surface area [of the expanding sphere] at 10-mm from the source with respect to beta 'particles' is then 1256.6-mm². The ratio of the surface area of the detector surface to the beta emissions is then 1:12.566.

The radioactive source is the aforementioned 1-*u*Ci, at 37-Kdps. The detector cross section [surface] will then see 37K/12.566 = 2944.5 dps, at 156.5 KeV. Which is 2.9445-KHz.

This is a sweet spot in the detector and electronic limitations because we do not yet know what these are and what to expect. We can comfortably sweep through the high MHz ranges looking for a Zeno Effect. Once we find a Zeno Effect, at that point we can begin doing some developmental work sweeping through the detection rates to quantify results. We would then have solid quantifiable and completely reproducible conditions to report.

A detector from TeraSource:

Details of Zeno Design in the Case of Radioactive Detection

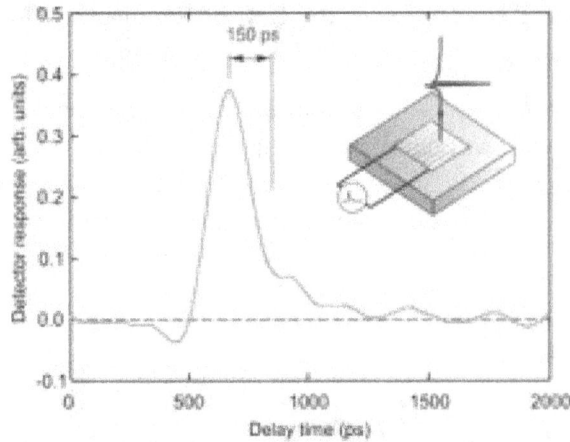

Taking the signal rise, then exponential decay-to-zero, without pushing any limits is 2 nanoseconds, thus the effective detection rate for this detector is 500-MHz. That seems like 2000 times less than the manufacturer's claim because it is. The detector will not effectively differentiate two separate signals in the 150 picoseconds depicted in the image. The peak is from the peak's midpoint to its half-height, only. The above detector is not a beta detector, just an example of a photon detector.

A good paper on detectors is: Radiation detector deadtime and pile up: A review of the status of science. Shoaib Usman, Amol Patil doi.org/10.1016/j.net.2018.06.014

In that paper dead time and pile up in radiation detectors is discussed. This is a topic that I cannot find a single reference to in the Zeno literature. It is not logical to assume the authors are knowledgeable of these concerns but do not discuss them because such should be regarded as common knowledge.

The design of a system that artificially alters spatial geometry obviously can only operate in the 4-observable dimensions of common spacetime. When constrained as such the principle is as simple as a rubber band and propeller; 2-clocks metering one another alters the path length between them as meeting the demands of General Relativity.

**Deadtime:**

Is the ability for a detector to separate one event from another and based upon:

- Intrinsic deadtime, drift time; for a gas detector a very long time. For a solid-state detector less, but there is luminescence inherent in such, not much less. Luminescence is a characteristic intrinsic to any form of scintillation. There is no detector void of both drift and/or luminescence, it will have one or both.
- Electronic digital to analog conversion. If this were as fast as the classic detector's claim your CD player wouldn't require oversampling, which is in the KHz range.
- Computer algorithm response time. This is a function of the language, which must be compiled, and processor speed. In some cases, the compiling may be serial, from for example LISP to C++ to Hex. C++ must be compiled down to C before going to hex, and so on. If the

Details of Zeno Design in the Case of Radioactive Detection

processor is in the GHz range, THz detection is not going to happen. If yu divide the above graph into individual data points, say, 2000, this is a limit in the MHz range.
- Amplifier characteristics. As we see in the TeraSource graph, the exponential shape of the peak may or may not be a function of the actual detector, is at least to some extent a function of the amplifier itself.
- Computer buffer time, which is slave to whatever laptop is being used.
- A system must have a preamplifier. This extends the exponential decay rate into the millisecond domain.

Usman and Patil point out: Charge collection time of the detector determines the rise time of the tailed pulse produced by the preamplifier. Amplifier's shaping of the pulse plays a critical role in preserving the spectroscopic and timing (or count rate) information. *A compromise between preserving the rate and the pulse height information is generally needed for any high-count rate application.*

When you compromise on this level you have the added burden that a lowered signal response capable of detecting at rates 'approaching infinity' cannot possibly have a detection efficiency of 100%. In turn, a higher efficiency results in greater dead time, which is the heightened inability to distinguish one event from another.

You thus have less than 100% efficiency, which spirals downward as you increase the detection rate. This will look like a Zeno Effect. As you increase efficiency, you heighten the amplifier and preamplifier effects, which results in a longer exponential decay rate, which can also look just like a Zeno Effect:

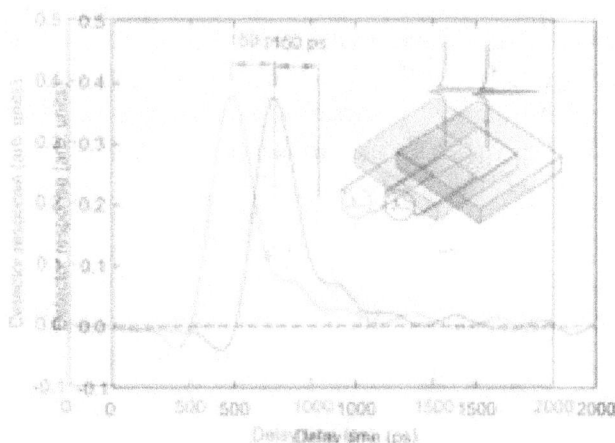

Few papers in the Zeno literature discuss these issues. Again, these appear to be limited to the Quantum Computing fields of study.

The next issue is the actual detector surface itself. The surface has a 'ballistic' map. This is the fact that locality on the detector surface results in different signal peak-to-peak amplification times. For example, if we look at an Electron Multiplier:

Details of Zeno Design in the Case of Radioactive Detection

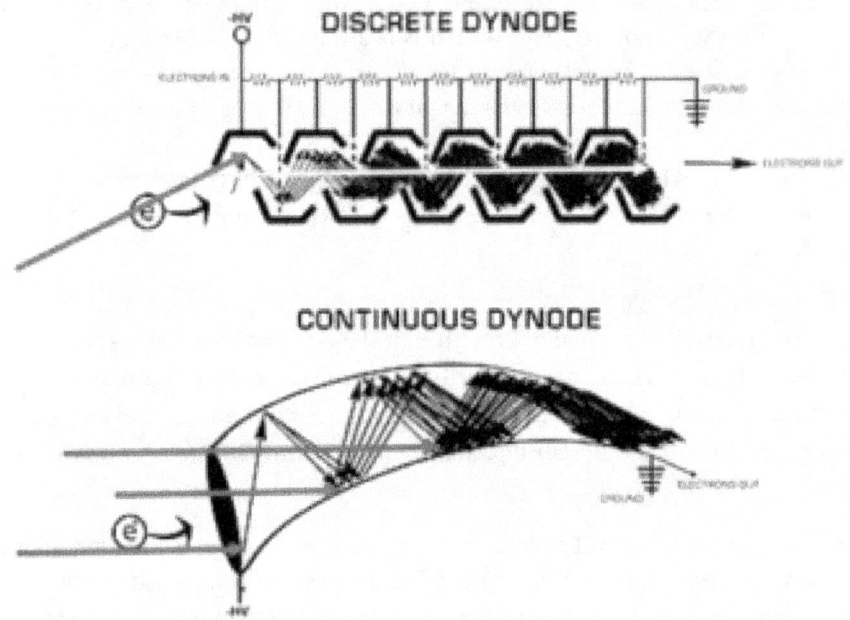

In the discrete dynode the yellow arrow indicates that an event can shoot clear through the dynode without being detected. In fact, it appears that if the detector is placed incident to the radioactive source or otherwise A-D converter this will usually be the case. The continuous or smooth dynode shows us that the signal amplitude varies by orders of magnitude depending on the angle of incidence of the source. The bottom arrow is maximally amplified, the top arrow may not be detected, depending on the cutoff. In order to detect the 'particles' that reach deep into the dynode that produce orders of magnitude less signal, you have to increase the efficiency, which in turn increases the dead time. If you do not increase the efficiency then you would have to take the statistical map of how many particles hit the physical map of the dynode surface at the rate you are claiming to detect at, at low efficiency.

Note that in the continuous dynode design, for example, the particle takes a different time of flight in striking each section of the dynode as we progress deeper into the detector. Over a distance of a centimeter this is in the 1-GHz range of differing detection times, assuming $v = c$, which cannot occur outside of a vacuum. Thus, a dynode for beta decay is limited by this practical limit alone.

- Signal rise time for a dynode is typically 10-ns to 100-ns [mostly Gaussian], equating to 100 to 10-MHz upper limit.
- Signal transfer to 1-GHz
- Latency 1-GHz
- computer speed in the GHz range

These add up to define an upper limit from 100 down to 10-MHz.

Details of Zeno Design in the Case of Radioactive Detection

**Pile Up:**

Signal pile-up is simply when two events occur that are not separated by all of the electronic limitations listed above. However, in the case of for instance a dynode, this can also be a surface chemistry phenomenon where the depleted orbitals cannot relax at sufficient speed to keep up. In a scintillation detector can be the same electron orbital deletion to relaxation time, which is more related to solid state scintillation. Then, given the actual velocity of the, for example, beta particle is unknown, pile up can literally be a population of beta particles emitted randomly that are physically indifferentiable from one another. In the case of a dynode, can be two or more particles striking the physical surface in close proximity that are not differentiated by the surface chemistry.

In the case of photon [gamma] detectors, the detection is similar, but in general less of an issue because the chemistry is strictly PN junction type. PN junctions will have a greater capacity to separate events. Of the various non-natural systems that appear in the literature, mostly in the Quantum Computing regime, ultimately refer back to PN junctions, as all detection technology is Electric and Magnetic boson dependent.

Given our maximum time issue is related to preamplification in the millisecond to microsecond domain, this pile-up issue has to be considered *in the KHz range*. This makes our 'Turing paradox' look more like:

$$\lim_{N \to 1KHz} P \to N = 1$$

As our detection rate is turned up into the KHz range the detector and associated electronics will only indicate a single event. I would be more confident If I could find any discussion of this sort in the Zeno literature, but I cannot. This particular phenomenon is characteristic of beta detectors, gamma detectors much less so. In systems not using radiative decays as the natural clock system these concerns are minimized. However, data transfer, digital to analog conversion rates, and computation times all figure into a lower value than researchers seem to be presuming their systems are capable of.

When we get into other natural clock systems, of which there are many, ultimately every type of detector is an ElectroMagnetic detector, because no other science exists. Nothing is different from the phenomenon related to radiation detection, only the exception that drift time and luminescence associated with scintillation isn't a concern. However, again, pile up can be a physical phenomenon of detector chemistry, of orbital depletion and relaxation times.

Then: We cannot detect native wave functions, only their associated Electric and Magnetic bosons. For example, we cannot detect the native wave function of the photon, only its associated Electric and Magnetic bosons. Even in 'field theories' these possess off-shell mass, which in turn demands they are limited to $v \ll c$. We plug that back into the century of ontology of the 2-slit experiments and effectively eliminate 99% of everything written on that subject.

Details of Zeno Design in the Case of Radioactive Detection

The reason wifi does not extend above roughly 5 to 10 GHz is simply, that is the limit of current technology, the ability to distinguish events at a rate above 10-GHz does not exist in the commercial market. It isn't the refractive index and scattering of air that is the problem. Fiber optic cable is limited to 10-Gbyte/sec. That is as fast as man can go. That includes dead time in the electronics of any computer attached to a detector. We cannot process information faster than this.

In the above description we have secured a C-14 source with precisely known decay energies and decay rate in decays per second, taken the cross section of the detector surface into account, and the electronics are well within the constraints of the proposed model for a Stable Zeno Effect; limited to:

> The radioactive source [C-14] is the aforementioned $0.07u$Ci, at 2620 dps. The detector cross section [surface] will then see $2620/12.566 = 208.5$ dps, at 156.5 KeV.

With that we can confidently perform an experimental setup in the high MHz range of detection and sweep from KHz through MHz searching for a stable Zeno Effect. Because we know the exacting details of the radioactive source, detector, and electronic limitations, we can thus derive exacting mathematical relationships describing the Zeno Effect, *which will be unique to this particular system.* If I change from C-14 to some other isotope, regardless of emission type, that will certainly *not have* the same mathematical relationships as C-14. This is likely true if I change the amount of C-14 isotope present, because the statistics change somewhat. If I change to an entirely different type of natural clock system, such as electron tunneling, that is also going to have different characteristics. The physical dimensions of the test apparatus are obviously important, as this defines cross section, particle travel times, *which are rather slow,* and so on. Even identical systems will have slightly different physical characteristics of this sort and have to be 'tuned,' to optimize and characterize any type of results.

No one has ever reproduced another researcher's experiment. No researcher has ever published sufficient information to do so. Few researchers have published *quantifiable results,* only summary graphs. All such graphs are normalized to some arbitrary, not quantifiable values that can be related back to any expectation value.

Every type of natural phenomenon follows a different probabilistic set and will thus have very different Zeno descriptions. The first steps involve exploring natural systems. The next steps should look at artificial systems, such as a simple clock circuit. When we get to that stage, we find a host of variables. However, if an artificial clock circuit *can be used*, this presents a host of superior qualities over that of a natural system. In all likelihood, two identical systems will have similar but not identical characteristics, requiring fine tuning of sorts.

A beta detector will have at best the same characteristics and will thus be limited to no better than 500-MHz. Electron multipliers will be significantly slower due to the much longer exponential decay rates of populations of events on the EM's surface. Photomultiplier tubes at least 2-orders of magnitude slower.

In this we see that we cannot take the manufacturer's claim of detection rate for face value. They after all, compete with one another based on these statements. The following is a published memo

Details of Zeno Design in the Case of Radioactive Detection

from DARPA: Photon Counting Detectors for the 1.0 - 2.0 Micron Wavelength Range. Michael A. Krainak. NASA Goddard Space Flight Center Greenbelt, Maryland 20771

In it the author describes a *single photon counter* operating in the 150-THz range, with explicit details on the detector's response time, etc., as well as the statistical analysis of how many photons are required to hit the detector surface that actually results in a signal. The technology discussed takes all of the issues and describes the detector type and associated circuitry for processing at the microcircuit level. The goal is to produce a true THz detector that operates in the true THz range of operation.

Most people have no search engines that will turn up DARPA memos, however, the noted paper is at the noted source and available. The more interesting aspect of the paper is that upon reading it one can see the Zeno Effect in a more skeptical light with respect to how many such experiments approach validity. It is simply evident that a portion of unknown weight, but considerable, of the Zeno literature are electronic issues, not Zeno Effects.

For example, in our Bell Loophole-Free test, we saw that the evidently unknown of the photons travelling at subluminal velocity, ~0.6c through fiber optic cable was not accounted for. The authors clearly describe $v = c$ as the transfer speed of the photons in the experiment, and no statement suggesting subluminal information transfer appears in the paper.

When we add on top that there is no discussion, probably because it is unknown, not considered, of the photon counter's probabilistic response, the entire experimental setup is now in total question. Both the passage of photons and the counting of photons is not accounted for, and there can be no relevance to the experiment. The type of detector and associated dead-time and pile-up issues cannot be ascertained because the authors have not addressed the subject at all. This is a highly cited paper, and it is evident that those citing the paper are equally unaware of this $v < c$ issue in the experimental setup. It is even treated as a landmark in Wikipedia.

The Zeno and Anti-Zeno Effect as Detector Dead Time and Pile Up

## The Zeno and Anti-Zeno Effect as Detector Dead Time and Pile Up

I like the Raizen et al graph because it has no hypothesis associated with it in the experimental description:

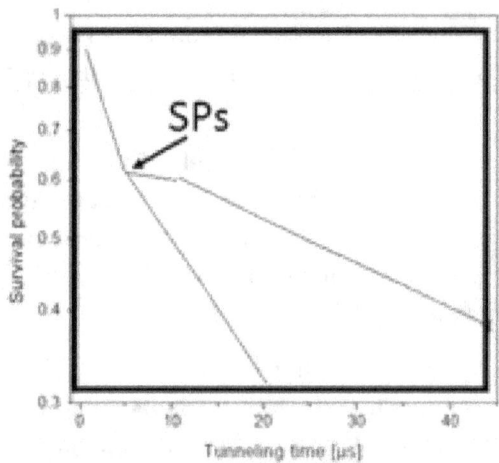

Note, as stated, we have an x-axis that appears superficially to be quantifiable, the y-axis is normalized only to 1. However, there is no information regarding the expectation values, e.g., the quantifiable expectation of tunneling rate under the conditions of the experiment, so there is nothing quantifiable about the graph. However, it does demonstrate a divergence between the Zeno and Anti-Zeno Effects, it does not however demonstrate that the experiment is reproducible even locally to the laboratory it was conducted in. This was the quantum tunneling experiment that has been reproduced most recently by [Y. S. Patil, S. Chakram and M. Vengalattore. Quantum Control by Imaging: The Zeno effect in an ultracold lattice gas. arXiv:1411.2678v1 [cond-mat.quant-gas] 11 Nov 2014]. That point marked $SP_s$ is what I refer to as the splitting point. Facchi et al produced this similar effect:

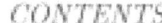

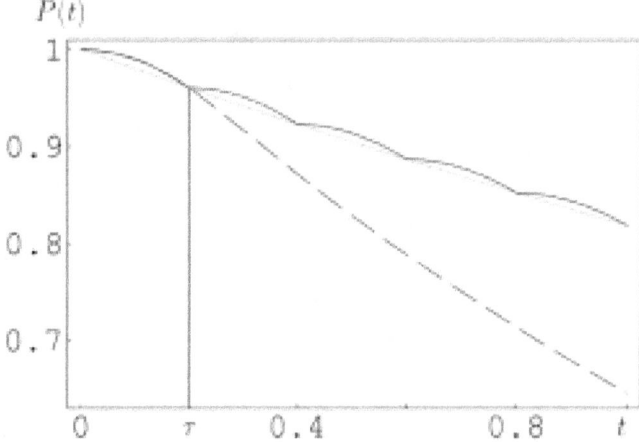

The Zeno and Anti-Zeno Effect as Detector Dead Time and Pile Up

The Fachii paper demonstrates how little information you can have and provided you generate a graph, publish it. There is nothing in the paper that describes the graph, it is just a set of arbitrary x an y axis. The paper evokes 'superspace' as the hypothetical rationale. The author does not present any x and y axis information because it is unknown. It is merely some penchant of numbers normalized to 1, then goes off into unobservable dimensionalities.

We look again at: **Pile Up:**

> Signal pile-up is simply when two events occur that are not separated by all of the electronic limitations listed above. However, in the case of for instance a dynode, this can be a surface chemistry phenomenon where the depleted orbitals cannot relax at sufficient speed to keep up. In a scintillation detector can be the same electron orbital deletion to relaxation time. Then, given the actual velocity of the, for example, beta particle is unknown, pile up can literally be a population of beta particles emitted randomly that are physically indifferentiable from one another.
>
> Given our maximum time issue is related to preamplification in the millisecond domain, this pile-up issue has to be considered *in the KHz range*. This makes our 'Turing paradox' look more like: [again]

$$\lim_{N \to 1 KHz} P \to N = 1$$

At the rate of processing where detector pile up is in the case, for example, of relaxation times, we have a population of excitations relaxing that are still in the Dead Time zone of the detectors physical surface, in this example [scintillation] luminescing. Because we have the processing power to differentiate these, but no true signal is coming from the C-14 source, we are seeing more counts than the C-14 is emitting.

The Divergence point in this case is as shown in the graph a direct function where one would expect the Dead Time and Pile-Up to overlap, because the Dead Time vs Pile-Up is the physical phenomenon that is occurring.

In a more straight forward approach, since it is suspect the authors do not know the actual value in decays per second to expect, the Anti-Zeno Effect being reported in any number of such papers may simply be a detection rate, in the above example, below 262 Hz. If the source is, as likely the case, hotter than $0.07 uCi$, detection rates in the KHz range can be the mere effect of detector vs dps.

This also is not a good sales pitch for the artificial alteration of spatial geometry hypothesis. However, it is a critical feature to be aware of in *any valid design* of a Zeno experimental setup. The idea is to note these issues, to nitpick them, to find every loophole and electronic issue, and deal with it on the front end. The number of papers that do not deal with such issues is, *almost all of them*.

Again, it is evident that quantifiable results within detector and processer issues are limited to Quantum Computing. This makes perfect sense, because each and all of these issues are common

153

## The Zeno and Anti-Zeno Effect as Detector Dead Time and Pile Up

knowledge to researchers in the filed of computer architecture design, electrical engineering with the focus on quantum computation, which must be validated by more conventional means.

For example, the 'Quantum Supremacy' paper that made headlines dictates that it would take 10,000 years for a conventional computer to render such processes in a Shannon type exponential form. However, this then demands that: *it would take 10,000 years to validate the experiment, which was not even done by the same procedure*. That is, there was no validation that the data was in fact as it claimed, not by conventional means, because it cannot be done. The system *cannot validate itself*, and no attempt was made to do so. Only the experiment within its own constraints was performed and analyzed. Thus, we have a hypothetical result with no means to validate it.

The goal is to produce a robust, quantifiable, and reproducible Zeno effect. Thus, all of these detector, data transfer, and computation rates have to be strictly understood and adhered to.

Summary of Quantifiable Zeno Experiment

## Summary of Quantifiable Zeno Experiment

In our example we are using C-14 $0.07u$Ci, at 2620 dps. The detector cross section [surface] will then see $2620/12.566 = 208.5$ dps, at 156.5 KeV.

Our detector can be a dynode or scintillator, each will perform at this rate without dead-time or pile-up. A 'pancake probe,' common to Geiger counters, is insufficient, as the detector characteristics of these are not consistent between manufactured units and they tend to decrease in efficiency with time and use.

Our information transfer should be LAN operating in the high MHz range, our processor operating at a minimum of 2GHz. Fiber Optic is acceptable, but introduces other variables associated with additional D to A conversion in the secondary detectors of the photons in the cables.

The expectation values for the decays per second give us a safe margin to sweep from the KHz into the high MHz range confidently, searching for a *quantifiable Zeno Effect* that will yield specific information in the form:

- *r:* $^{14}$C at 1-$u$g @ 37Khz events.
- *R:* the detection, data transfer, and computation rates in observations per second limited to the range ~1KHz through 500MHz.

There is no prior art of this sort. There are quantitative results in many Quantum Computing papers, but in radiative sources, none. After a stable Zeno Effect is achieved under these operating conditions the obvious next step is to change the dps by changing the number of micro-Curies present, alter *r*, so as to comparatively characterize the effect of changing the temporal progression rate of a natural system as per the Zeno Effect. This will be dependent on the results. For example, if we find that we are at the maximum high end of our ability to *[observe]*, approaching our practical maximum data acquisition rate of 500MHz, then increasing the quantity from 1$u$Ci is not much of an option. We find that in order to remain within our practical data acquisition constraints we must decrease the amount of C-14 present in the source.

If, for example, C-14 at 1$u$Ci represents the upper end of what we can effectively observe as a Zeno Effect with current technological limitations, then isotopes with significantly shorter half-lives will not be practical. On the other hand, we can determine what the upper end is for the sample in question and extrapolate out to other elemental isotopes. In this, we keep in mind that a beta detector will have some noticeable disadvantage to a gamma detector in the detection rate scheme, but probably only up to about 1GHz as a practical observation rate that accounts for all of the other electronic, computer processor, and information transfer technological imitations. Any system will only be as fast as its slowest component.

## A Faster Than Light Discontinuous 'Jump' Drive Spacetime Manifold with Zero Mass-Energy Requirement and Zero Negative Energy Density: Part I

### The Zero Energy FTL Spacetime Manifold

There will be no motion derivation because again, there is no notion of continuous motion associated with this manifold. The Alcubierre derivation cannot possibly be correct, again, as the requirement of motion for the original manifold is 10E38 solar masses worth of mass-energy, *prior to setting the thing in motion*, there can neither be any sense of motion for the original derivation by Alcubierre, although conceptually, the notion of a localized spacetime manifold and the initial mathematical attack was an act of brilliance.

There will also be no 'contraction' derived as there is no 'length contraction' associated with the drive mechanism. Length contraction only occurs in nature under General Relativity in the most extreme conditions of the one millisecond 'kick' of two coalescing black holes. This is the root to our 10E38 solar mass issues in GR, and under Lorentz transform, 'length contraction' is the problem that renders the original manifold a futile effort, albeit again, was an act of admitted brilliance.

When we look at the Schwarzschild transformations, what was not taken into account by Alcubierre, evidently, was that the Preferential Frame of Reference must be that from *the inside of the manifold, looking out at flat spacetime*.

As for 'length contraction' under Lorentzian conditions, this was described in prior papers and texts and will not be a subject of focus here.

The manifold is purely a discontinuous 'jump' drive that works by extending the manifold out via *dilation* to great distances. The dilation requires no energy as this is associated with the phenomenon of observation, only.

Quantizing then over the localized domain:

1. $R$ is the observing system.
2. $r$ is the observed system.
3. $k$ is the divergence between the Zeno and Anti-Zeno Effects.
4. The primes are fractal designations.
5. The term, $Zt'$ is the change in temporal rate of progression as a causal result of the Zeno Effect.

- The term $c^4$ becomes $(L_p^4/t_p^4)$. This will not be primed as a fractal; it is a constant.

The Zero Energy FTL Spacetime Manifold

- The term G becomes G' (relativistic), then in relation to **R**, designated as G''. G-prime is not a suggestion that the gravitational constant of the universe is changing, it is an artifact of Preferential Frame of Reference.
- **The velocity $v$ becomes quantized as nLp/xtp.** Note that these are integer values, $n$Lp and x$t_p$ as describing velocity $< c$, provided $n < x$.
- The term $v_s^2$ can take the form $((nLp/xt_p)/(1Lp/1t_p))^2$. At $v > c$, $n$Lp $>$ $x$tp. That is all that is required.
- The term $(y^2 + z^2)$ becomes $(Lp'_y{}^2 + Lp'_z{}^2)$. These are primed fractals.
- The determinant metric $g^2$ will be treated like T' (the Trace Matrix*) and simply referred to as $g'^2$. This is an extremely lengthy description of the Tensor in a fractal conformal scaling relationship and is a separate lengthy text from this paper.
- The term $r_s$ (radius of the center of the phenomenon) is represented by $r_s(t) = [(x-x_s(t))^2 + y^2 + z^2]^{1/2}$ essentially replaced by quantized values such that:

$$[(Lp''x - Lp''x_s(Zt'))^2 + Lp'_y{}^2 + Lp'_z{}^2]^{1/2}$$

The highlighted t' term is the determining factor for the fractal scaling of the unitary intervals, Lp', Lp'', tp', and tp'', along with G'. For the term:

$$\theta = v_s \frac{x_s}{r_s} \frac{df}{dr_s}$$

From the original Alcubierre Metric, as a function of $x$ and $\rho = (y^2 + z^2)^{1/2}$ which he plotted as:

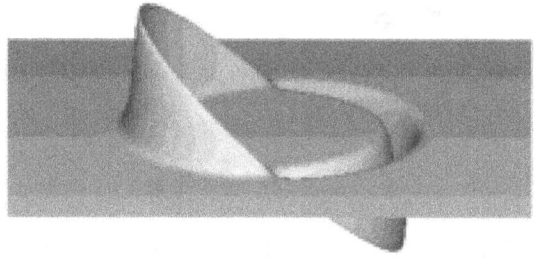

He has spacetime dilating to the left [behind the ship] and contracting in front [to the right]. This is backwards and is the reason 10,000-trillion universes worth of exotic mass-energy are required to produce the manifold. What we need to do is:

The Zero Energy FTL Spacetime Manifold

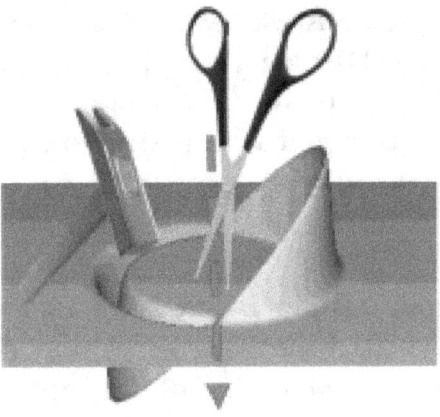

We need to preserve the front, which was the rear, of the manifold and ditch what is the rear, which was the front, of the manifold. Meaning, the original manifold proposed by Alcubierre is upside down and backward. We need to flip it around and discard what is now the rear of the manifold, the 'length contraction' portion, leaving that portion totally flat. I believe I pointed out in prior papers that 'length contraction,' particularly in a Schwarzschild transform, only occurs in the rarest and extreme instance in nature when two black holes coalesce, for just 1-millisecond, regarded as the 'kick,' referred to as 'chirp mass.' This only is observable if we are in the direct position where one of the coalescing bodies is 'kicked' directly towards us, causing spacetime to ripple, like kicking a carpet toward the observer. The Alcubierre Manifold requires this extreme condition along each point-for-point of the 'length contracting' portion of the manifold. In a sense, this 'negative energy' can be manifest by $10E38 M_\odot$ 'kicked' toward the manifold's rear end and does not require 'exotic matter universes.' The upward spire in the Alcubierre Manifold is 'chirp-mass,' and occurs in nature, it is neither negative nor is it exotic.

Slicing off that portion of the manifold is currently a work in progress and may not be included in this paper, as I have to write a shaping function from scratch, which will probably change the energy Tensor. For now, we will proceed with the manifold as is and I will update it later.

This is done via the equations that follow. Note that the origin of the fundamental shaping function is double primed, this will be explained later.

$$R\alpha \pm k$$

Where $k$ is defined as the Splitting Point, where the Zeno and Anti-Zeno Effects diverge. This is a value purely derived via experimental data from the specific system being used. Every type of system will have a different Splitting Point where the Zeno and Anti-Zeno Effects diverge.

And

$$R\alpha t''$$

## The Zero Energy FTL Spacetime Manifold

Where t'' is the second order fractal from:

$$L_p'' \rightleftharpoons L_{p0} / \sqrt{1 - \frac{2G''M}{n'L_p(\frac{1Lp'}{1tp'})^2}}$$

$$t_p'' \rightleftharpoons t_{p0} / \sqrt{1 - \frac{2G''M}{n'L_p(\frac{1Lp'}{1tp'})^2}}$$

$$\mathcal{B} \rightleftharpoons \sqrt{1 - \frac{2G''M}{n'L_p(\frac{1Lp'}{1tp'})^2}}$$

$$L_p' \rightleftharpoons L_{p0} / \mathcal{B}$$

$$t_p' \rightleftharpoons t_{p0} / \mathcal{B}$$

*It is important to keep in mind that the above form is from the Preferential Frame of Reference of the 'engine,' looking out into flat spacetime.* Each point on the spacetime manifold will have a different value t' and t'', thus l' and l''. The working principle of the jump drive is to dilate the unitary Planck interval Lp in front of the 'ship.' This means that the Alcubierre Manifold will essentially be 'cut in half,' there will be no region of 'length contraction.'

As stated, the negative mass-energy problem is the 'contraction,' specifically, it is the need to dilate time and 'contract' space, we get into this unobservable sqrt(-1) issue, which is also a violation of Preservation of Information Conservation.

Also note that the inside term $c^2$ takes the quantized form $(1Lp'/1tp')^2$ and is primed, this is an artifact of observation that must be taken into account.

The Zero Energy FTL Spacetime Manifold

And the Corrected Metric then begins to take the quantized form:

$$\varphi f_{R:r}(\pm k) \mapsto \left| \frac{\left(\frac{Lp^4}{tp^4}\right)}{8\pi G'} \frac{\left(\frac{nLp'}{xtp'}\right)^2 (Lp'^2{}_y + Lp'^2{}_z)}{4g'^2[(Lp''_x - Lp''_{xs}(Zt'))^2 + (Lp'^2{}_y + Lp'^2{}_z)]} \left(\frac{df}{dr}\right)^2 \right|$$

$\varphi$ is the Maric Operator, which is the reason for the absolute value brackets. This is a natural consequence of:

$$\varphi(\pm v) \equiv |v|$$

This **dr** highlighted in yellow is not the Zeno Rate Equation, $f_{R:r}$, but the shaping function as Alcubierre presented in 1994. At this time, I am going over a paper [*1] that treats the original metric in terms of Conformal Gravitation. It may be months before I work out the hybridized metric and quantize it. Therefore, I am leaving this particular term as is, and just highlighting and noting that it will be edited at some future date. The shaping function is the graphics you see of the Alcubierre metric. The paper noted below is highly recommended reading, as it analyzes and elucidates the original metric very nicely.

*1: Conformal Gravity and the Alcubierre Warp Drive Metric. Gabriele U. Varieschi and Zily Burstein, Department of Physics, Loyola Marymount University, Los Angeles, CA. http://dx.doi.org/10.1155/2013/482734

The new shaping function will have no 'contraction,' as this is an absurdity. The principle of the new metric is to extend spacetime out to great distances, e.g., purely dilation. There is no instance in all of nature where spacetime 'contracts,' save for the very millisecond when two black holes coalesce. If we are lucky enough to be aligned with this 'kick,' we can see the upward spire of 'length contraction,' under such extreme conditions, for exactly 1-millisecond. There is no Lorentz condition where length 'contracts.' That absurdity has never been observed simply because it does not happen.

The Zero Energy FTL Spacetime Manifold

This is imaging by LIGO of GW150914:

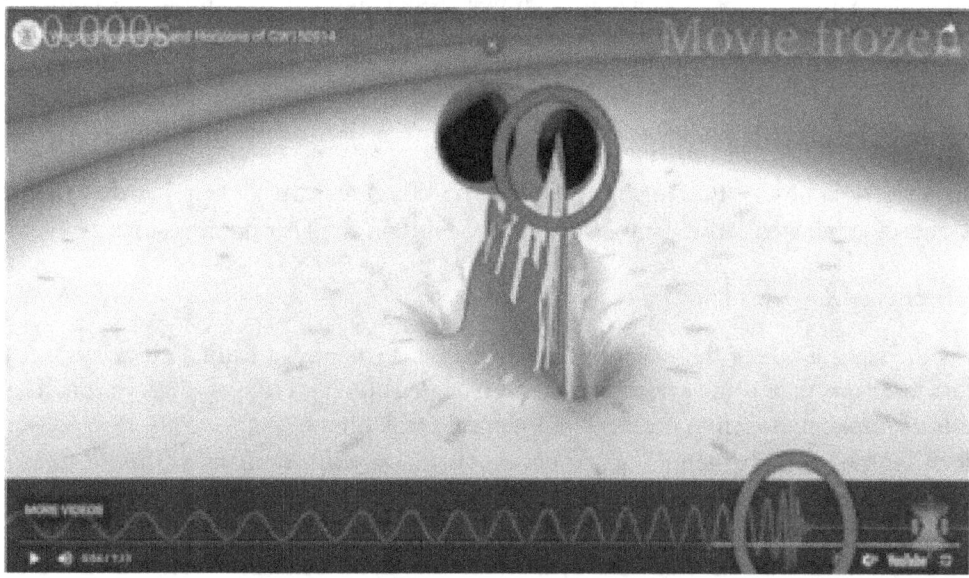

That peak is 'length contraction.' It requires two massive black holes to occur. The circle at the lower right is the 'chirp-mass' signal. The two gravity wells on either side are typically regarded as 'length contraction.' However, this is incorrect, as the amount of time it would take you or I to journey down into one of those holes would be *infinity*. The idea of time dilation co-occurring with 'length contraction' is the reason physics has achieved absolutely nothing tangible since the bomb in 1945. If you 'believe' in Minkowski spacetime, you will never achieve this engine. Time dilation co-occurring with 'length contraction' is a gross violation of Preservation of Information Conservation, because we have a permanent record of temporal rate change via the permanent disparagement between clocks, but our 'length contraction' is regarded conventionally as artifact of observation; thus, an accounting of the temporal elements but bankrupt in the spatial elements, violation of Preservation of Information Conservation.

It is critical to point out here that the term highlighted in blue is $(v/c)^2$, originally [Alcubierre] just $v^2$. This is illogical in the Alcubierre manifold as there is no 'velocity' within the local constraints of the manifold, nor is there any velocity associated with any notion of motion of the entire manifold itself. Alcubierre's derivation is a static entity, there is no dynamic quality to it whatsoever. It cannot move, because *tangible* motion is described by energy, and 'negative energy' does not occur in this universe, thus, the Alcubierre manifold does not move in this universe, or any other, it is simply incorrect, brilliant, but incorrect. To put the incorrectness into perspective, the authors of the above paper [*1] did the calculation to work out to 6.12E92 erg which works out to $3.42E38 M_o$. However, it is 3.42E38 *exotic solar masses, of negative energy*. Last week's value for our cosmos was about 2E22 solar masses. Thus, the Alcubierre manifold requires 2E16 *non-existent 'negative' universes* worth of mass-energy, *and that is not even moving, much less superluminal*.

Thus, the entire numerator in Alcubierre's derivation [of the energy Tensor] is *zero*. As such, the entire Alcubierre Spacetime Manifold Tensor *is zero*, the velocity is zero, and the negative sign is consequently

161

## The Zero Energy FTL Spacetime Manifold

non-sequitur, as it only states, negative zero. The only thing that can survive is the shaping functions, and those have to be changed radically such that there is no 'length contraction,' only dilation.

Note the use of the Maric Operator highlighted in red on the left and the resulting absolute values on the right. This is key, as there is no sequitur 'negative energy density,' nor is there a description for such an aberration.

Again, this is via observation rate only, e.g., Zeno Effect and has no sequitur description of energy nor mass-energy equivalent. The 'Bray' Spacetime Manifold requires no mass-energy equivalent.

The function, $f_{R:r}$ is explained as follows:

- $r$ is the natural rate of the system being observed. Typically, a natural clock system such as a radioactive source is the system being interrogated at a high rate of observation. The system, in this example, radioactive source, has a natural clock rate, $r$.
- $R$ is the 'observing system,' e.g., detector, what have you. By manipulating $R$, the detection rate, we produce a stable Zeno Effect, which affects the natural clock system, $r$.
- $f_{R:r}$ thus becomes a self-similar, iterative process, where the natural clock system $r$ is affected by observation rate, $R$, yielding a new [in this example] decay rate, which is then affected by $R$, and so on:

$$f_R \rightleftharpoons f_r$$

Note: Again, this is *not* the shaping function $df/dr$, but the Zeno Rate Equation.

The value, $(\pm k)$ is the 'Splitting Point,' where the Zeno and Anti-Zeno Effects *diverge*:

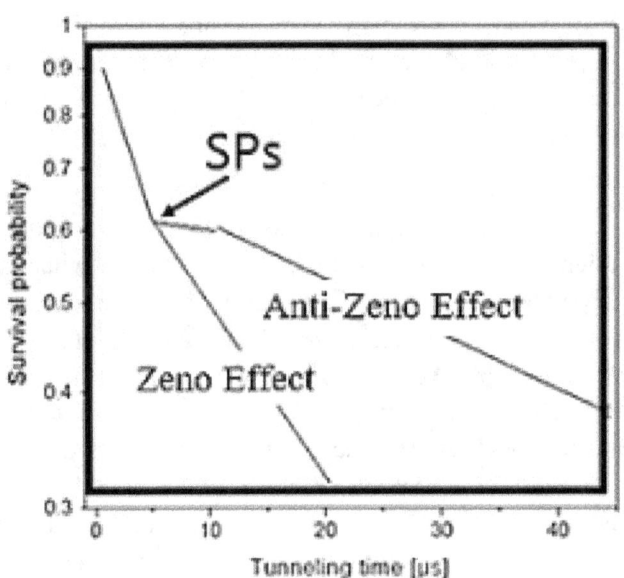

This was the quantum tunneling experiment that has been reproduced most recently by [Y. S. Patil, S. Chakram and M. Vengalattore. Quantum Control by Imaging: The Zeno effect in an ultracold

The Zero Energy FTL Spacetime Manifold

lattice gas. arXiv:1411.2678v1 [cond-mat.quant-gas] 11 Nov 2014]. That point marked $SP_s$ is what I refer to as the splitting point.

And $G'$ is:

$$G' = \frac{\daleth nLp'(\frac{Lp'}{tp'})^2 - \beth^2 \daleth nLp'(\frac{Lp'}{tp'})^2}{2M}$$

$$G' = 6.67384(80) \times 10^{-11} \text{ m}^3/\text{Kg (s)}^2$$

$$G' = 6.67385E - 11 \frac{Lp^3}{Kg * (\frac{t_0}{t'})^2_\beth}$$

Where:

$$\beth_{Sc} = \sqrt{1 - \frac{2G'M}{nLp(\frac{1Lp}{1tp})^2}}$$

And/or

$$\beth_{Lo} = \sqrt{1 - \left(\frac{\frac{nLp}{xtp}}{\frac{1Lp}{1tp}}\right)^2}$$

The self-similar relationship regarded above as $G'$ is the quantitative result of treating the subspace below the AdS Horizon Surface as superluminally fixed but altering the normal light distance between points figuratively 'below' the Horizon Surface. This results in a 'gap,' as a result of the vanishing value epsilon at a fixed scope and rate defined by $c \equiv 1Lp/1tp$ not being equated with the observation rate of the Information as it traverses the distance between points. This was all described and derived in detail in the prior series of papers. It is not the gravitational constant of the universe that is changing, but again, the *artifact of* observation from that Preferential Frame of Reference of the 'engine' looking out into flat spacetime.

The Zero Energy FTL Spacetime Manifold

Again, the value (±k) is a point of divergence between an alteration in the rate of progression of unitary time as [+k] increasing rapidity and [-k] slowing of progression of time. This is observed in the history of Zeno experiments:

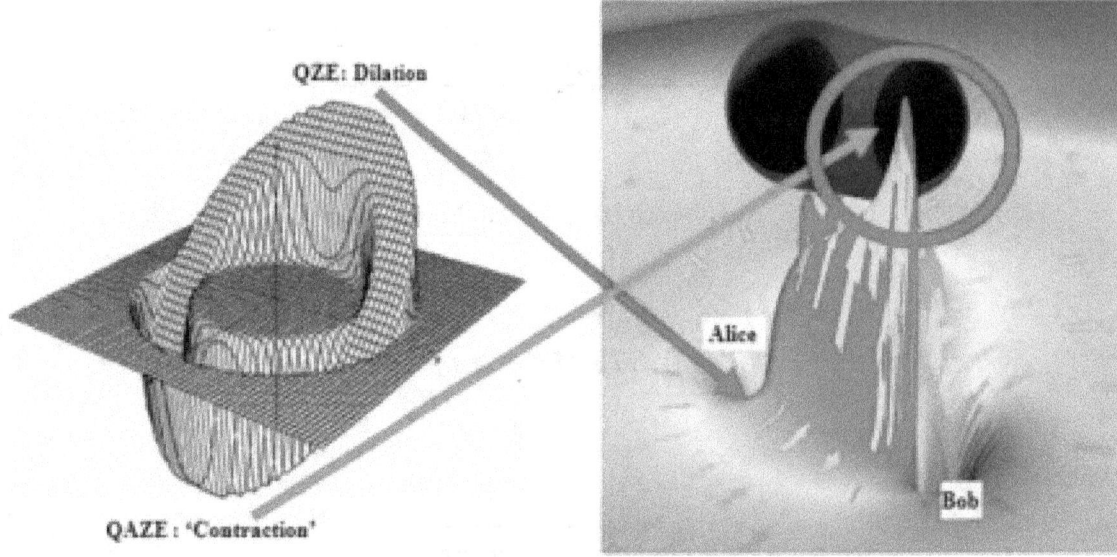

The image to the left is the original Alcubierre manifold, with length dilating to the right, as I took the liberty to flip the image. As described and derived in prior [Bray] papers and texts, this upward spire, the length dilation is associated with the Zeno Effect. On the right we have LIGO GW150914 modeled at the LIGO computer labs, as per the image a few pages back.

The question is, that no one seems to ask, what do the black holes, Alice and Bob see as they look out at flat spacetime? The answer is clear, the see a growing infinite distance to get out of their situation. What Alice and Bob see from *inside the manifold* is *length dilation, not contraction*. Perhaps you see something different, that is irrelevant. What the engine sees is length dilation, and this is necessary to make the engine produce a working manifold that dilates the unitary Planck intervals out to great distance in the form of a discontinuous jump.

This in turn means that what that upward spike in the LIGO images *sees when it looks out at flat spacetime is thus that it is 'contracted.'* For some bizarre reason convention never deals with what the system sees, but only what we see of the system, which is utterly illogical. Eulerian, Lagrangian, neither apply to this nor any other manifold, *that could work.*

The right image is this same upward spire appearing in nature as the upward spike, regarded as 'kick' or chirp-mass, not unlike a carpet bunching up toward the detector as a result of a sudden change in vector of a 30-solar mass black hole for about a millisecond. On the left is the modeling of a conventional Alcubierre type spacetime manifold. Here the downward spire that invoked this notion of negative mass-energy is equated to the Anti-Zeno Effect, which is identical in spatial geometry to the upward 'kick' circled in red in the LIGO image, with the inner flat region of the Alcubierre manifold being equated with the 'Splitting Point,' ±*k* or divergence from Zeno to Anti-

The Zero Energy FTL Spacetime Manifold

Zeno Effect. The QZE [not QAZE] defines length dilation in the Bray Manifold via the fractal relationships on the prior pages.

Note that via the fractal relationships there is no change in the *quantity of* Planck intervals, which represent energy, thus, there is no logical sense of requiring energy to produce the manifold. The fractal relationships are all temporally drive by $Zt'$, which is the temporal change in rate due to the Zeno Effect, not any other factor. Meaning, as the observed system $r$ is observed at some phenomenal rate by $R$, the temporal rate, $Zt'$, changes, which feeds back into $G'$ and all of the fractal relationships. *This is a change in conformal scaling only*, not, again, a change in quantity of unitary Planck intervals. And again, $G'$ is the artifact of observation from the Preferential Frame of Reference of being *inside the manifold, looking out at flat spacetime*. I really do not understand why no one has considered this crucial frame of reference issue. In general, convention continues to flip-flop frames of reference and comes up with absurdities such as the negative energy tensor of the Alcubierre manifold, which is a simple thing, like a rubber band and a propeller: *two clocks, observing one another, will produce this [Bray] manifold*.

The details of the actual engine design is in a forthcoming paper that describes the Phased Array network which steers and 'wiggles' charged beta particles to produce magneto-Bremsstrahlung radiation, which is *non-destructively* detected via the RF emitted. The physical engine design is technically challenging but not un-doable.

## Making Sense of the Spacetime Manifold

The first step is to produce a quantifiable, stable, and reproducible Zeno Effect. In this example we use a bit of carbon-14 which has a Specific Activity of 165K decays per second per microgram.

We place the 1μg-$^{14}$C 1cm from the opening of a beta detector whose orifice is 1cm in diameter. Thus, 1K decays per second enters the detector orifice under 'normal' observation rate conditions. The detector in this case will be a simple electron multiplier.

*The following numbers are purely fictitious for demonstration argument's sake.*

We say at a detection rate of 10MHz the Zeno Effect is 0.5, meaning, the $^{14}$C is yielding 500 decays per second. The Zeno Effect in this case *should be a linear, not second order response.* That will take far too long to explain. However, refer back to the Raizen et al plot and you can see that even in a semi-quantifiable mode, the Zeno Effect is linear.

Thus, at 20MHz the $^{14}$C yield is 250 dps, 30MHz is 125 dps, and so on.

These values feed into the first function:

$$f_R \rightleftharpoons fr$$

Note: Again, this is not the shaping function, but the Zeno Rate Equation.

At 500KHz, the Zeno yield is 1.5, e.g., Anti-Zeno. The [fictitious] linear values are then 250Khz 2.0, 125KHz 62.5, and so on.

$$\varphi(f_R \rightleftharpoons fr) \pm k$$

Note that fractals cannot be treated as equalities. There is no way to back calculate, for example, how many iterations have occurred at any given point in a fractal. Simply, if you look at any classic Mandelbrot, there is no possibility of determining how many iterations have occurred at that point in the result: [Below: *10-million iterations*]

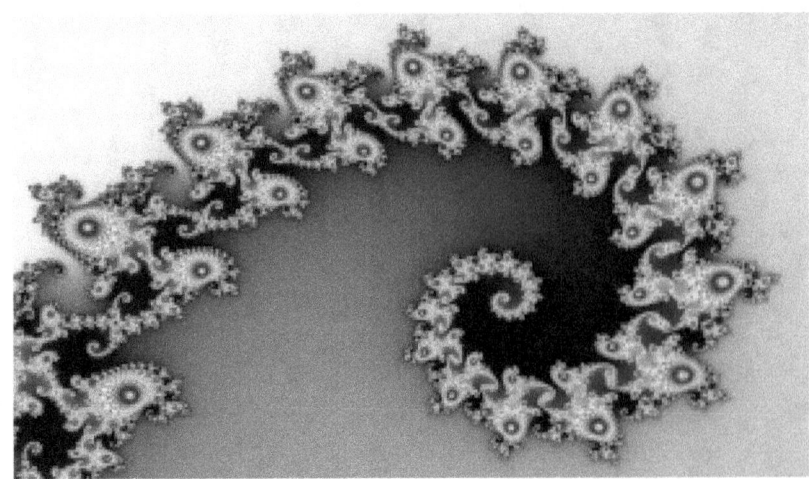

Making Sense of the Spacetime Manifold

Nor is there any means by which we can pick a point or portion and reverse engineer the equation.

The Maric Operator yields only positive values, e.g., no sqrt(-1) values, as:

$$\varphi(\pm v) = |v|; \; \varphi(\pm l) = |l|; \; \varphi(\pm t) = |t|;$$

$$\varphi(\pm 1) = |1|$$

This, then, is mapped into the Spacetime Manifold:

$$\varphi f_{R:r}(\pm k) \mapsto \left| \frac{\left(\frac{Lp^4}{tp^4}\right)}{8\pi G'} \frac{\left(\frac{nLp'}{xtp'}\right)^2 \left(Lp'^2{}_y + Lp'^2{}_z\right)}{4g'^2[(Lp''_x - Lp''_{xs}(Zt'))^2 + (Lp'^2{}_y + Lp'^2{}_z)]} \left(\frac{df}{dr}\right)^2 \right|$$

The $dr$ highlighted in blue is not the Zeno Rate Equation, $f_{R:r}$, but the shaping function as Alcubierre presented in 1994. I will update this paper when I have worked out and quantized a new shaping function.

Note the absolute value brackets surrounding the Spacetime Manifold, highlighted in red and yellow. There is no 'negative energy,' nor is there *any energy or mass-energy equivalent* to this manifold, as there is no sequitur description of energy associated with observation rate.

Thus, we had the rules:

- When mass-energy is not coupled to the local spacetime geometry, the *quantity* of N-Qubits must remain fixed, else violate preservation of Information conservation.
- There is no sequitur coupling of mass-energy to spacetime in this system as there is no energy involved in the transitions or transformations.
- In cases where spatial and temporal geometries are *not* coupled to their local source of mass*energy, Preferential Frame of Reference *does not apply* for General Relativity under Schwarzschild transformations. Tus, a system may detect its own change of state under a Schwarzschild transformation, or otherwise *create its own state.*\*
- As such, it is the *magnitude of* the unitary Planck intervals that changes, rather than the *quantity of* unitary Planck intervals.
- There is no energy nor mass-energy equivalent sequitur to a description of the magnitude of the unitary Planck intervals changing.

*This is the failing point of the Alcubierre and all subsequent attempts at localized spacetime manifolds. Regardless of the energy requirements, because they employ mass-energy in an attempt

Making Sense of the Spacetime Manifold

to alter spatial geometry, the spatial geometry is thus coupled to the mass-energy. As a result, no such system can alter its change of state under a Schwarzschild transform.

We turn to the values highlighted in green for this *change in Conformal Scale* [magnitude] of the unitary Planck intervals:

$$L_p'' \rightleftharpoons Lp' \rightleftharpoons L_{p0}/\sqrt{1 - \frac{2G''M}{n'L_p(\frac{1Lp'}{1tp'})^2}}$$

$$t_p'' \rightleftharpoons tp' \rightleftharpoons t_{p0}/\sqrt{1 - \frac{2G''M}{n'L_p(\frac{1Lp'}{1tp'})^2}}$$

$$\mathcal{A} \rightleftharpoons \sqrt{1 - \frac{2G''M}{n'L_p(\frac{1Lp'}{1tp'})^2}}$$

$$L_p' \rightleftharpoons L_{p0}/\mathcal{A}$$

$$t_p' \rightleftharpoons t_{p0}/\mathcal{A}$$

Note the seeming inversion of these metrics, the reciprocals highlighted in red and yellow. This is because the Preferential Frame of Reference is that of *being inside the Spacetime Manifold, looking out at 'flat' spacetime*. This is a critical point.

Likewise, the other unitary Planck intervals must change in Conformal Scale:

$$\varphi f_{R:r}(\pm k) \mapsto \left| \frac{\left(\frac{Lp^4}{tp^4}\right)}{8\pi G'} \frac{\left(\frac{nLp'}{xtp'}\right)^2 (Lp'^2_y + Lp'^2_z)}{4g'^2[(Lp''_x - Lp''_{xs}(Zt'))^2 + (Lp'^2_y + Lp'^2_z)]} \left(\frac{df}{dr}\right)^2 \right|$$

Same note for the red highlighted *df/dr* term. This is the original shaping function. Note that *t'* is highlighted in red and yellow. This *t'* brings about an effect, artifact of observation, given the entire manifold is built upon observation rate, of the sort:

Making Sense of the Spacetime Manifold

And $G'$ is:

$$G' = \frac{\frac{1}{2}nLp'(\frac{Lp'}{tp'})^2 - \theta^2 \frac{1}{2}nLp'(\frac{Lp'}{tp'})^2}{2M}$$

$$G' = 6.67384(80) \times 10^{-11} \, m^3/Kg \, (s)^2$$

$$G' = 6.67385E - 11 \frac{Lp^3}{Kg * (\frac{t_0}{t'})_\theta^2}$$

If the artifact, $G'$ is not taken into effect, this will in turn affect the observation rate $R$, and affect thusly $r$. Thus, we see $G'$ highlighted in red and yellow in the denominator of the Einstein-Schwarzschild metric.

For example, we had an observation rate $R$ of 10MHz yield a Zeno Effect of 0.5, reducing $r$ from 1KHz to 500Hz. The $t'$ is 0.5, thus, $G'$ *observed* is then, in the simplest sense:

$$G' = 6.67385E - 11 \frac{Lp^3}{Kg * (\frac{t_0}{t'})_\theta^2}$$

$$G' \, [observed] = 26.7E\text{-}11$$

And so on. This 'observed' $G'$ value figures back into the metric as shown above.

## The Stress Energy Tensor

As a set of permutations describing the Tensor connectivities and stress energy. If we look at:

$$(T^{uv})_{u,v}|\{0,1,2,3\} = \begin{pmatrix} T^{00} & T^{01} T^{02} & T^{03} \\ T^{10} & T^{11} T^{12} & T^{13} \\ T^{20} & T^{21} T^{22} & T^{23} \\ T^{30} & T^{31} T^{32} & T^{33} \end{pmatrix}$$

On a Planck scale, the matrix above is senseless, not because of any conventional argument, but because the 'worldsheet,' $A_\Omega$ of such a system is defined by:

$$A_\Omega = \hbar N_Q = \hbar \frac{Lp^2}{tp^2} = c^2 \equiv \hbar 1$$

And can be depicted, [highlighted in red above] *metaphorically*, as:

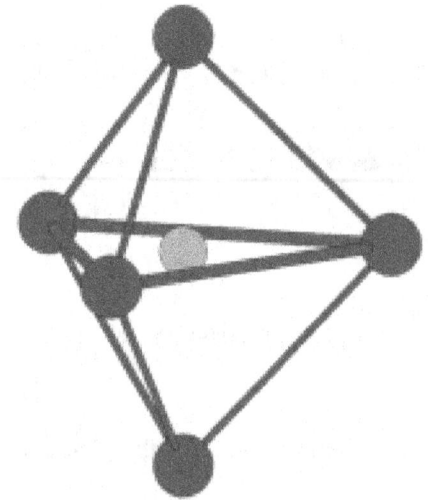

Consequently, the simple fix is to eliminate the diagonal:

$$(T^{uv})_{u,v}|\{0,1,2,3\} = \begin{pmatrix} T^{00} & T^{01} T^{02} & T^{03} \\ T^{10} & T^{11} T^{12} & T^{13} \\ T^{20} & T^{21} T^{22} & T^{23} \\ T^{30} & T^{31} T^{32} & T^{33} \end{pmatrix}$$

I tried to explain this in prior papers and texts, but I don't think I got it off clearly enough. The diagonal of the matrix simply makes no sense with geometry of this sort, keeping in mind that any Platonic shapes are inherently impossible and meaningless on a Planck scale. Nonetheless, the geometry of spacetime on a Planckian scale is not 'boxlike.'

The Stress Energy Tensor

The problem is that one can describe the trigonal-bipyramid as Stress-energy as it points *outward* from the center of the trigonal-bipyramid, but there is no way to describe Stress-energy pointing *inward* from some arbitrary distance to the center of the trigonal-bipyramid. The Stress-energy Tensor as is, a Combination, is a one-way description, from the Preferential Frame of Reference of *looking outward from the center of the trigonal-bipyramid*, but totally lacks any description where I can surgically apply that Stress-energy *looking inward* at the trigonal-bipyramid; it cannot be done.

Furthermore, that is, visual metaphors aside, the fine structure of spacetime on a unitary Planck scale. By eliminating that diagonal, we can begin to quantize the Stress-energy tensor on a Quantum Scale, which follows.

The other thing to know and keep in mind is that the AdS Horizon Surface is essentially a 2-dimensional construct. There are literally tens of thousands of papers by authors trying to make sense of this and at this time there is no agreed upon description of the actual Horizon Surface, and certainly not on a Planckian scale. The greatest amount of effort seems to be trying to communicate our 3-dimensional world into a 2-dimensional 'worldsheet.'

The AdS Horizon Surface value epsilon [$\varepsilon$] is thusly translated, derived in prior [Bray] papers and texts and I am not going into detail in this paper:

$$4\varepsilon' \equiv c^2 = \frac{E}{m} = \frac{Lp'^2}{tp'^2} = 4A'_\Omega = 4N'_Q = \frac{L'}{e^{\ln(\frac{L}{\varepsilon})}} = 4_{path}L'$$

$$\therefore$$

$$4\varepsilon' = \frac{E}{m} \equiv T^{00}$$

Since the energy density is defined by dimensions that are of the normal light distance between points on all scales, regarded then as [Ryu-Takayanagi] Path $l$, for the Matrix:

$$T_{uv} \rightleftharpoons S \rightleftharpoons [\frac{E}{\varepsilon}]_{uv}$$

Again:

$$\varepsilon = \frac{Lp^2}{tp^2} = c^2 = \frac{E}{m} = 4N_Q = A_\Omega = 1HUP = Le^{-\frac{3l}{\varepsilon}}$$

The Stress Energy Tensor

So, this quantizes as:

$$T_{uv} \rightleftharpoons S \fallingdotseq [\frac{E}{\frac{Lp^2}{tp^2}}]_{uv}$$

Then in terms of Quantum Information Dynamics

$$T_{uv} \rightleftharpoons S \fallingdotseq [\frac{E}{\tfrac{1}{4}N_Q}]_{uv}$$

$$T_{uv} \rightleftharpoons S \fallingdotseq [\frac{E}{\tfrac{1}{4}A_\Omega}]_{uv}$$

This alteration in the scope of the worldsheet, $A_\Omega$ is the key to the manifold's structure and qualities.

$T_{uv}$ is in terms of unitary intervals that define Path $l$. Again. All scaler and vector quantities must be in magnitude values in terms of Path $l$. The permutations are then:

$$[T_{uv} \rightleftharpoons S \fallingdotseq l \equiv \frac{e}{3} Ln \frac{L}{\varepsilon}]$$

$$[T_{uv} \rightleftharpoons S \fallingdotseq l \equiv \frac{e}{3} Ln \frac{L}{\frac{Lp^2}{tp^2}}]$$

$$[T_{uv} \rightleftharpoons S \fallingdotseq l \equiv \frac{e}{3} Ln \frac{L}{c^2}]$$

$$[T_{uv} \rightleftharpoons S \fallingdotseq l \equiv \frac{e}{3} Ln \frac{L}{\tfrac{1}{4}N_Q}]$$

$$[T_{uv} \rightleftharpoons S \fallingdotseq l \equiv \frac{e}{3} Ln \frac{L}{\tfrac{1}{4}A_\Omega}]$$

**The Bell Loophole-Free Test**

### The Bell Loophole-Free Test

When I discussed the Bell Loophole Free test, this is what I was referring to in the case where the photons, by forcing them to travel in fiber optic at subluminal velocity 0.6c because if the refractive index, forces them to remain during the entire journey *on* the Horizon Surface. This is because in a medium with such a refractive index, the photons cannot be in an eigenstate, they must be eigenvalues, because they are actively interacting with the medium's electrons. Furthermore, a portion of them is deflected, which is another factor Hensen et. Al. did not account for.

In any case, the Bell Loophole Free test as it was thusly performed cannot be valid just by virtue of the velocity of the photons at 0.6c.

That may be interpreting the system a bit too literally, however, it is in essence the case, and all of the pervading papers more or less agree in math to that extent. This was the issue I was trying to get across with the Hensen et al experiment [which is actually discussed in a later section in detail]:

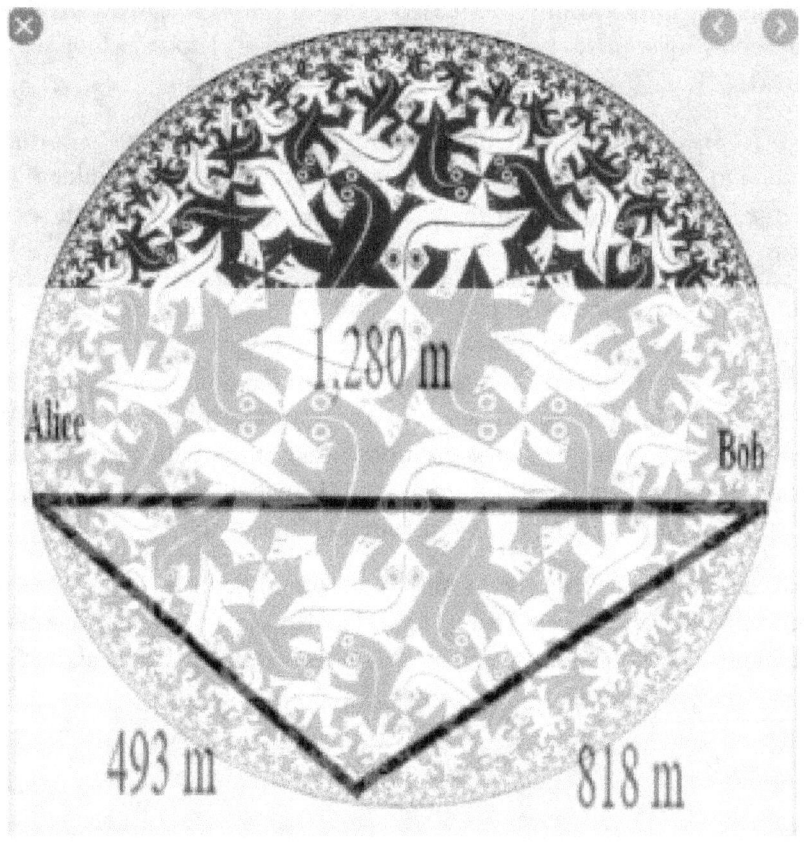

**The Bell Loophole-Free Test**

In the Loophole-Free Bell Test, Hensen et al, by passing the photons through fiber optic cable, which has a refractive index [average] of 1.5, cause the photons to travel at $0.6c$, *not* $v = c$. They clearly do not account for this, and all of the math is clearly treating the system as $v = c$, as per their stated calculations. Information in the experiment is being transferred via fiber optic cable, with a non-zero refractive index, which means the photons are interacting with matter *on the Horizon Surface* as they travel.

Keep in mind that eigenvalues cannot exist in the region regarded as 'below' the horizon surface, because it does not exist, it is only a record of past states of the horizon. Eigenstates come by a different set of rules, namely, the static domain of the conventional Ryu-Takayanagi Path $l$.

Photons cannot be in an eigenstate when interacting with the medium of fiber optic cable. Any interaction at all precipitates native wave functions from eigenstates to eigenvalues, that is in fact the dilemma. If the native wave functions of the photon are not traveling at $v = c$ then it is conclusive that they are interacting with the electrons in the medium. Furthermore, this interaction cannot be non-destructive, else the photons would not be impeded in motion. Consequently, what Hensen et al got was 'noise.' And in fact, the 'noise' fit the coin-flip scenario nicely. We have 2-bit [binary] noise, which will always be a perfect 50/50.

The problem is, photons in the AdS description seem to increase in Conformal Scale as they travel, and at some halfway point, begin decreasing to the local value $\varepsilon$ where they will 're-emerge.' That is a problem that I have not seen addressed, and albeit seems a literal interpretation, is the case.

However, in the Loophole-Free Bell test, the photons, by 1) being limited to $v = 0.6c$ and 2) because they are physically and electrodynamically interacting with matter during the entire time of Flight, are thus being forced to travel between points *on the Horizon Surface*. Keeping in mind that the Horizon Surface is by convention superluminal, this seems to go along with the quanta-for-quanta trip as each photon interacts with each atom in the fiber optic cable; a go-stop-go type of quantized motion. However, we can regard the lessening of velocity of the photons as fitting this superluminal limitation.

The idea that the fiber optic cable goes through Conformal Scale Invariance has to be taken into account; *as not being the case.* That is, does the fiber optic cable increase in Conformal Scale as we move toward the center, as depicted in the visual metaphor above?

**The Bell Loophole-Free Test**

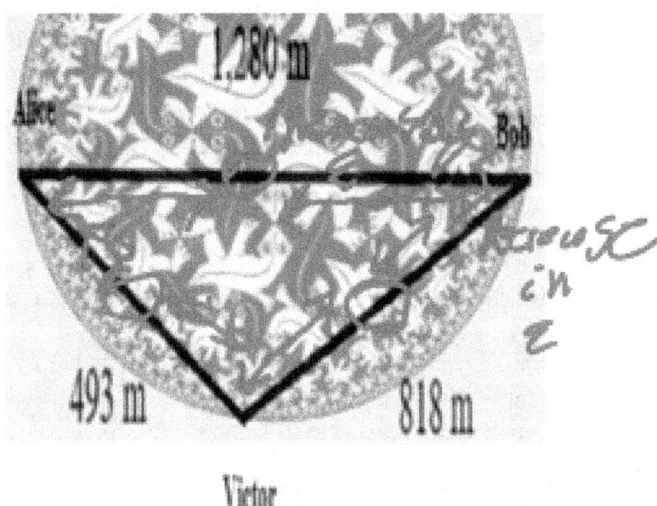

This isn't a subject that has been addressed to the best of my knowledge. The value epsilon regards the Horizon Surface value. However, the prior states of epsilon are represented as the larger Conformally Scaled 'lizards.' As Information travels from Alice [upper left] to Bob [to the right], in the form of *single quantized photons*, it would seem they gain in scale as they approach the center-point, the red circle, then at exactly that point begin decreasing in Conformal Scale until they [each] reach Bob on the Horizon Surface, where the current state of epsilon describes their scaling. I'm not certain if I find this troubling in the description. It is again a bit of a literal interpretation, but when you do the various species of maths that describe it, it seems to invariably work out to be the case.

Nonetheless, this change in Conformal Scale cannot be true for the length of the fiber optic cable. However literally I am interpreting this, is non-sequitur, the convention regards Ryu-Takayanagi Path *l* in a manner that as hypothesized in this text, can only apply to entangled systems. The Bell-Loophole-Free test fails to measure entangled photons because of the refractive index of the fiber optic cable, demands that the photons are interacting with the medium. As such, they are forced to travel via Ryu-Takayanagi Path $L$, the superluminal horizon, which is why they travel at $v \ll c$ at $0.6c$, rather than the speed of light. This is an interesting proposition. That is, regarding the superluminal horizon as the causal component for measuring the photons traveling at velocities less than light.

We can thus regard another interesting hypothesis:

- Traversing the superluminal AdS Horizon Surface, for native photon wave functions, results in the photons travelling at $v \ll c$.

Digital noise is binary and a 'lot of' digital noise is always going to be 50/50. This test does not validate any Bell postulates because of the transfer medium. Again, they clearly regarded the photons, looking at the math, as traveling at $v = c$, which is definitively not the case and not possible.

## The Bell Loophole-Free Test

Rather or not, this is the case, the problem with the Loophole-Free Bell test is that the *fiber optic cable* cannot in either case change Conformal Scale from one end, getting greater in magnitude to the middle, then lesser in magnitude to the opposite end. A cognitive argument of the sort, 'this change in Conformal Scale does occur in the fiber optic cable, cannot be validated.

The fundamental properties are:

1. *A change in Conformal Scale represents by convention a change in the coordinates of time.*
2. That is, we assign a time-zero. At time-zero epsilon has the value $\varepsilon_0$. At time-zero $\varepsilon_0$ is associated with the unitary values $\{Lp, tp\}_0$.
3. If the Conformal Scale is greater than $\{Lp, tp\}_0$, then time is less than time-zero [e.g., causal past], e.g., $t_{0-n}$. Epsilon's value was greater than time-zero in the causal past. Thus, $\varepsilon_{t0-n} > \varepsilon_0$ and epsilon is equated as the unitary Planck length Lp and time tp such that $\{Lp, tp\} \varepsilon_{t0-n} > \{Lp, tp\} \varepsilon_0$
4. If the Conformal Scale is less than $\{Lp, tp\}_0$, then time is greater than time-zero [e.g., causal future], e.g., $t_{0+n}$. Epsilon's value is less than at time-zero in the causal future. Thus, $\{Lp, tp\} \varepsilon_{t0+n} < \{Lp, tp\} \varepsilon_0$

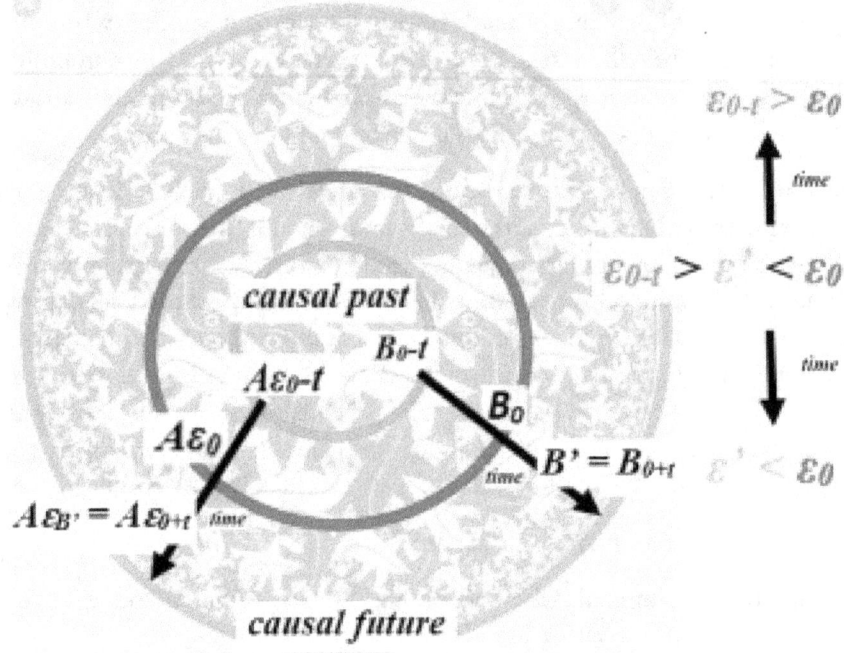

The above diagram [hypothetically] represents time, as indicated, at time-zero, time-zero minus *t* [causal past], and time-zero plus *t* [causal future], with respect to the Conformal Scale of epsilon. The standard visual metaphor:

**The Bell Loophole-Free Test**

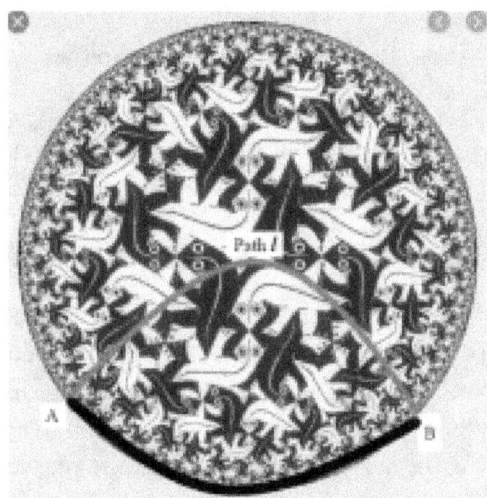

**Path L**

Can only be relevant in the rarest cases where we discuss this one relationship of Path L vs Path *l*, and only in the case of entangled systems.

## Alice, Bob and the Horizon Surface

First, we have to define a few parameters of the systems. The next section will go into the details of why and how Shannon Entropy is limited to greater than one causal path and conventional $\Delta S$ entropy is limited to exactly one causal path. This is a key feature in understanding the nature of systems that are observable vs unobservable.

Again, it is important to consider that the region figuratively below the horizon cannot change physical geometry as this would then demand physically changing the past. Meaning, the geometry of the region figuratively below the horizon does not actually exist, but is merely a record, if you will, of past states of the horizon. Any physical change to the geometry figuratively below the horizon then would require physically changing the past.

We tie Alice and Bob together via a tether that is 1-Planck interval in length. The first thing we find is that this is not possible, because the lower such boundary [Hawking-Bekenstein] is defined by the Heisenberg Uncertainty Principle [HUP] as 2-Planck lengths. Thus, we let out some slack and now Alice and Bob are tethered together by a 2-Planck length tether. We begin lowering Bob closer and closer to a Schwarzschild *Surface*, e.g., black hole Surface. Note that I avoid the term horizon as there are numerous descriptions for this that differ rather widely.

We find that we can only lower Bob by intervals of 2-Planck lengths because of our HUP lower boundary constraint, 1-Planck length is a finer resolution than the HUP will allow.

At 4-Planck lengths Bob has not yet coalesced with the horizon and is observable to Alice, and Alice is still observable to Bob.

- The system {Alice, Bob} is described by conventional $\Delta S$ entropy.
- The system {Alice, Bob} is thusly defined by one and only one causal path.
- The system {Alice, Bob} is {observable}.

At 2-Planck lengths:

- Bob coalesces with the horizon.
- Bob is now unobservable to Alice.
- Alice is now unobservable to Bob.

And the entropy description:

- The system {Alice | Bob} is now described by Shannon entropy.
- As Shannon entropy, the system can only be described by $> 1$-*causal path*.
- The system can no longer be described by conventional $\Delta S$ entropy, because conventional $\Delta S$ entropy only applies to systems that are described by only 1-causal path.

### Alice, Bob and the Horizon Surface

- The system, regarded as both Alice | Bob is no longer {observable}, but regarded as {unobservable} with respect to each other.

This represents

- A *divergence* from conventional ΔS entropy [limited to an upper constraint of 1-causal path] to Shannon entropy [lower constraint of >1-causal paths].
- The divergence occurs at a proximity to a Schwarzschild Horizon [Surface] of exactly 2-Planck lengths, as defined by the Heisenberg Uncertainty Principle [HUP] on a quantized scale.
- At such time the system {Alice, Bob} is described by 1-causal path all elements of the system are {observable}.
- At such time the system {Alice, Bob} is described by >1-causal path all elements of the system are {unobservable}.

The conditions {observable, unobservable} are inextricably interdependent such that

- Conventional ΔS entropy [upper] limited to 1-causal path, {observable}.
- Shannon entropy [lower] limited to >1-causal path, {unobservable}.

As a result, as will be discussed in more detail, the evolution of the DeSitter Domain is of a Shannon type, not a conventional 19th century gas law ΔS type. Conventional ΔS entropy cannot be used to describe the AdS Horizon Surface. That is, the AdS Horizon Surface is {unobservable} because each element $\{Lp^2, tp^2\}$ is superluminally isolated from each other element $\{Lp^2, tp^2\}$. In order to observe some other element on the Horizon Surface, Information must traverse some Path type $l$, regardless of what form that Path takes, be it regarded in the conventional Ryu-Takayanagi sense or otherwise. The Ryu-Takayanagi Path L is superluminal, and Path $l$ is the normal light distance between points.

- Ryu-Takayanagi Path $L$ is described by Shannon entropy.
- Ryu-Takayanagi Path $l$ is described by conventional ΔS entropy.

As a consequence of the second bullet point, Path $l$, the region figuratively 'below' the Horizon Surface does not actually exist, but only represents a collection of past states of the horizon. Thus, conventional ΔS entropy only describes *past states*, and cannot have any meaning with respect to the present. Keep in mind that the present is a two Planck unit value, a very small slice of time.

Shannon entropy, then, describes the Horizon Surface, but because the slice of time is only two Planck units of time, cannot evolve past $n = 1$ iterations.

The normal light distance between points is defined by subspace, the region figuratively 'below' the AdS Horizon Surface. This, then, is a Path that is constantly changing, as the

### Alice, Bob and the Horizon Surface

Conformal Scale of the AdS Horizon Surface is vanishing. However, derived later, the vanishing rate that is conventionally regarded on cosmological time scales is actually vanishing at a scope and rate that is defined by $c \equiv 1Lp/1tp$, on a quantized scale. This brings up several issues regarding Conformal Scale Invariance, as this CSI is changing during the measurement process.

This transfer of Information via the Ryu-Takayanagi Path $l$ is limited in scope and rate to luminal velocities and takes place figuratively 'below' the Horizon Surface, regarded as subspace, which does not tangibly exist, but is only a record of past states of the Horizon Surface. Thus, all elements $\{Lp^2, tp^2\}$ are {unobservable} with respect to their immediate state of function, provided the exist *on the AdS Horizon Surface*.

We can regard each element as {observable} in the sense of a limited, finite time span. However, until such time detection occurs on the Horizon Surface, and it cannot occur anywhere else [cannot occur below the horizon Surface], the system is said to be {observable, unobserved}. This is a *non-ontological* approach to the logical set:

{observable, unobservable, observed, unobserved}

Observation herein is regarded as any electromagnetic interaction.

This logical set of states defines time, as stated in other sections, *by Bob*, as he is the system that is the causal component of Information being *forced to* the Horizon Surface. There is no other description that has been observed. Meaning, Bob 'forces' the Information to the Surface is a tangible rationale for causality, any other rationality has never been observed, else, it would be regarded as *observation*. There is no ontological content in this, any electromagnetic interaction can fit the definition of *observation*. Observation, as should be obvious, precipitates some eigenstate to some eigenvalue.

This means that *observation*, which can take the form of any electromagnetic interaction, is a conditional operator that defines some eigenstate precipitating to some eigenvalue. Again, albeit there are countless hypotheses put forth that describe this in a myriad of different ways, *that is the only observed phenomenon*. Observation, be it electromagnetic phenomenon, definitively results in some eigenstate precipitating out to some eigenvalue.

It is very important at this point to clearly differentiate:

{observable, unobservable} | {observed, unobserved}

Albeit mundane to list, the necessity will be clearer shortly. A system can take on the following logical states:

- {observable} | {observed, unobserved}
- {unobservable} | {unobserved}

At 1 Planck length there are only two (2) possible things a thing of one (1) dimension can do, {0, 1}. Heads | tails. We can think of this metaphorically as a genuine Heisenberg

## Alice, Bob and the Horizon Surface

eigenstate of the result of being only one Planck unit. Quantum Temporal and Information Dynamics is an evolution of this notion that will be expressed in some later volume of text.

Wherein all convention of Quantum Theory fails at the fundamental level of a superposition of the sort spin up spin down for our listless characters, Alice and Bob. Whereas we can disregard the eigen*state* as having any range between 0 and 1, e.g. the coin of type heads | tails has no edge that represents some fractional range between the two values, the eigen*value* heads *or* tails, a priori, is in itself non-sequitur. Before such time *that the coin lands*, e.g., *detection*, there is zero certainty (not HUP type) *that the coin even exists*. If we are suggesting that the coin exists because we set it in motion, *then it has been observed*, else we would have no information about it a priori.

This is in terms of pure eigenstate. We may know that the system {Alice, Bob} exists because we created it in the lab. However, maybe the device was not plugged in or malfunctioned [metaphor]. We have no certainty that the system exists and certainly no notion of locality, aside from predictions a-priori, based on past experience with similar phenomenon. But there is no certainty of 1) the system exists 2) any notions regarding locality; other than presumptions based on prior behavior, not detection, which is observation [non-ontological].

We keep in mind the former description, an eigenstate can only exist [figuratively] 'below' the AdS Horizon Surface, an eigenvalue can only exist *on* the Horizon Surface. This should be clear, otherwise, the eigenvalue must be in the literal past, where nothing actually exists *in the past*. It either exists in the present or it does not exist.

Also keep in mind here that I treat the electric and magnetic bosons as discrete and massive 'virtual' particles that are entangled with the native wave function. For example, a native wave function of a photon has a sea of virtual electric and magnetic bosons that arise out of the vacuum or otherwise, that because they arise out of the HUP possess both discreteness and mass. As a consequence of this, the electric and magnetic bosons cannot travel at $v = c$, but are mass limited to $v < c$. When we discuss issues with the 2-slit experiment and variations thereof, we have to keep in mind that the photon, at $v = c$ has no sequitur description of time nor locality, only the discrete and massive electric and magnetic bosons experience time and locality, and travel at the mass limited velocity of $v < c$. Thus, we trace the history of the 2-slit back a century and find ourselves chasing luminal velocity native photon wave functions via subluminal velocity electric and magnetic bosons that are essentially left in the wake of the native photon wave function.

### Shannon Entropy vs. Conventional ΔS Entropy and the Number of Causal Paths they Define

#### Shannon Entropy vs. Conventional ΔS Entropy and the Number of Causal Paths they Define

##### Why is Shannon Entropy a Description of Greater than One Causal Path?

If we look at the most fundamental description of Shannon Entropy:

$$H(X) = -\sum_{i=1}^{n} P(x_i) Log_2 P(x_i)$$

When $i = 1$, the entire term falls to zero.

$$H(X) = -\sum_{i=1}^{n=1} P(x_i = 1) Log_2 P(x = 1) = 0$$

This refers specifically to a Lone Temporal Endpoint, such as $u_0$ **or** $u_{0+1}$. If we have an eigenstate of $<u_0|u_{0+1}>$ then the Shannon entropy is 1. However, this describes one causal path. If we regard the Shannon entropy in terms of 1-causal path, again:

$$H(X) = -\sum_{i=1}^{1} P(x_i = 1) Log_2 P(x = 1) = 0$$

Specifically, if 1-causal path is described by Shannon entropy, from $i = 1$ to $n = 1$, then the Shannon entropy is zero. Furthermore, if the Shannon entropy is in the macroscopic form of *nats:*

$$S_{nat} = -\sum_{i=1}^{1} (P_i) Ln(P_i)$$

Again, the Shannon entropy is zero and cannot describe any tangible thing.

Finally, if Shannon entropy is determined by way of Hartleys, which is the base 10, again, the Shannon entropy is zero and cannot describe any tangible thing.

### Shannon Entropy vs. Conventional ΔS Entropy and the Number of Causal Paths they Define

It can only be concluded that Shannon entropy cannot describe 1-causal path. Meaning, zero entropy cannot describe any tangible thing, as all systems must be observed as undergoing some degree of entropy.

Furthermore, if Shannon entropy cannot describe any tangible thing, then it is conclusive that Shannon entropy in all of its forms can only describe *unobservables*. Specifically, given Shannon entropy cannot describe 1-causal path it must therefore be conclusive that Shannon entropy can only describe greater than one causal path. [>1-causal path].

In later sections it will be described in greater detail how there is a divergence from conventional ΔS entropy to Shannon entropy as a function of distance from a Schwarzschild 'horizon,' where the term horizon specifically refers to a Schwarzschild Surface. Here, I am using the term Surface rather than horizon because there are too many conflicting descriptions of what a 'horizon' is. Another example is the entry to an Einstein-Rosen Bridge, which cannot be regarded as a 'horizon' under any of these definitions but can only be regarded as a Surface.

Briefly, we tie Alice and Bob together on a tether of some unitary Planck interval(s), specifically 2 Planck units of time and length 2{Lp, tp}, so as to adhere to Heisenberg descriptions on that scale [HUP]. We begin to lower Bob onto the Schwarzschild Surface. At such time that he coalesces with the Schwarzschild Surface, he is then *unobservable* to Alice, cannot be described by conventional ΔS entropy because he now answers to Shannon Entropy, where he is becoming infinitely multiply connected on the other side of the Schwarzschild Surface. Since conventional ΔS entropy can only describe 1-causal path, there is no conventional ΔS entropy description for him, and he can only be described by greater than one causal paths [>1-causal paths] and is therefore only described by Shannon Entropy. It is at the exact moment that he coalesces with the Schwarzschild Surface that conventional ΔS entropy, limited to only 1-causal path and Shannon Entropy, having a lower limit of >1-causal path, diverge.

This leads to the postulates:

- Conventional ΔS Entropy, which is limited to 1-causal path {observable}
- Shannon Entropy, which is defined as >1-causal path, always {unobservable}

And furthermore, the terms {observable, unobservable} are defined purely by proximity to a Schwarzschild Surface, such as a black hole, Einstein-Rosen Bridge, or what have you. This is a purely non-ontological approach to the terms *observable* and *unobservable*. This will become of greater value when we begin to discuss the 'observer effect' as it pertains to the Delayed Choice Quantum Eraser. We then have a logical set, which begins to form as:

- *observable: observed, unobserved*
- *unobservable: unobserved*

## Shannon Entropy vs. Conventional ΔS Entropy and the Number of Causal Paths they Define

Which then extends to the postulates:

- *observable: observed, unobserved:* Conventional ΔS Entropy, which is limited to 1-causal path, *always observable not always observed*.
- *unobservable: unobserved:* Shannon Entropy, which is defined as >1-causal path, always *unobserved*.

This may seem mundane, but it is not. When we revisit these fundamentals later in the DCQE it will become clearer how these logical definitions take the form:

*{observable, unobservable, observed, unobserved}*

In many cases, some regard the asymmetric progression of common time as obeying entropy and in fact regard conventional ΔS entropy as somehow *governing* the asymmetric forward progression of common time. I do not agree with this, however, it is exemplary of how any progression of time or some other tangible and observable aspect of the progression of time is coupled with entropy.

Nonetheless, I believe it is agreeable that zero entropy cannot describe a tangible system, either as existing in an eigenstate or having evolved to some eigenvalue or set of eigenvalues. Thus, I believe it is conclusive that conventional ΔS entropy can only describe tangible systems, which in turn must be limited to no more and no less than exactly 1-causal path. No causal path [0-causal paths] cannot be regarded as tangible nor observable. And conventional ΔS entropy cannot describe greater than one causal path [>1-causal path].

In turn, Shannon Entropy [is a proper noun] can only describe systems of greater than one causal paths [>1-causal paths] and as such cannot be regarded as either tangible or observable.

Meaning, where we regard a causal path as consisting of two Lone Temporal Endpoints, $(u_0 \rightarrow u_{i\pm 1})$, there are two conditions where $i = 1$.

- $u_0$ referring specifically to a Lone temporal Endpoint.
- $(u_0 \rightarrow u_{i\pm 1})$ regarded as the unitary Planck intervals {Lp, tp}, referring specifically to 1-causal path.

In the first case, $u_0$ is regarded as the Lone Temporal Endpoint and thus $i = 1$ and the Shannon Entropy is equal to zero. However, this is not a tangible, nor does it qualify as any eigenstate. It is sufficient to say it has no logical meaning.

In the second case, $(u_0 \rightarrow u_{i\pm 1})$ is regarded as 1-causal path, $i = 1$ and the Shannon Entropy equals zero. Shannon Entropy cannot be regarded as having any meaning with respect to 1-causal path, but such a system then answers to conventional ΔS entropy.

## Shannon Entropy vs. Conventional ΔS Entropy and the Number of Causal Paths they Define

When ($u_0 \to u_{i\pm 1}$) is regarded as two distinct temporal endpoints, for whatever reason, then $i = 2$, *with each p being 0.5* and the Shannon Entropy is not zero, but one. In such case then each $u_0$ and $u_{0+1}$ are p(0.5) and the Shannon entropy is 1. However, this is not a description of a tangible thing, as this can only be regarded as a system that is in some eigenstate and thusly not observable.

The reason I am describing a Lone Temporal Endpoint, such as $u_0$ and its progression to $u_{0+1}$ at all is because I want to derive how it is that the progression of the unitary intervals occurs.

Here, each $u_0$ and $u_{0+1}$ are being regarded as Lone Temporal Endpoints, which is the beginning of an eigenstate with a false vacuum state. The 'false' vacuum state refers to the notion that a true vacuum state cannot actually exist, nor can it be regarded as tangible, nor observable. A true vacuum state, in simple and archaic definitions in QED, would lead to [hypothetically] a Bubble Nucleation, resulting in another universe, as an example of the improbability of such a state. A false vacuum state is regarded conventionally as entangled with some other state.

Therefore, I designate $\psi_m$ as representing any arbitrary origin [single point] on or below the DeSitter domain, and fracking to the Conformally Scale Invariant Transformation value for Lp by the process:

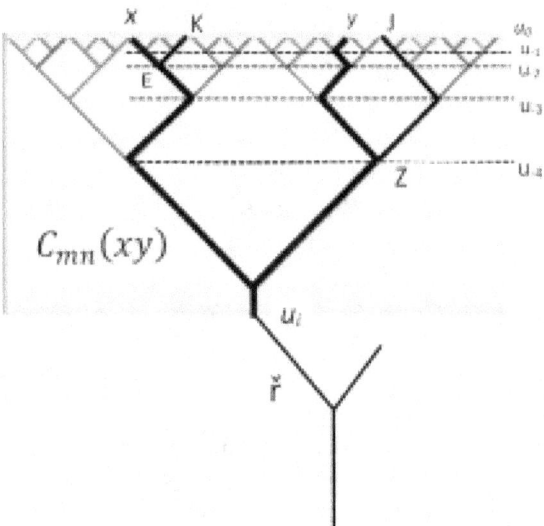

I assign the *frack*, which is described in detail in a later section, from the point $u_i$ to ($u_i$ + 1) such that the Temporal Endpoint $m$ evolves to Length [L, not $l$] Lp [epsilon] on the Horizon represented by nm, considering $m$ as a Temporal Endpoint and $n$ as a False Vacuum State [does not exist] leading to *{m', m}* is the superposition for the binary false vacuum state of $m$ as any progression for any 1-dimension has only two possible values, (+/-) [that is, the designation of $S_p$ as a go|no-go *{m', m} state* of only type 'go' *{m}*]. In

### Shannon Entropy vs. Conventional ΔS Entropy and the Number of Causal Paths they Define

addition, we can regard the false vacuum state of $m'$ as 'precipitating' to an eigenvalue as event, $n$.

Meaning, we have the Lone Temporal Endpoint $m$ which is entangled with the False Vacuum State $m'$. The state $\{m, m'\}$ evolves from a False Vacuum State to an eigenstate of $m'$ to $n$, which is then $nm$. If I haven't already said so, Markov Mechanics is 'reverse Polish,' such that the progression of $m$ to $n$ is written $nm$, e.g.,

$$\psi_m \rightarrow \psi_{nm}$$

Thus:

$$X_{m',m}$$

$$\psi(m)$$

$$\sum_{m',m} \psi^*(m') X_{m',m} m$$

$$<\psi_m|\chi|\psi_{m'}>$$

For the Lone Temporal Endpoint $m$, [my prior reference to the Trace Matrix for $T_m$ as a permutation rather than combination] which has a dual state $\{m', m\}$, a Vacuum Sate $n$ *does not exist*. However, the Vacuum State is *entangled* with $m$ as a result of being *unobservable*. This is the demand of Shannon Entropy of any state function for $\psi_{\{m', m\}}$. Here, we can regard the dual state for $\{m', m\}$ as our go|no-go state, otherwise, neither term, m' nor m have any sequitur meaning. As such, P ≡ 1. Because P > 0, there exists some entangled state. Here for argument, I will arbitrarily assign m' as P = 0 and m as P = 1. Since the P = 0 {m'} state has no sequitur description, and $P_m$ =1, as per the definition in the Shannon Entropy, the following process must occur:

Given:

- Conventional ΔS Entropy, which is limited to 1-causal path {observable} [but is not necessarily observed].

### Shannon Entropy vs. Conventional ΔS Entropy and the Number of Causal Paths they Define

- Shannon Entropy, which is defined as >1-causal path, always {unobservable} [thus, never observed]

The states are inextricably linked in such a way that:

- When a system is *observable*, it is thusly defined by conventional ΔS entropy and defines exactly 1-causal path. This is true rather the system is observed or if it is unobserved.
- When a system is *unobservable*, it is thusly defined and described by Shannon Entropy and defines *always* >1-causal paths. It can never be observed.

This term *unobservable* is not ontological, but refers to, for instance, any Information Threshold [novel term discussed later] as it exists *below or behind* A Schwarzschild Horizon. That is, because $n$, our 'spectator' is *unobservable*, as the demand for the Shannon Entropy defines Tensor Connectivity with the *unobservable* as a result of being *unobservable*. In simpler terms, the [true] Vacuum State $n$ is entangled with the lone Temporal Endpoint $m$ as being unobservable from $m$ at the temporal origin $u_0$, which is the temporal locality of the Lone Temporal Endpoint $m$. As a result, where for the progression of some arbitrary lone Temporal Endpoint $m$ at proper time interval $u_0$, the state ($u_0 \to u_0 \pm 1$) defines the unitary proper time interval where it would progress to nm, *even if such interval approaches infinity*, it is nonetheless defined by ($u_0 \to u_0 \pm 1$), which is our Fractal Value on the DeSitter Horizon (shown below) which defines the Conformally Scale Invariant

$$Y_m \to Y_{nm}$$

Would describe some probability of $m$ evolving to $nm$ as some range from zero to infinity, the Shannon Entropy reduces this to $ln(2)$, which is > ½, meaning, P = 1, *it will occur*, and the Density Probability Matrix for:

$$Y_m \to Y_{nm}$$

$$Y_n \to Y_{mn}$$

is

$$P[(Y_m \to Y_{nm})|(Y_n \to Y_{mn})]_{(m,n)} \equiv 1$$

### Shannon Entropy vs. Conventional ΔS Entropy and the Number of Causal Paths they Define

As a result:

$$\sum_{n,m',m} \psi^*(m') \chi_{m',m} \varphi_{(mn)}$$

$$\sum_{n} \psi^*(m') \varphi_{(mn)} = P_{m,m'}$$

$$|\psi_i\rangle$$

$$\langle \chi \rangle = \sum_{m,m'} P_{m,m'} X_{m,m'}$$

Here, P is the Probability of the Trace Matrix for the process $P_m \to P_{nm}$ [by convention, nm refers to the evolution of n from m]

$$P_{m',m} = 1 = P_{n',n}$$

$$P_{m',m} = 1 = P(N=1)_{n'}, (N=0)_n = 1$$

$$0 \leq P_{m',m} \leq 1$$

for the progression:

$$\Psi_m \to \Psi_{nm}$$

## Shannon Entropy vs. Conventional ΔS Entropy and the Number of Causal Paths they Define

And the Density Probability Matrix for

$$Y_m \rightarrow Y_{nm}$$

$$Y_n \rightarrow Y_{mn}$$

is

$$P[(Y_m \rightarrow Y_{nm})|(Y_n \rightarrow Y_{mn})]_{\{m,n\}} \equiv 1$$

However, again, the above statement does not regard some range of arbitrary units defining 'between zero and one.' As a go|no-go eigenstate, where 'no-go' has no sequitur outcome, $P \equiv 1$. To simplify, there is no arbitrary 'slices' of the range 0 to 1. The unitary interval ($u_0 \rightarrow u_0 \pm 1$) defines the 'slice,' in this case, tp [Epsilon]. This leaves only two choices, $P = 0$ and $P = 1$, with no range of arbitrary slices in between. If ($u_0 \rightarrow u_0 \pm 1$) does not occur, the universe exists in some frozen (no change of sate) the has no description. Therefore, empirically, $P \equiv 1$. The unitary step of the sort ($u_0 \rightarrow u_0 \pm 1$) does occur. Furthermore, P is always $P \equiv 1$, else, the domain does not change state.

The statistical mechanics for this are given then [again], as a result of the above description for the [False] Vacuum State (not yet *function*)

$$N_m(t+1) = 2N_m(t+1) - \sum_n Y_{nm}N_m + Y_{mn}N_n$$

Here, $N$ refers to any one of an arbitrary lone Temporal Endpoints of type $m$. The (t+1) term is the unitary step, ($u_0 \rightarrow u_0 \pm 1$) in terms of time. As a result of the unitary step, this demands a second Temporal Endpoint, designated as $n$. The term, 2N refers to the fact that for any unitary step, ($u_0 \rightarrow u_0 \pm 1$), requires two Temporal Endpoints, namely, the origin, $u_0$, and the endpoint, $u_0 \pm 1$. The designations which otherwise seem arbitrary, $Y_{nm}$ and $Y_{mn}$, refer to any $u_0 \pm 1$ as arbitrary with respect to sign.

### Shannon Entropy vs. Conventional ΔS Entropy and the Number of Causal Paths they Define

Obviously, any progression of the type m → n requires, given $P \equiv 1$, that a second unitary interval of the type ($u_0 \to u_0 \pm 1$) occurs from the endpoint n, such that n → m. Thus, for every:

$$Y_m \to Y_{nm} \; \exists \; Y_n \to Y_{mn}$$

$$Y_n \to Y_{mn} \; \exists \; Y_m \to Y_{nm}$$

For every unitary step ($u_0 \to u_0 \pm 1$) of the type $y_m \to y_{nm}$ *there exists* (because P is always $P \equiv 1$, without ceasing) the interval $y_n \to y_{mn}$. Else, the progression ($u_0 \to u_0 \pm 1$) stops, and the (DeSitter Horizon evolution) ceases. Furthermore, note that ($u_0 \to u_0 \pm 1$) *does not designate any asymmetric linear 1-dimensional choice* for the progression. That is, the requirement that ($u_0 \to u_0 +1$) *only*, is non-sequitur.

There is no condition where 1-causal path, such as is the case with conventional ΔS entropy, where the above description can render only 1-causal path, e.g., yield a value of one. Keeping in mind that by causal path I am specifically referring to the definition, ($u_0 \to u_{i \pm 1}$). Meaning, two temporal endpoints make one causal path. *A lone temporal endpoint is non-sequitur and can only exist as an eigenstate, not an eigenvalue.* Thus, we are regarding:

$$(u_0 \to u_{i \pm 1}) \text{ as } i = 1.$$

As 1-causal path. Conventional ΔS is limited, somewhat conventionally, to one causal path and only exists as a scalar value. It is insufficient to describe numerous causal paths.

In the equation above it is agreeable that our summation is:

$$H(X) = -\sum_{i=1}^{n=2} P(xi) Log_2 P(xi)$$

And our two choices are:

$$n=1. \; u_0 \therefore p(1) = 0.5$$

$$n=2. \; u_{0+1} \therefore p(2) = 0.5$$

## Shannon Entropy vs. Conventional ΔS Entropy and the Number of Causal Paths they Define

Whose internal structure is:

$$u_0 \to u_{0+1}$$

In terms of our Markov Process:

$$u_n \to u_{nm}$$

In terms of a visual:

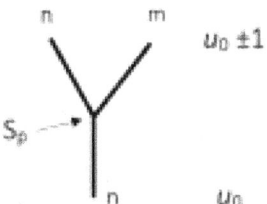

And in terms of a continuous process:

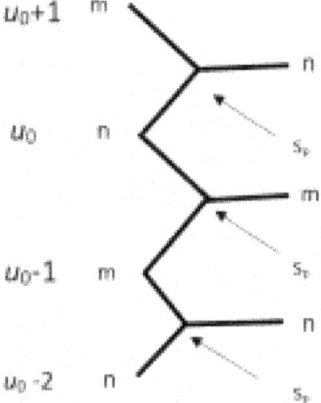

## The AdS Horizon Surface and the Delayed Choice Quantum Eraser

For all of the following relationships,

- the unitary Planck interval of length is designated $Lp$
- the unitary Planck interval of time is designated $tp$
- The Heisenberg Uncertainty Principle is designated **HUP**.
- The quantized lower limit of the HUP is 2 unitary Planck intervals of length and time.
- Koppa [ϟ] is the police pseudo-operator that prevents the human penchant to change the quantities of things, typically via the Heisenberg Uncertainty Principle [HUP]. It is a place holder, for the most part, that polices holding a quantity as constant.

A rearrangement of prior equalities:

$$\varkappa\varepsilon' \equiv \frac{L'}{e^{\ln(\frac{L}{\varepsilon})}} = c^2 = \frac{\varkappa Lp'^2}{\varkappa tp'^2} \equiv \varkappa N_Q = \varkappa A'_\Omega$$

$$\varkappa\varepsilon' < \varepsilon_0$$

The koppa in front of the epsilon above should be regarded as the fractal relationship:

$$\varkappa\varepsilon' \rightleftharpoons \varepsilon_0$$

The more detailed derivations of these fractal relationships appeared in prior [Bray] papers and is summarized in the section of this text on fractal Relationships.

First, we look at the difference in Information content, which would suggest a violation of Information Conservation, *if* Path $l'$ was not of different Conformal Scale from Path $l$. Meaning, we must change the *magnitude of* the unitary Planck intervals in order to obey Conformal Scale Invariance without violation Preservation of Information Conservation.

This has been a major issue I have dealt with over the years. For example, we look at a Lorentzian time dilation; convention has not regarded that the *quantity of* unitary Planck intervals cannot change, as this would indeed violate Preservation of Information Conservation. It would seem that I have never seen anyone address the issue, convention would simply have it that the *quantity of* unitary Planck intervals of time *increases*. This, again, cannot be, else violate Preservation of Information Conservation.

If one is thinking, the change in *quantity of* unitary Planck intervals does not represent a violation of conservation laws, the entire premise of Quantum Electrodynamics is based on this principle, specifically, the vacuum energy. I am not going to get into a separate text on QED and vacuum energies. The basic premise is that the Heisenberg Uncertainty

## The AdS Horizon Surface and the Delayed Choice Quantum Eraser

Principle in all of its conventions demands a vacuum state to be energetic. Only the amount of energy is not clear.

Furthermore, when we regard the fact that [most] every system is in some sort of proper motion with respect to every other system, there is some Lorentzian change in temporal rate between [most] every system. This would mean, as most convention shrugs this off as being artifact of the observing system, which makes no sense because there is a real disparagement between clocks, that every system has a set quantity of unitary Planck intervals of time that make up its worldsheet, $A_n$ and every system has a near infinite number of other *quantities of* unitary intervals according to every other system in the cosmos. Meaning, the *quantity of* unitary intervals is different in every space relative to every other space in the cosmos. This is the quintessential 'loss of information regarding the microstates of a system,' which cannot by that convention of altering the quantity of unitary intervals be the same between any two Lorentzian frames of reference. Then we extend this to Schwarzschild metrics, and every point in cosmological space would differ in information content.

Preservation of Information Conservation must be observed, and this is the reason for placing the pseudo-operator ⊬ *in front of those values* that are identified as needing to be held constant under certain types of transformations.

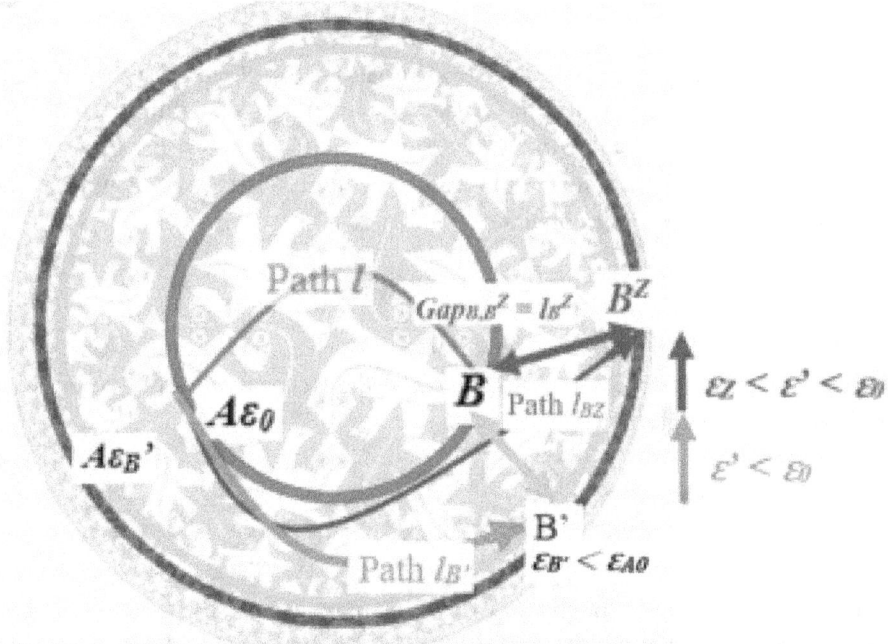

Both Paths $l$ and $l_B'$ are set by ⊬$nL_p$/⊬$xt_p$ because the *quantity* of Information cannot change between frames of reference. Path $l_B'$ is always greater in magnitude than Path $l$. And again, Path $l$ is postulated here as only being relevant in the case of entangled systems. Because Path $l_B'$ is greater in magnitude than Path $l$, this is the causal component of that

### The AdS Horizon Surface and the Delayed Choice Quantum Eraser

'Gap,' $Gp$, that describes Preferential Frame of Reference as established purely by distance and no other factor as well as defining *why* what is true in one system may not be true in another, e.g., Lunar Landing.

*This is also true of both Lorentz and Schwarzschild transformations.* The red arc is that which is typically used in these descriptions. The green arc extending from Alice at $A_\epsilon$ to $B'$ is a more accurate representation. The rigid red arc of convention extends from Alice at time-zero to Bob who is also at time-zero. Therefore, the conventional Ryu-Takayanagi Path $l$ cannot represent any Lorentzian nor Schwarzschild transformations, as it is *atemporal* in nature, extending from and to time-zero in all frames of reference. Motion cannot be described on Ryu-Takayanagi Path $l$, with the exception of course of entangled systems.

$$\hbar\varepsilon' \equiv \frac{L'}{e^{\ln(\frac{L}{\varepsilon})}} = c^2 = \frac{\hbar L p'^2}{\hbar t p'^2} \equiv \hbar N_Q = \hbar A'_\Omega$$

$$\hbar\varepsilon' < \varepsilon_0$$

Epsilon-prime is always lesser in scope than epsilon-naught. Again, this is generally regarded on cosmological time scales, however, the scope and rate are defined locally by c.

Thus, Bob's value $B'$ at point $\varepsilon'$ is less in Conformal Scale [magnitude] than it was when he was at point $B$ at point $\varepsilon_0$. Again, this is because epsilon is vanishing in real time at a scope and rate defined by $c = Lp/tp$. And again, typically epsilon is only regarded on cosmological time scales, which also renders the statement as true. However, the vanishing scope and rate is immediately significant.

$$\hbar B'_{\varepsilon'} < B_0 \varepsilon_0$$

epsilon at point $B'$ is less than epsilon at time-zero, $B_0\varepsilon_0$. The term, $A'_\Omega$ defines the worldsheet of the local system, as is convention of the Bekenstein-Hawking relationship. The temporal distance is given by:

$$\hbar t p'^2 = \frac{\hbar L p'^2}{\hbar \varepsilon'} = \frac{e^{\ln(\frac{L}{\varepsilon})}}{L'} = 1/c^2 = 1/\hbar N_Q = 1/\hbar A'_\Omega$$

The color and highlighting is merely to draw your attention to these terms, as I will be further manipulating them as we proceed. In non-quantized terms looks like:

$$\hbar t^2 = \frac{\hbar d^2}{c^2} = \frac{e^{\ln(\frac{L}{\varepsilon})}}{L'} = 1/c^2 = 1/\hbar N_S = 1/\hbar T_{uv}$$

And the vanishing rate is given by the self-similar relationships:

### The AdS Horizon Surface and the Delayed Choice Quantum Eraser

$$*G' = 6.67384(80) \times 10^{-11} \text{ m}^3/\text{Kg (tp')}^2$$

*The current literature value G. The value itself is not of importance.

From:

$$t', l' = t_0, l_0 \sqrt{1 - \frac{2G'M}{rc^2}}$$

Quantizes as:

$$\varphi \pm \hbar(t'_p, L'_p) \rightleftharpoons |\hbar(t_{p0}, L_{p0}) \sqrt{1 - \frac{2G'\hbar M'}{\hbar n L'_p \left(\frac{1 L p'}{1 t p'}\right)^2}}|$$

$m = E/c^2$, and $c^2 \equiv \varepsilon$ thus:

$$\varphi \pm \hbar(t'_p, L'_p) \rightleftharpoons |\hbar(t_{p0}, L_{p0}) \sqrt{1 - \frac{2G'\hbar(\frac{E}{\varepsilon})}{\hbar n L'_p \hbar \varepsilon'}}|$$

Where the term $nLp$ is r, the distance from the center of mass. And $\varepsilon$ and $\varepsilon'$ are not the same value, $\varepsilon'$ represents the evolutionary state over time of epsilon.

The value epsilon was equated above with $c^2$:

$$\hbar \varepsilon' = c^2 = \frac{\hbar L p'^2}{\hbar t p'^2} \equiv \hbar N_Q = \hbar A'_\Omega$$

The rationale for the constancy of the speed of light would be moderately explained by the evolution of epsilon, given [Ryu-Takayanagi] Path *l* is the normal light distance between points.

Going back to the Lunar Landing for a moment, we could have the LEM 'explode' on the moon in Bob's frame of reference but does not explode according to Alice on Earth, as the

## The AdS Horizon Surface and the Delayed Choice Quantum Eraser

result from running out of fuel in Bob's [lunar] frame of reference but with 1.2 seconds of fuel remaining in Alice [CAPCOM] frame of reference. Eventually the information that Bob exploded on the moon would reach Alice, but for 1.2 seconds, Alice's reality is different from bob's reality. If you are thinking, that is true for all things over distance, that is the point. What is true in one system may not be true in another system. This can be a temporary state, such as Bob's demise on the moon, or a permanent state, such as our 2-train scenario, where the train crash *never occurs in Alice's Preferential Frame of Reference.*

This sounds like the stuff of science fiction; however, the ongoing 2-train scenario is recorded in both Preferential Frames of Reference with Bob crashing into Alice but Alice not crashing into Bob. If the LEM had run out of fuel above the lunar surface, there very likely would have been no explanation, because CAPCOM had fuel remaining in their Preferential Frame of Reference.

In the simplest sense, we can regard the train wreck [CAPCOM cutting it] as some quantity of Information and assign it $jN_Q$, where j is some integer value. In Bob's reality [Moon] the value j has a positive integer value, in Alice's [Earth] reality $j = 0$. If we take it one step further and assign that Information $jN_Q$ mass*energy equivalence, then there is a violation of classic thermodynamics $\Delta S$, where j implies some positive integer amount of mass*energy in Bob's frame of reference but is zero in Alice's frame of reference. The example of the LEM running out of fuel is this $\Delta S$ issue in the scalar quantity, *j*, differing between frames of reference.

There is no conventional resolution of this. Only, 'what is true in one system may not be true in another system,' can extend to scalar quantities, as the prior example describes.

Purely for a common reference's sake, we can look at histograms of the Earth vs Moon recordings:

**Earth [8.7 seconds of Information]**

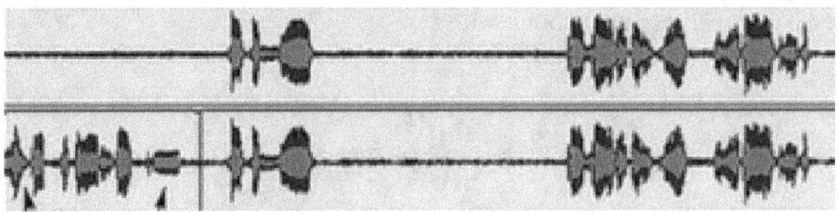

**Moon [7.5 seconds of Information]**

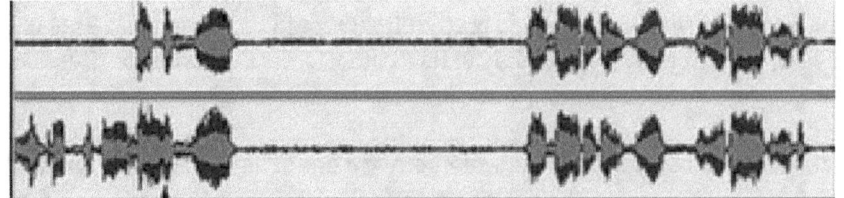

### The AdS Horizon Surface and the Delayed Choice Quantum Eraser

With the difference in length squeezed down to *appear* equal, as representing Information content, we have a different set of events, regardless of type, between the two frames of reference. Events occur in a different causal order. There is no conventional resolution for this either.

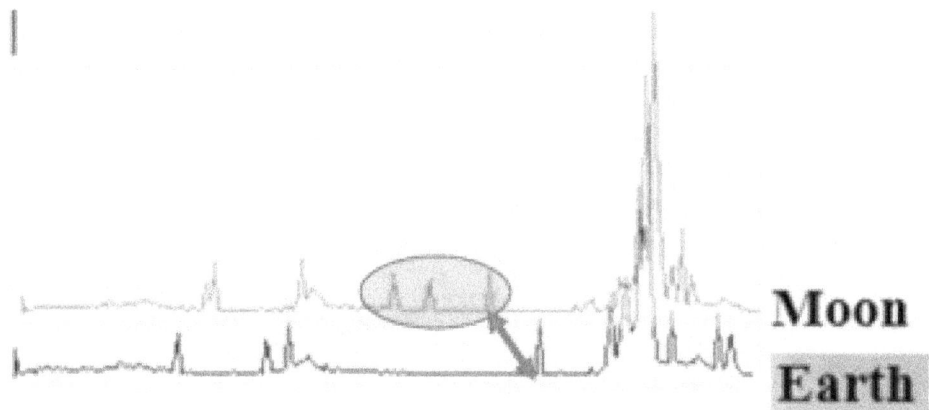

This is not a quantitative analysis; I am merely showing an example of the real difference in Information content between frames of reference established purely by distance. There are peaks in the Earth frame of reference that are buried in the lunar frame of reference and vice versa. The placement of the peaks would [initially] attempt to be explained off as simple signal delay issues. However, the content differing in the two frames of reference cannot be explained by signal delays and have no conventional $\Delta S$ solution because of that, else, the circled peaks would appear but offset. There is, in fact, different entropies clearly visible between the two systems in terms of quantitative information. However, because the volumes of the two recordings in their original form were not exactly equal, we cannot reverse engineer what the difference would be.

The three circled peaks are of a relative scale that should appear in the Earth recording but do not. It could be internal signal noise of the instrument, but the scale indicates that should not be the case, they are quite large in scale. If they were internal signal noise the amplitude is sufficient it would appear across both systems.

All of the other peaks are shifted in time and not of consequence in this argument. Those three peaks, even if it is shot or Johnson noise should appear in the Earth based recording because of their magnitude. I have no answer for what they are, only that they are a complete anomaly of the sort, $jN_0$. Here, the koppa pseudo-operator cannot apply, only, 'what is true in one system may not be true in another system,' which can include scalar quantities and total Information content [$jN_0$]. There are *bits* of information that appear in one Preferential Frame of Reference but not the other.

This may seem insignificant, but these *bits* could potentially be of any magnitude. When we discussed the destruction of information as per LIGO detection of gravity waves, the *bits* that are missing are the two black holes, Alice and Bob, as there is no causal path that

## The AdS Horizon Surface and the Delayed Choice Quantum Eraser

can lead back to them. The only causal path is to the product black hole, Victor, whose evolution is the cause of the chirp-mass and gravity wave. As for the chirp mass, it is the only example of genuine 'length contraction' in a natural system. I am not going into those details as they were discussed at length in prior texts.

In particular, the three peaks in the lunar recording do not appear in the Earth recording at all, regardless of this being an open channel. It isn't that the peaks are buried elsewhere in the recording, as they are too distant from any CAPCOM peaks. The third peak could be that peak down and to the right, which is important because this is an indication of the actual original volumes of the two recordings. Meaning, they are not just noises on the LEM that aren't being broadcast, as the third peak clearly indicates the magnitude is sufficient between the two systems.

This *quantity of peaks* is violation of conventional $\Delta S$ entropy, as well. It is not, however, a 'paradox.' It merely implies that conventional $\Delta S$ [not Shannon] entropy is not equal between the two systems.

This is less perplexing than one may think. In the Delayed Choice experiments, all conventional $\Delta S$ holds true, oddly, there is no difference in entropy between detector zero and the other detectors:

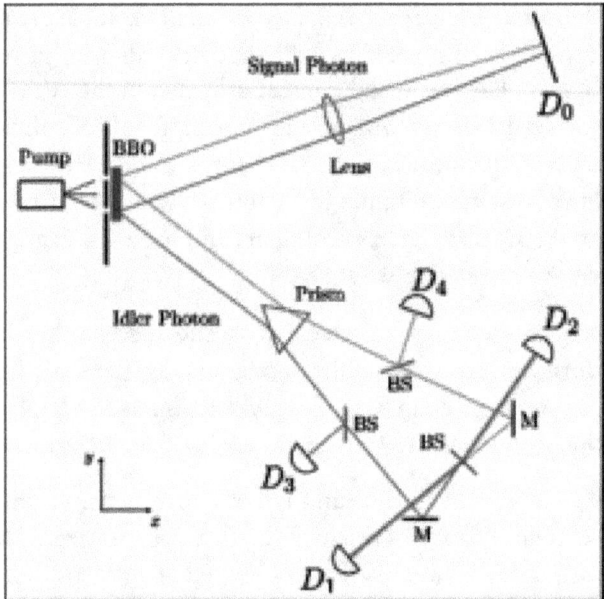

This would actually appear to be the dilemma, oddly, that the conventional $\Delta S$ entropies are equal between all detectors, rather than differing. That is, if we detect an interference pattern [train wreck] at detectors {1, 2, 3, 4} we also get an interference pattern at detector zero, although detector zero is further down [progressed] in time than detectors one through four. There is no difference in the information content or type between the systems.

### The AdS Horizon Surface and the Delayed Choice Quantum Eraser

Likewise, if we detect 'particle-like' behavior at {1, 2, 3, 4} we also get a particle-like pattern at detector zero, although detector zero is further down [progressed] in time than detectors one through four. The sum of the information is the same between all entangled systems when that sum should differ at detector zero. That is not to say that there is a difference in mass-energy rather it is an interference pattern or particle-like behavior, energy is conserved, but the arrangement of energy should differ, and it does not.

- Again, the supposition is that, for entangled systems, Preferential Frame of Reference *does not apply, and thus:*

*What is true in one system is true in all systems.*

This of course, is also referring to the conventional $\Delta S$ quantity of information between systems, that should differ but do not, which is again, the dilemma.

This is a rather bold and odd postulate. However, if one studies the phenomenon with respect to the AdS Horizon Surface descriptions from the last section very carefully, it becomes clear what Preferential Frame of Reference is and what it is not. This is where and why I use the syntagm, Preferential Frame of Reference rather than *frame of reference*.

Entangled systems by convention would seem to ignore the region figuratively 'below' the AdS Horizon Surface altogether and seem to exist solely on the AdS Horizon Surface, the superluminal Path $L$. However, I want to present that this [conventional] notion is incorrect, shortly. I am not implying that they are indeed superluminal, merely that Information travelling through the region figuratively 'below' the surface, the normal light distance between points, is not representative of the eigenstates of the entangled information. There can be no actual locality, if you will, of the Information along the arc, only that it starts with Alice at time-zero and ends with Bob at time-zero, traveling the red arc Ryu-Takayanagi Path $l$, proceeding inward toward the larger Conformal values epsilon [causal past] then shifting at midpoint toward the larger Conformal value epsilon [causal present]. The notion of 'motion' gets rather hazy. I don't think it is correct to think of information or such as 'travelling' in any sense of the word whatsoever, as the conventional Ryu-Takayanagi system is static, atemporal. There is no progression of time as per convention in that system.

At such time they evolve or precipitate to eigenvalues, it would seem [by convention] as though they have not travelled via Path $l$ at all, but have in fact, not traveled at all. That supposition is consistent with the conventional $\Delta S$ solution of all of the detectors detecting the same thing. However, the explanation I want to put forward is very different:

## Preferential Frame of Reference in Entangled and Non-entangled Systems

When a system obeys Preferential Frame of Reference, the Information takes Path $l'$ [$l$-prime] as the normal light distance between points, the Conformal Scale of the actual Information content changes *in magnitude but not quantity*, preserving Information Conservation. [Note again that I highlight the 'primes' because of the penchant to read these things on pads and even phones].

In prior papers I put forward that it is the *magnitude of* the unitary Planck intervals {Lp, tp} that change, not the *quantity of* unitary Planck intervals, under the conditions of a Lorentzian or Schwarzschild metric. I will summarize that in an appendix of this text.

Preferential Frame of Reference is, looking again at:

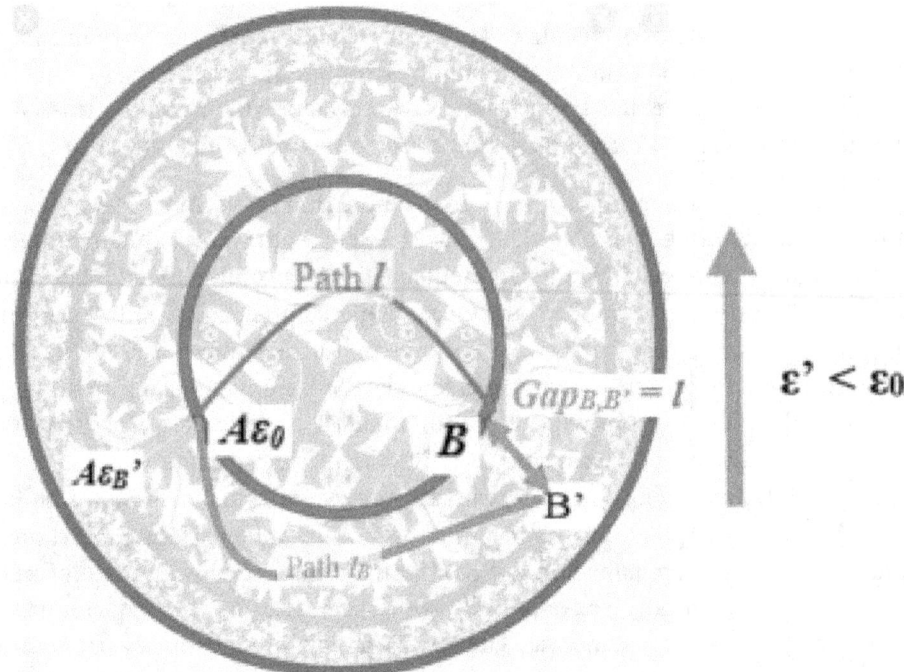

As stated, the green arc represents the effect of Preferential Frame of Reference with respect to the causal series of events between systems. This is what results in, 'what is true in one system may not be true in another system.' Convention would have the Information traverse [Ryu-Takayanagi] Path $l$, however, this is impossible because points $\{A_\varepsilon, B\}$ represent *the same instant in time, namely, time-zero, no time has evolved in this system.* This is true regardless of the scope and rate of the vanishing value, epsilon.

However, when Preferential Frame of Reference *does not apply*, which is specific for entangled systems, it seems to me that in this case the Information does actually take the rigid red arc of convention. Furthermore, it is clear that typical causality *is not represented*

**Preferential Frame of Reference in Entangled and Non-entangled Systems**

by information taking the rigid red arc of convention, meaning, entangled systems do not represent typical causality because:

- The Information leaves Alice at time $A_{\varepsilon_0}$ and is observed by Bob at time $B$. Both of these intervals represent time-zero, not time that has evolved. What is true in one system [Alice] is true for all systems [Bob] because time has not evolved or progressed. Preferential Frame of Reference *does not apply* because epsilon [Ryu-Takayanagi $\varepsilon$] has not progressed in time or scope.

*Keep in mind that this is true regardless of the scope and rate of the vanishing value epsilon, even on cosmological time scales.*

Thus:

- When Preferential Frame of Reference applies, Information takes the [Ryu-Takayanagi] Path $l'$ *[l-prime]* which is the green arc. The Information leaves Alice at time $A_{\varepsilon_0}$ [time-zero] and is observed by Bob at $B'$ [causal future from $A_{\varepsilon_0}$]
- When Preferential Frame of Reference *does not apply*, Information traverses the rigid red arc of convention. Information leaves Alice at time $A_{\varepsilon_0}$ and is observed by Bob at time $B$ *[not B-prime]*. Both $A_{\varepsilon_0}$ and $B$ represent the same instant in time. This is actually the convention.

Keep in mind here, again, convention has the red arc going 'down into' the 'field of lizards,' where the lizards [unitary Planck values] are of greater Conformal Scale, meaning, toward the past, backward causality. And at exactly the midpoint begin burrowing back up toward the present. This can neither be true nor correct. Only in the case where the entire system is completely frozen in time, static, non-dynamic, can the conventional Ryu-Takayanagi Path $l$ be valid, and only entangled systems demonstrate this characteristic.

*What is true in one system is true in all systems because all systems are fixed at time-zero.*

Thus, when we go back to the **Delayed Choice Quantum Eraser**. In the image below, note that:

$$\text{Path } l_{B'} > \text{Path } l = \varepsilon_0 > \tfrac{1}{2}\varepsilon'$$

- Path $l_{B'}$ is greater than Path $l$ [defined by convention as proportional to $\varepsilon_0$] because $\varepsilon'$ is always less than $\varepsilon_0$].
- The 'Gap' equals the temporal distance from $B$ to $B'$ and is always equal in magnitude to Path $l = \varepsilon_0$. The Gap, again, is the result of the horizon evolving *in scope*.

**Preferential Frame of Reference in Entangled and Non-entangled Systems**

If Bob at $B'$ sends a signal back to Alice at $A\varepsilon B'$ she will be at temporal index $A\varepsilon B''$ [double-prime] at such time the Information reaches her. *This is the Lunar Landing Anomaly, the Gap, in two directions:*

$$Gp \text{ of } \{B, B'\} \text{ to } \{A\varepsilon B', A\varepsilon B''\}$$

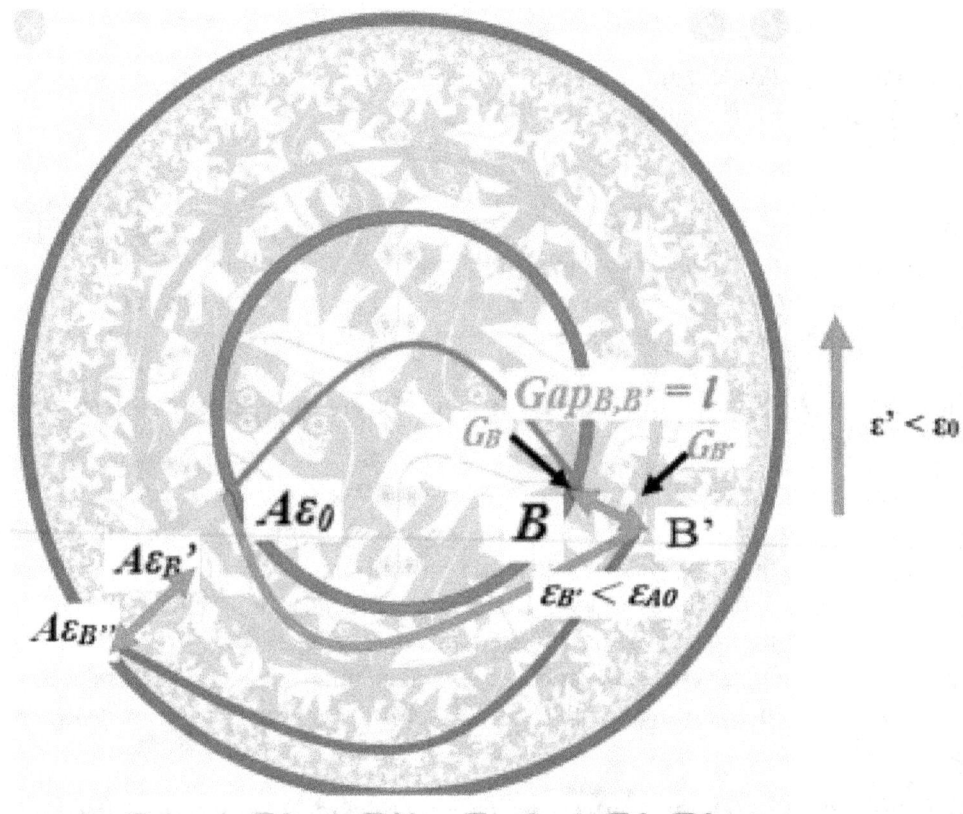

$$Gap\ A\varepsilon B', A\varepsilon B'' = Path\ A\varepsilon B', B'$$

1. Alice sends a signal at time $A\varepsilon_0$ to Bob who is at point $B$ at time-zero.
2. In the time it takes for the signal to reach Bob, he is now at point $B'$ and Alice is now at point $A\varepsilon B'$. We will call this *time-two*.
3. At time-two, Bob, who is at $B'$ and Alice at $A\varepsilon B'$, sends a signal back to Alice, who again, is at $A\varepsilon B'$.
4. In the time it takes for Bob's signal to get to Alice, she is no longer at $A\varepsilon B'$, but is now at $A\varepsilon B''$ [double-prime]. This is *time-three*.
5. This is the second Gap in green [small double headed green arrow] stretching from $A\varepsilon B'$ to $A\varepsilon B''$ [double-prime].

**Preferential Frame of Reference in Entangled and Non-entangled Systems**

6. Thus,
   - the first Gap is from $B$ to $B'$
   - the second Gap is $A\varepsilon B'$ [prime] to $A\varepsilon B''$ [double-prime]
   - time-one is Alice at time-zero $\{A\varepsilon 0\}$ and Bob who is also at time-zero $\{B\}$
   - time-two is from Alice at time-zero $\{A\varepsilon 0\}$ to Bob who is now at $B'$.
   - time-three is from Bob at $B'$ to Alice, who is now at $A\varepsilon B''$ [double-prime]

These two 'Gaps,' are the Preferential Frame of Reference we see in the Lunar Landing.

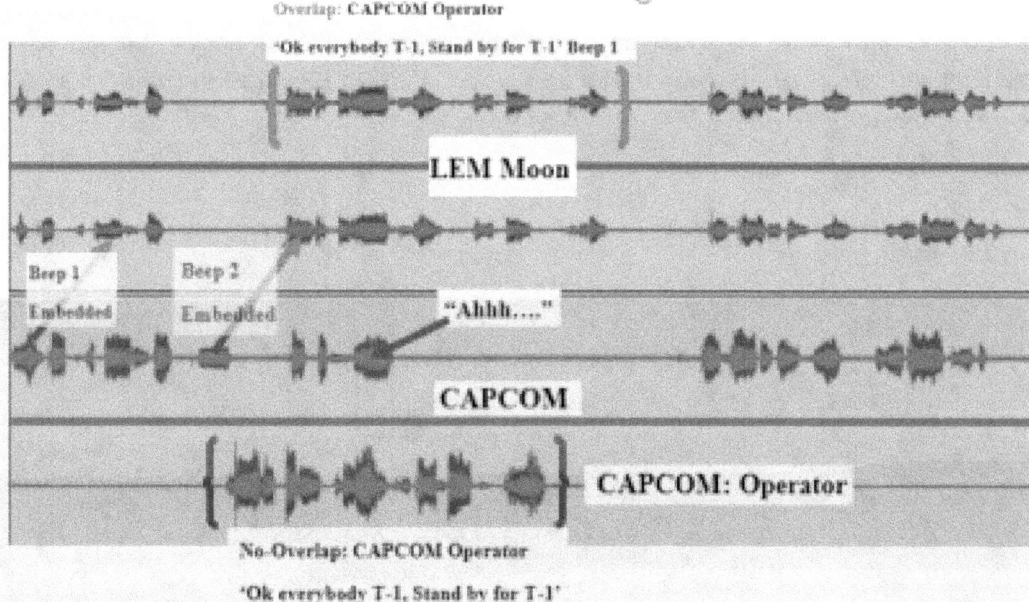

We have information that is bouncing back and forth between Earth and moon. However, what is happening here is that:

- **Time-zero A:** Armstrong begins to speak, intending to say, "Tranquility Base here, the Eagle has landed."
- **Time-zero B:** Unknown CAPCOM operator does speak, saying, "Ok everybody, T-1, standby for T-1"

These two time-zeros occur because neither speaker is aware of the other speaker or any sequence of events regarding who is speaking and/or what they are saying.

- **Time-one A:** *Beginning of* Unknown CAPCOM Operator's message reaches the moon.
- **Time-one B:** Armstrong is cutoff mid-sentence by Unknown CAPCOM operator.

These two events occur A because of distance and B is the simple train-wreck.

## Preferential Frame of Reference in Entangled and Non-entangled Systems

- **Time-three A:** The end of the Unknown CAPCOM operator reaches the moon.
- **Time three B:** Armstrong hears that the air is clear and resumes speaking.

This may all seem like detail overkill, however, to the best of my knowledge no one has attempted a detailed analysis of this set of recordings. Keep in mind that the notion that this is not a signal timing delay issue but one of questionable causality is not novel. In fact, I think it was Feynman who first brought the subject up in lecture not months later in 1969. And if I recall I think there may be a note or footnote on this in one of the very early Feynman grad texts, since they were based on his lectures.

If you look at the wave form images, you can piece this together and begin, hopefully, to realize that there is no means by which to reverse engineer the argument back to signal delays. I think we left that behind several sections ago.

## What exactly is the 'Gap?'

### What exactly is the 'Gap?'

Time and distance were equated by Einstein over a century ago. We typically regard time and distance as a factor when we are discussing large scales, lightyears, for example. However, we don't typically have much regard for much smaller scales. This is a bit puzzling, as the DCQE is a benchtop experiment where the distances between detectors is perhaps on a meter scale, at best:

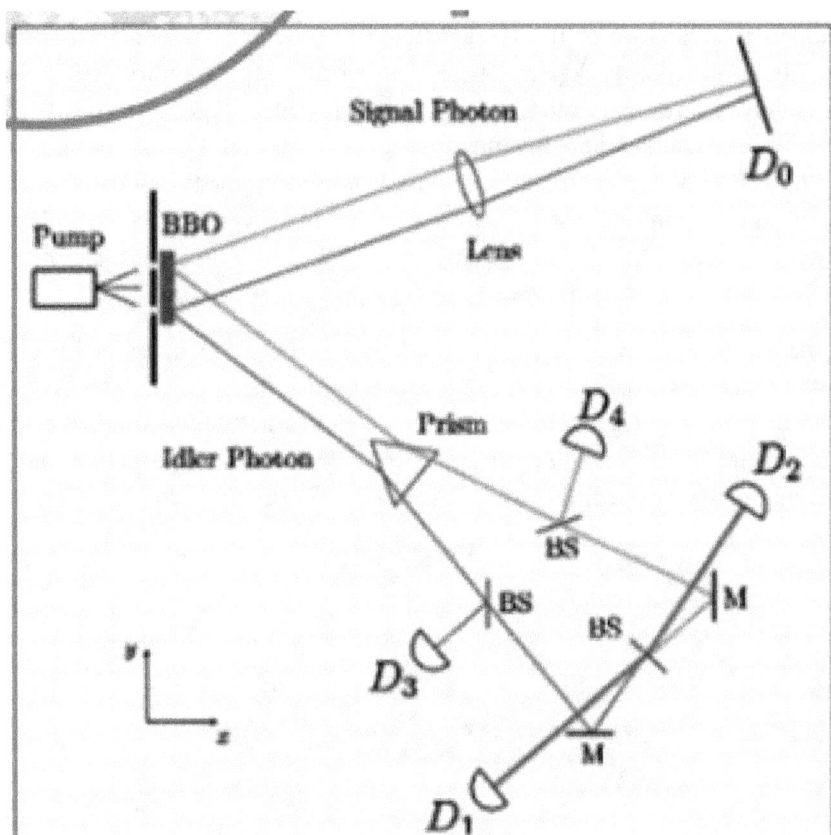

We puzzle over the observation that detectors one through four see the same behavior, e.g., interference pattern vs particle-like behavior, over distances less than an arm stretch apart.

The moon is about 1.2 light-seconds away. This again, is oddly the furthest man has ever ventured in any direction.

Clearly, I think I have established that the events in the recordings of the first Lunar Landing cannot be explained off as simple time delays. The astronauts were as distant in time as they were in meters. This is true all the way down to, if you regard quantization, a single Planck length.

### What exactly is the 'Gap?'

Alice's reality [CAPCOM, Earth] is 1.2 light-seconds removed from Bob's reality [Armstrong, moon]. If Bob were 1.2-billion lightyears distant it would not be any different, just greater in magnitude. Bob can be born and die, his civilization rises and falls, and although we could perhaps witness those events in the form of observing the light from that extinct civilization, we could not interact with it; *there could be no train crash.*

It is the *inter*acting with systems, which is at this time only relevant on smaller scales, light-seconds, that brings about seemingly odd results that are different between systems, e.g., train-crash vs no-train-crash.

As the AdS Horizon Surface evolves, regardless of scope and rate, epsilon diminishes in magnitude, always, non-stop. Again, where we typically regard epsilon on cosmological scales, the quantization demands that this is true on all scales. Thus, our Locally Quantized Metersticks are constantly changing magnitude with respect to one another. I am not suggesting that this change in magnitude is the causal component of the Lunar Landing Anomaly, [what is true in one system may not be true in all systems], I am suggesting that the evolution of epsilon on local scales 'moves' Information around in a way that is not consistent between Preferential Frames of Reference.

Meaning, the 'movement' of Bob from $B$ to $B'$ is a real change in his position from one horizon value to the next, with respect to his relative position to Alice, who is also moving from one horizon value to the next from $A\varepsilon_0$ to $A\varepsilon B'$. The actual localities of things on the horizon surface are not an agreed upon description, but many hypothetical descriptions. The entire thing is hypothetical.

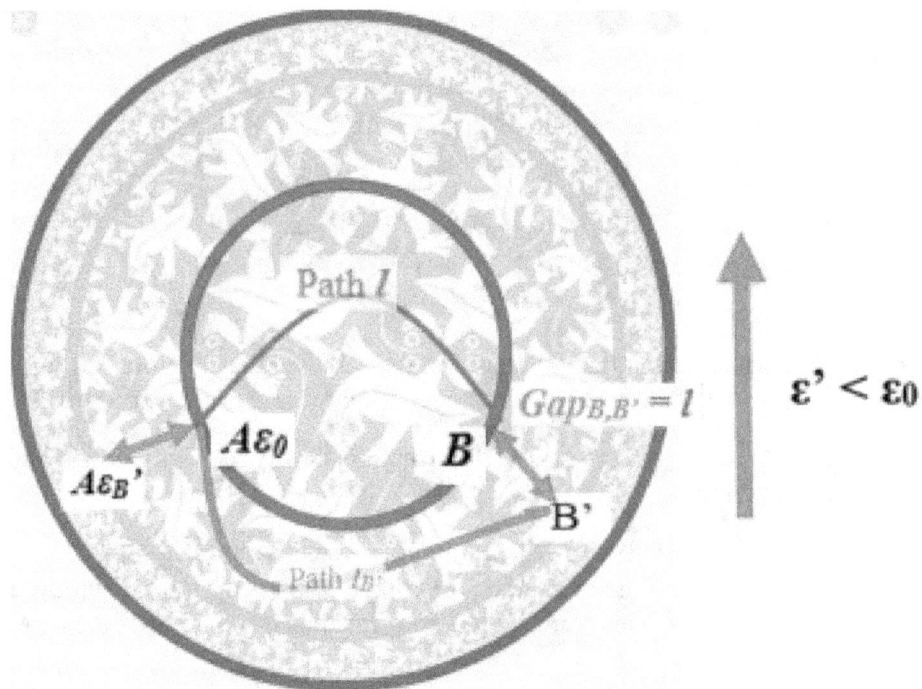

### What exactly is the 'Gap?'

The two double-headed green arrows depict Bob moving from the innermost red horizon to the yellow horizon, and Alice from the innermost red horizon to the yellow horizon. They are not moving 'on the horizon,' the horizon itself is changing in real time in the manner depicted above. Eventually, the horizon will pseudo-expand to the outermost *blue horizon* [no pun Jeff].

If Alice and Bob were pseudo-moving 'on the AdS Horizon Surface' that would be rather impossible to describe. To date, I haven't seen two papers that agree on what that would be characterized as, and nothing I have seen is satisfying. I think the problem goes back to what I put forth in earlier papers, that, given each Qubit is superluminally isolated from each other Qubit, there is characteristically *only one Qubit that defines the entire AdS System*. I understand that is preposterous to put forth, however, mathematically there is no way to describe relationships between Qubits of Information that are superluminally isolated from each other Qubit of Information in some continuous way. This is because in the Einstein universe, we regard space and time as equal and defined by the speed of light as its limiting factor. Meaning, superluminal isolation evades a description in Einstein's universe. This is where people suggest transcendentals, such as the sqrt(-1) and so on, which becomes impossible to clearly define.

Again, the sqrt(-1) is a state that is {unobservable}, and thus defined and described by Shannon, not conventional $\Delta S$ entropy.

Furthermore, the evidence to date given the history of the 2-slit seems to validate the notion that the conventional Ryu-Takayanagi Path $l$ is not a motion, but a pseudo-motion, because temporal progression does not exist in that description. That would be characteristic of Alice and Bob not having a clearly defined spatial relationship to one another *on the AdS Horizon Surface*. We cannot be certain at this time.

Whatever the case may be, it is not a motion in any sense of the word that defines Bob's transition from his point $B$ on the horizon to his later locality, $B'$ [prime]. It is the horizon itself that is regarded as changing, in some sense of the word pseudo-expanding, in some sense starting at the center of the image above and pseudo-shifting figuratively 'outward.'

As the AdS Horizon Surface pseudo-expands figuratively outward, there is this 'Gap' that is left between Bob's former locality at $B$ to his new locality at $B'$ *[prime]*. The train crash issues come from the fact that CAPCOM cannot directly communicate with Bob at his former pseudo-locality figuratively 'below' the horizon, because the region figuratively 'below' the horizon does not exist, it is only a record of past states of the horizon.

Thus, CAPCOM sends a message to Bob at point $B$, which is a pasts state of the horizon but cannot reach Bob at point $B$ as this would require the transfer of Information to be instantaneous. It reaches future Bob, who is looking at Information from the past, not the present.

I think the trick is not to make it more complicated than it is.

**What exactly is the 'Gap?'**

The summary of math derivations that describe all of this are again, moved to the Appendixes of this text. I chose to do this because of the modern human penchant to read papers on pads and even phones. The math is nearly impossible to read under those conditions and makes the readability of the entire text worse than it already is.

If we regard the placement of 'things' on the horizon surface in some quantitative way, and regard the postulates presented in this text, there are much less holes in understanding, and figuratively no unobservable dimensionalities. Keeping in mind by unobservable dimensionalities we treat the AdS system as a 2-dimensional rendering of our 3-dimensional world, or a 3-dimensional rendering of our 4-dimensional world. I am not comfortable with unobservable dimensionalities. As stated in this and prior texts, there can never be any validation of these things, and claiming that the data that invoked the unobservables validates the unobservables is completely illogical and unscientific.

If you regard the prior statement that a system that is in the state, {unobservable} is at least hypothetically described by Shannon entropy, and not conventional $\Delta S$ entropy, then this would demand that an unobservable dimensionality can only be described by Shannon Entropy and has no relationship with conventional $\Delta S$ entropy. At such time a system is unobservable and described by Shannon Entropy, the causal connectivity between each system drops into infinite causal connectivity.

The 'Gap' is exactly what it appears to be, a change in relative position in time of systems on the horizon surface. It is not a phenomenon related to the position in space, save for it is defined as some normal light distance.

When those things are entangled, as the postulate goes, they exchange Information via the conventional Ryu-Takayanagi Path $l$. When those systems are not entangled, it seems clear to me that they cannot traverse Path $l$ but must traverse a different path that does not violate causality and Preservation of Information Conservation. That path seems to be clearly defined in this text.

Moving on and expanding the image out a bit: [following page]

**What exactly is the 'Gap?'**

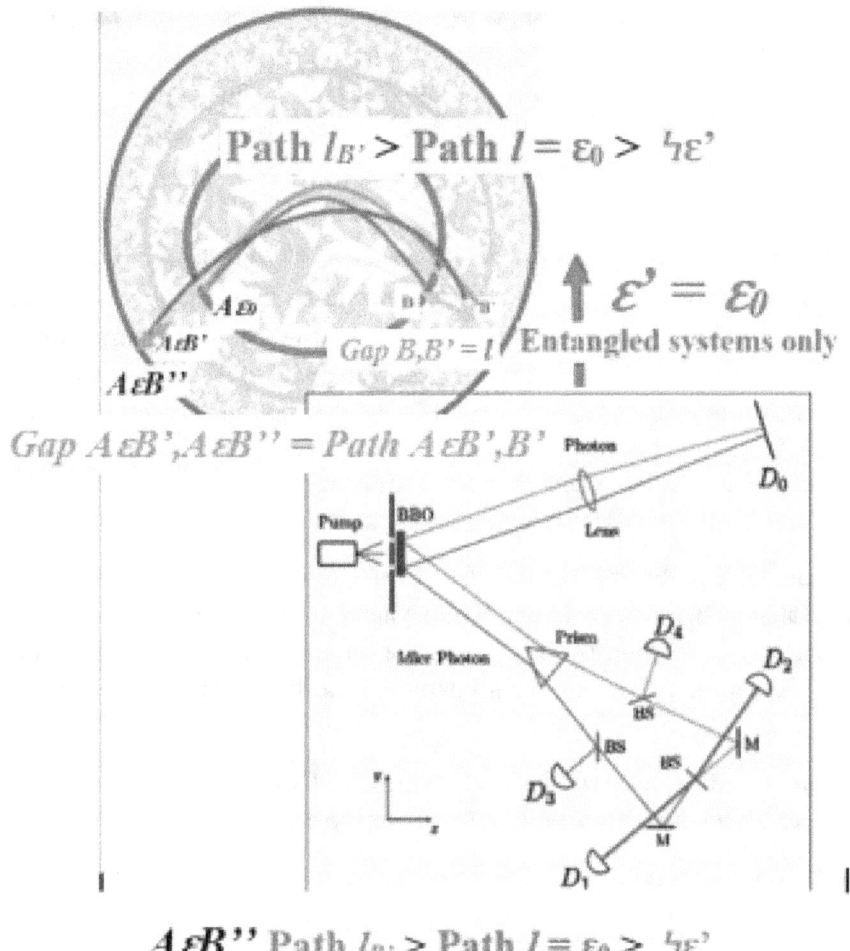

Note that above, the red indicating $\varepsilon' = \varepsilon_0$ applies to entangled systems only.

- $A\varepsilon$ to $B$ [not-prime] is the conventional Ryu-Takayanagi Path $l$, and again, only relates to entangled systems. Both $A\varepsilon$ and $B$ are at time-zero, and *no causal progression, no temporal progression can be associated with this system*. This is consistent with observed entangled Information, e.g., DCQE.
- $A\varepsilon$ to $B'$ [prime] is the 'normal' path Information takes for non-entangled systems. This is represented above by Path $l_{B'}$ [prime]. Path $l_{B'}$ is always greater than Ryu-Takayanagi Path $l$. Entangled systems do not travel via Path $l$.
- The Gap, $Gp$ is represented above by the double-green arrow, and is always equal in magnitude to the conventional Ryu-Takayanagi Path $l$. The Gap, $Gp$ is always in the causal future to that of the conventional Ryu-Takayanagi Path $l$, meaning that what *was true* in system A is no longer true in system B.
- Conventional Ryu-Takayanagi Path $l$ is set proportionally to epsilon-naught [$\varepsilon_0$]. epsilon-naught represents
    o time-zero for all systems
    o therefore, no temporal progression for any system.

**What exactly is the 'Gap?'**

- o this is why 'what is true in one system is true in all systems,' because no temporal progression has taken place.
- The AdS Horizon Value $\varepsilon'$ is always *less than* $\varepsilon$, because epsilon-prime represents temporal progression, which in turn means epsilon has evolved [diminished in magnitude].

We can think of the BBO crystal as Alice at time-zero, $A_\varepsilon$ and *all detectors* {D1, D2, D3, D4} as Bob at $B$. Note that now we have $\varepsilon' = \varepsilon$ [in red upward pointing temporal arrow], not $\varepsilon' < \varepsilon$ because Information taking the rigid red arc of convention is doing so in a manner where the evolution of epsilon has not occurred, which is actually also by convention. What is true for Alice at $A_\varepsilon$ is true for Bob at $B$ because the two points, {$A_\varepsilon$, $B$} both represent the initial conditions at time-zero. This seems to defy typical causality, but entangled systems behave in this manner because they are taking the rigid red arc of Ryu-Takayanagi convention, as such, Preferential Frame of Reference *does not apply*.

This is not a superluminal transfer, it is the transfer of Information by the rigid red arc of convention, which demands that epsilon is static, not dynamic. The entire system is static, no progression of time is taking place. Thus, the notion of Information *travelling* from Alice to Bob is also not correct. There can be no motion. Exactly what it is, is a causally frozen system.

**A causally frozen system does not defy the Preservation of Causal conservation, it is a different phenomenon and for the most part *an artificial one*.**

I honestly cannot think of any natural system that has this causally frozen characteristic.

1. Again, the notions that electrons in an orbital [opposite spins] are entangled is incorrect because the electrons in an orbital are in *eigenvalues, not eigenstates: eigenvalues cannot be entangled*.
2. The notion that motions, such as an electron's orbital momentum is entangled with the force that altered the angular momentum is also incorrect: the forces of nature cannot be entangled, there is zero evidence of this, and the hypotheses are naïve.

There are no conditions in nature that demonstrate entanglement. Entanglement is purely an artificial phenomenon. And, like the Zeno Effect that was used in prior chapters to define it:

1. The Zeno Effect is an artificial alteration of the progression of time, and as such, the local geometry of space, else violate General Relativity in some way that has no prior art.
2. The Zeno Effect can artificially slow the progression of time.
3. The Zeno Effect can artificially accelerate the progression of time.
4. The DCQE and 2-slit variations artificially *freeze time altogether*.
    a. The DCQE and 2-slit variations are the only examples of systems that exchange, for lack of a better term, Information via the conventional Ryu-Takayanagi Path $l$.

**What exactly is the 'Gap?'**

> b. Thus, only artificial systems can pseudo-traverse the conventional Ryu-Takayanagi Path *l*.

Convention to the best of my knowledge makes no provision for the time of flight for the Information as being in any way equated with the evolutionary scope and rate of epsilon because epsilon is by convention only regarded on cosmological time scales. Thus, convention in that sense treats epsilon as non-dynamic [static], on such small-time scales. Again, regardless of the scope and rate of the progression of epsilon, it is not static, but is dynamic, and the above diagrams represent the demands of a dynamic AdS Horizon Surface, regardless of epsilon's vanishing rate.

Thus,

- When Preferential Frame of Reference *does not apply*, which is specific to entangled systems, Information takes the rigid red arc [of convention] of Ryu-Takayanagi Path *l*. In this case, epsilon has not progressed [evolved] because the information leaves at time-zero for $A\varepsilon$ and arrives at point $B$, which also represents time-zero.
    - Since the rigid red arc of convention represents the normal light distance between points, it is clear that the information has not travelled in a manner consistent with the normal progression of time. It has not, however, travelled via [Ryu-Takayanagi] Path *L*, which is superluminally forbidden.
    - What is Temporally True for one system is true for all systems.
- When Preferential Frame of Reference *applies*, which is true for all non-entangled and macroscopic systems, Information takes the green arc, from $A\varepsilon$ [time-zero] to $B'$ [time has progressed forward]. Epsilon [$\varepsilon$] has progressed [evolved] in its vanishing scope and rate defined by $c \equiv 1Lp/1tp$. As a result of the vanishing scope and rate of epsilon, $\varepsilon' < \varepsilon$ [epsilon-prime is less than epsilon at time zero] which leaves the 'Gap' indicated in the diagram, *which is always equal in magnitude to the normal light distance between points.*
    - What is Temporally True in one system may not be true in all systems.
    - The 'Gap' represents what *may seem like* atypical causality violations.

Going back to my statement above:

> Since the rigid red arc of convention represents the normal light distance between points, it is clear that the information has not travelled in a manner consistent with the normal progression of time.

Entangled systems travel Ryu-Takayanagi Path *l*, which connects two systems both at time-zero. Information travels the red arc, but the red arc does not represent the evolution or progression of time, it is a static value of a static system. Time, to the best of my knowledge, was not figured into Ryu-Takayanagi initial description of this sort. And to the best of my knowledge, no one has addressed the quality or quantitative aspect of temporal evolution in the Ryu-Takayanagi system. Thus, the postulates are a bit bold, but nonetheless, seems to make sense.

**What exactly is the 'Gap?'**

We go back to the DCQE setup:

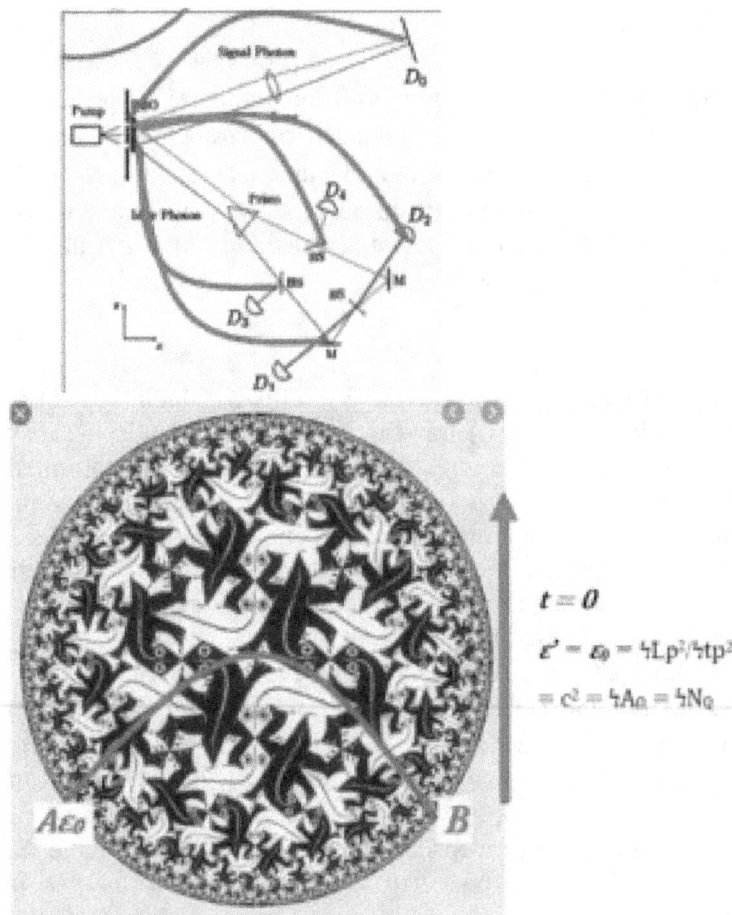

In the top portion of the image, the red arcs all represent the conventional Ryu-Takayanagi Path *l*, which is indicated in red throughout this text. Alice at $A\varepsilon_0$ and Bob at *B* are entangled systems. There can be any number of causal paths connecting them, as would be the case in sending some quantity of photons around the DCQE setup.

The key factor is that the system in the first image of the DCQE is identical in nature to the fractal lizards image. Time is static, frozen, non-existent. This is not the same as say, infinitely dilated, it is not at all related to such phenomena. It is an *artificial state* where the progression of time hasn't changed in any way, e.g., Lorentzian or Schwarzschild transforms, it is simply, *frozen*.

The reason time has *frozen* is because nature has no provision for entanglement. Again, I am not going to argue with sub-sufficient notions of naturally entangled systems, those notions are insubstantial and naïve. What we have done is introduced a phenomenon that nature has no provision for, entanglement, thus, time freezes because nature has no clue

## What exactly is the 'Gap?'

how to proceed. Because time takes a siesta space becomes meaningless, and our observations are weird because nature is stumped.

In the bottom portion of the image, the red arc Ryu-Takayanagi Path *l* begins with Alice at time-zero at $A_0$ and ends with Bob also at time-zero at point *B*. The red upward arrow to the right indicates time, however, the value t = 0 is indicating that zero time has elapsed. The koppas are placed in front of the worldsheet, $A_0$ and $N_0$, the *quantity of* Qubits that make up the worldsheet.

The koppas in front of the Planck length Lp and Planck time, tp are there to indicate that *no change in Conformal Scale has taken place, because no time has elapsed throughout the system*. Because Conformal Scale has not changed for the conventional Ryu-Takayanagi Path *l* system, that is thusly entangled, it seems to defy our common sense of time, which is founded upon a constant epsilon change in Conformal Scale as describing and defining nature.

In effect, all of the red arcs in the upper portion of the image, the DCQE setup, are the same path, Path *l*. This does not break any convention, it in fact observes convention in every respect. We can connect any number of things in this way, provided they are all systems that have some inherent entangled quality, such as spin. Again, the native wave functions are not entangled, only the, in these examples, spin states are entangled. If that disagrees with your thinking, so be it. However, there is far too much Information that describes a native wave function for the entire thing to be entangled in that way you may be thinking of. There is no agreed upon description of entire wave functions being entangled and there is no convention for it.

Nature has no provision for entangled spins [in this example], if the entire native wave function were entangled with some other native wave function, say, the vector and momentum, then changing the vector of one would appear to spontaneously 'deflect' the other. We do not see that. What we see is that we bounce them off of mirrors all over the lab and nothing else is spontaneously changing direction.

New Rule:

> *Only one property of a native wave function can be entangled.*
>
> *Entire native wave functions cannot be entangled.*
>
> *Only EigenStates of native wave functions can be entangled.*
>
> *EigenValues cannot be entangled.*

And the reason I specific 'Native Wave Function' is because in many schools of thought the Electric and Magnetic bosons are treated as both discrete and massive, arising via the Heisenberg Uncertainty Principle [HUP].

However, no temporal progression, as a property, has to the best of my knowledge never been discussed. The fact that by convention the Information travels figuratively 'inward'

### What exactly is the 'Gap?'

toward those values of epsilon of greater Conformal Scale, which then demands into the causal past, has also never been addressed. Again, perhaps I am interpreting this to literally, but that seems to be the case. And again, at midpoint the Information begins moving again figuratively 'upward' toward those values of epsilon of lesser value, toward the causal *present*. However, the math that I have seen seems to describe this literal interpretation.

*There is no causal future relationship in the Ryu-Takayanagi conventional description. There is no causal past description, either.* Perhaps the greatest per chance awakening in physics was Ryu-Takayanagi drawing the thing as it is, a static image depicting, of course, a static thing. What else could a drawing be?

Again, in order for Information to travel toward the causal future, the path would have to take this form:

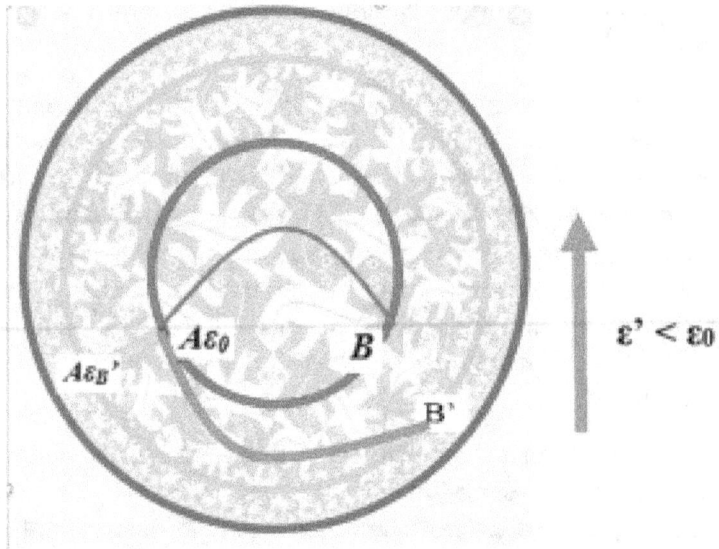

Again: The red arc:

- The red arc connects two points, both are at time-zero, there is no temporal progression.
- The red arc would seem to violate Causality Preservation by Information being required to move inward, toward the unitary values of greater Conformal Scale, which is the definitive past. At exactly midpoint the Information begins 'burrowing,' if you will, back toward the present, *but arrives also at time-zero*. However, this is not the case, in some sense, because the entire system is temporally static. There is no time and thus no sense of 'motion.'
- There is no possible way to pass through [via the red arc] the least number of 'lizards.' This is because there are two changes in vector, the Conformal Values of epsilon first increasing in magnitude, then decreasing in magnitude, symmetrically. The demand then is that the passage must always travel completely into the dead center, which is the 'Big Bang.' Hence, motion in the conventional Ryu-Takayanagi Path *l* manner must violate causality all the way back in time to the 'Big Bang' and then back to the temporal present.

**What exactly is the 'Gap?'**

The green arc:

- The green arc, however, is moving only toward the causal future. The green arc begins at time-zero [$A\varepsilon_0$] and traverses a distance that only progresses forward in time, toward the Conformal Values of epsilon that are ever decreasing in magnitude. This is true regardless of the scope and rate of the vanishing value, epsilon.
- The green arc does not violate Causality Preservation in any way.
- The green arc is passing through the least number of fracs, values epsilon, e.g., 'lizards.'
- The green arc will typically pass through less unitary Planck values ['lizards'] than any path via the red arc.
- The green arc is always greater in magnitude than the red arc, Ryu-Takayanagi Path $l$. However, the green arc passes through less lizards [Conformal Values of epsilon] then is possible on the red arc, Path $l$. Path $l$ is deceivingly 'shorter' because it connects two points at the same instant in time. The green arc is deceivingly of greater physical length because it connects two points with typical causality.
- The green arc connects Alice at time-zero, $A\varepsilon$, to Bob, who in this case is not at time-zero, but the causal future, at $B^+$, at time $\varepsilon^+$.
- Epsilon has evolved, meaning it must diminish in scale, thus $\varepsilon^+$ is less than $\varepsilon$.
- As a result of $\varepsilon^+$ being lesser in magnitude, Bob's Locally Quantized Meterstick is thus lesser in magnitude. Consequently, just like a Lorentz or Schwarzschild transform, he measures Information coming to him as being quantitatively *larger in magnitude, because of his shorter local meterstick.*

I think it is safe to say that traversing Ryu-Takayanagi Path $l$ is not actually of zero distance, but of zero time. If one imagines that we are traveling that red arc, starting with Alice at time and coordinate $A\varepsilon$, we are moving toward the larger Conformal values of epsilon, which means the past. We are travelling some distance, but that distance is nullified by the fact that the 'interior' of the AdS system does not actually exist but is merely a record of past states of the AdS Horizon Surface. Thus, qualitatively, we are not travelling any *real distance*, as this would suggest the interior literally exists. At midpoint we begin travelling up toward the larger Conformal values epsilon, *only toward the causal present, not future.*

*This is the path that entangled Information travels. This is why entangled systems appear to display pseudo-causality violations.*

We have thus traversed a distance in our 'frame of reference,' but according to the AdS system, we have only travelled toward a figurative past and '*back to the present,*' not future, and thus have not travelled at all.

In the DCQE, let us regard electrons rather than photons, such that they have mass and so on. The electrons, in their own 'frame of reference,' are moving. However, they are moving

### What exactly is the 'Gap?'

toward a causal past that no longer exists and turning around midpoint to the causal present, thus, as we observe, are traversing zero time.

Do not confuse this with the electrons individually are or are not in motion. Collectively, which is always the case if they are entangled, they are connected to their causal source at time-zero, $A$, and to their destination, also at time-zero, at $B$. Thus:

What is true for the electrons at point $A$ is true for all of the electrons at point(s) $B$. They do not observe Preferential Frame of Reference.

I see Youtubes with titles like, 'The Delayed Choice Quantum Eraser Debunked.' Oddly, these are physicists presenting this information. The ontology and confusion regarding temporal Causality Preservation is profound. I think looking at the system in this way is satisfying intellectually. The math [Appendix] is rather straight forward, because there are no transcendental aberrations of unobservables and unobservable dimensionalities.

Interestingly, the source of all of the unobservable dimensionalities in all of physics is purely the result of the history of the 2-slit experiment. If we regard the AdS description as valid, then the resolution to the history of the 2-slit, void of unobservables and unobservable dimensionalities is clear. There are no transcendental hypotheses in this, perhaps with the exception of the AdS system itself.

I think this a bit more satisfying than suggesting 'higher dimensionalities' and so on in an attempt to explain the behavior of the DCQE system. There is nothing on that tabletop apparatus that would invoke a set of 'higher dimensionalities.' Ontologically, this description does not exist, there is nothing ontological in any of this.

In prior papers I described how it is Bob, by detection, which can be any electromagnetic interaction, who forces the Information content from figuratively 'below' the AdS Horizon Surface to the AdS Horizon Surface. There is no empirical evidence otherwise. The electromagnetic interaction 'forces' the system which is in some eigenstate to precipitate to some eigenvalue or values. Again, the common rules are:

- Eigenstates can only exist figuratively 'below' the AdS Horizon Surface.
- Eigenvalues can only exist 'on' the AdS Horizon Surface.
- Entangled information can only exist in eigenstate and cannot exist as an eigenvalue or values.

By convention, Bob makes the choice rather the system remains in an eigenstate or precipitates to some eigenvalue or values. When we talk about Delayed Choice Entangled Swapping, it becomes much clearer that observation, which can be any electromagnetic interaction, does in fact 'force' Information from eigenstates to eigenvalues. That much is rather elementary.

Keep in mind frame of reference only refers to what Alice and Bob see when they observe one another, there is no paradox nor anomalies associated with this. Preferential Frame of

## What exactly is the 'Gap?'

Reference is when some Conformal Value(s) differ between Alice and Bob. This can be the result of, again:

- Lorentz transformation
- Schwarzschild transformation
- Distance only

Distance is equated with the Zeno Effect. This was discussed in detail in the section 'Preferential Frame of Reference Issues.'

The postulates are:

- In all systems that can be regarded as in a sum of *eigenvalues*, such as macroscopic systems [e.g., Alice and Bob on Earth and moon], Preferential Frame of Reference as a result of pure distance applies and seeming causal violations can occur; what is true in one system [Bob, moon] may not be true in another system [Alice, Earth].
    - Information travels via [Ryu-Takayanagi] Path *l'* *[l-prime]* the fluid [green] arc in the images above, from $A$ to *B'* *[B-prime]*.
- In entangled systems [in *eigenstates*], Preferential Frame of Reference *does not apply* and what is true in one system is true in all systems. The Delayed Choice Quantum Eraser demonstrates that when Preferential Frame of Reference does not apply, what is true in one system [train crash, interference pattern] is true in all frames of reference. For example, it seems a causal violation for a particle-like pattern [no train wreck] to occur at detector-zero as well as detectors one and two, or otherwise an interference pattern [train wreck] occurs at detector-zero as well as detectors one and two. What is true in one system is true in all systems that are entangled.
    - Information travels via Ryu-Takayanagi Path *l* *[not prime]* the rigid red arc of convention from $A$ to *B* [not prime].

Thus, as Information is taking the path figuratively 'below' the AdS Horizon Surface, it remains in an eigenstate, and for the most part that is conventional. At such time Bob detects the Information, which can be any type of electromagnetic interaction, the Information is then an eigenvalue. This gives a rule set:

- An eigen*value* can only exist *on the AdS Horizon Surface*.
    - Preferential Frame of Reference applies.
    - What is true in one system may not be true in another. [e.g., what is true for Alice may not be true for Bob]
- Information figuratively 'below' the AdS Horizon Surface can only be in an eigenstate or eigenstates: *provided the Information is travelling the rigid red arc of convention*.
    - Preferential Frame of Reference *does not apply*.
    - What is true in one system is true in all systems. [e.g., what is true for Alice is always true for Bob] Or, what is observed at detector-zero is observed at all detectors.

## The Delayed Choice Quantum Eraser and Non-Preferential Frame of Reference

## The Delayed Choice Quantum Eraser and Non-Preferential Frame of Reference

In this description the interference pattern is regarded as the train wreck and the particle-like behavior is *no train wreck*. It doesn't matter which is which, we simply have to make a decision. The interference pattern as the train wreck should suffice.

I will use the term, Non-Preferential Frame of Reference, to mean 'Preferential Frame of Reference does not apply.'

The basic premise is that in this experiment, Preferential Frame of Reference as it is defined in this text as a function of pure distance\* *does not apply* to entangled systems. Thus, where we had the Lunar Landing anomaly described as a train wreck that occurs in the Lunar Preferential Frame of Reference between Armstrong and CAPCOM, which are not entangled systems. In that description a train wreck in the form of CAPCOM cutting Armstrong [Bob] off as he begins to speak occurs in the recording as events took their normal causal relationship on the moon, Lunar Preferential Frame of Reference. In the Earth [CAPCOM, Alice] Preferential Frame of Reference no train wreck occurs.

\*Note again that Preferential Frame of Reference can be the result of:

- Pure distance
- Lorentz conditions
- Schwarzschild conditions

Preferential Frame of Reference, then, is what would be regarded as 'normal,' as one takes the meaning to apply perhaps on a macroscopic scale. However, it is not scaling that can account for this, as there is no definitive line that differentiates the quantum from the macroscopic. We can only assert that the macroscopic scale is a composite, a collective sum of a large number, that is also undefined, of quantum systems.

In terms of Information content, we have the values $A_\Omega$ of $N_Q$. I put forth the notion in some prior work that the scale $A_\Omega$ seems to have an upper boundary condition of 10E20 unitary Planck lengths [1-d], 10E40 Planck areas [2-d], and 10E60 Planck volumes [3-d]. And the upper constraint for the cosmos appears to be about 10E80 $N_Q$. I don't want to get into that here and fill another appendix with yet more derivations.\*\* However, these numbers not only seem to be inherent in describing nature, but they also appear to be the same values that drove poor Dirac crazy with his 'large number hypothesis.' I think the derivations were in Quantum Information Dynamics Volumes I and II.

\*\*Also note that I am moving as much of the derivations to the appendix as possible to lend readability to the text as a whole.

In short, there are two distinct and true realities that are not paradox, because they are *real*. Two trains leave their respective stations, train Alice and train Bob. In Bob's [Armstrong,

### The Delayed Choice Quantum Eraser and Non-Preferential Frame of Reference

moon] Preferential Frame of Reference he collides into train Alice. In Alice's [CAPCOM, Earth] Preferential Frame of Reference she does not collide with Bob.

Both realities are *real, are recorded, and are macroscopic.*

Again, there is no conventional ΔS solution to this scenario, as conventional ΔS can only deal with one causal path. Shannon Entropy *may possibly define* this condition because Shannon entropy is not limited to one causal path but applies only to systems where the number of causal paths are greater than one.

- Conventional ΔS Entropy, which is limited to 1-causal path {observable}
- Shannon Entropy, which is defined as >1-causal path, always {unobservable}

Any *observable* system must be limited to 1-causal path, this is by convention. Shannon Entropy is defined as >1-causal path, always, and is thus *unobservable*. The hard dividing line from conventional ΔS [1-causal path] to Shannon Entropies [>1-causal path] would be a Schwarzschild Horizon surface. It is at the point of coalescing with a Schwarzschild Surface, such as a black hole, that Shannon Entropy takes over, spacetime becomes multiply connected, and conventional ΔS entropy can no longer describe the system. Consequently, as black holes Alice and Bob coalesce, *the causal paths back to Alice and back to Bob cease to exist.* We can see a chirp-mass that results a literal millisecond before they coalesce, but no information back to Alice nor to Bob.

First, a metaphor. We tie Alice and Bob together via a tether that is 1-Planck interval in length. The first thing we find is that this is not possible, because the lower such boundary [Hawking-Bekenstein] is defined by the Heisenberg Uncertainty Principle [**HUP**] as 2-Planck lengths. [That is by convention]. This 2-Planck length criteria is important, so keep it in mind.

Thus, we let out some slack and now Alice and Bob are tethered together by a 2-Planck length tether. We begin lowering Bob closer and closer to a Schwarzschild Surface, e.g., black hole horizon. We find that we can only lower Bob by intervals of 2-Planck lengths because of our HUP lower boundary constraint, 1-Planck length is a finer resolution than the HUP will allow. At 4-Planck lengths Bob has not yet coalesced with the horizon and is observable to Alice, and Alice is still observable to Bob:

- The system {Alice, Bob} is {observable}.
- The system {Alice, Bob} is described by conventional ΔS entropy.
- The system {Alice, Bob} is thusly defined by one and only one causal path.

At 2-Planck lengths from the Horizon Surface:

- Bob coalesces with the horizon.
- Bob is now unobservable to Alice.
- Alice is now unobservable to Bob.

## The Delayed Choice Quantum Eraser and Non-Preferential Frame of Reference

And the entropy description:

- Bob is becoming multiply connected [to his environment] by multiple causal paths.
- Bob is unobservable.
- The system {Alice, Bob} no longer exists, as conventional $\Delta S$ entropy cannot describe a system where Alice, in 'normal' spacetime thus limited to 1-causal path can be connected to Bob, who is dropping into infinite multiple connectivity. Shannon Entropy, also, cannot describe such a system; *thus, the system has no causal path or paths leading back to it.*
- As Shannon entropy, the system can only be described by >1-causal path. [see the internal link]

Then, Alice also falls into the black hole:

- Alice is becoming multiply connected [to his environment] by multiple causal paths. She will eventually be connected to Bob an infinite number of times as well.
- Alice is unobservable.
- Alice cannot be causally connected back to any system in 'normal,' conventional $\Delta S$ entropy spacetime.

This represents

- *A divergence from conventional $\Delta S$ entropy* [limited to 1-causal path] to Shannon entropy [lower constraint of >1-causal paths].
- The divergence occurs at a proximity to a Schwarzschild Horizon [Surface] of exactly 2-Planck lengths, as defined by the lower limits [Beckenstein-Hawking] of the Heisenberg Uncertainty Principle {HUP}.
- At such time the system {Alice, Bob} is described by 1-causal path all elements of the system are {observable}.
- At such time the system {Alice, Bob} is described by >1-causal path all elements of the system are {unobservable}.

The conditions {observable, unobservable} are inextricably interdependent such that

- Conventional $\Delta S$ entropy limited to 1-causal path: {observable}.
- Shannon Entropy *lower limited to* greater than 1-causal path: {unobservable}.

In an entangled system, such as Alice and Bob, we see anomalies in the interference pattern or particle-like behaviors because albeit they are separated by distance, Preferential Frame of Reference *does not apply*. In this case, we have:

> A train wreck *or* lack of train wreck is true in all *frames of reference*.

### The Delayed Choice Quantum Eraser and Non-Preferential Frame of Reference

Here, I use frames of reference for the obvious reason that they are not preferential as established by pure distance, they are in fact not yet defined. Preferential Frame of Reference applies to what was at one time the convention of frame of reference. However, that term has become so clouded over the years that it has become essentially unusable. Again, Preferential Frame of Reference can result from distance, a Lorentzian description, or a Schwarzschild description.

In this case, we take a brief look at a diagram of the DCQE:

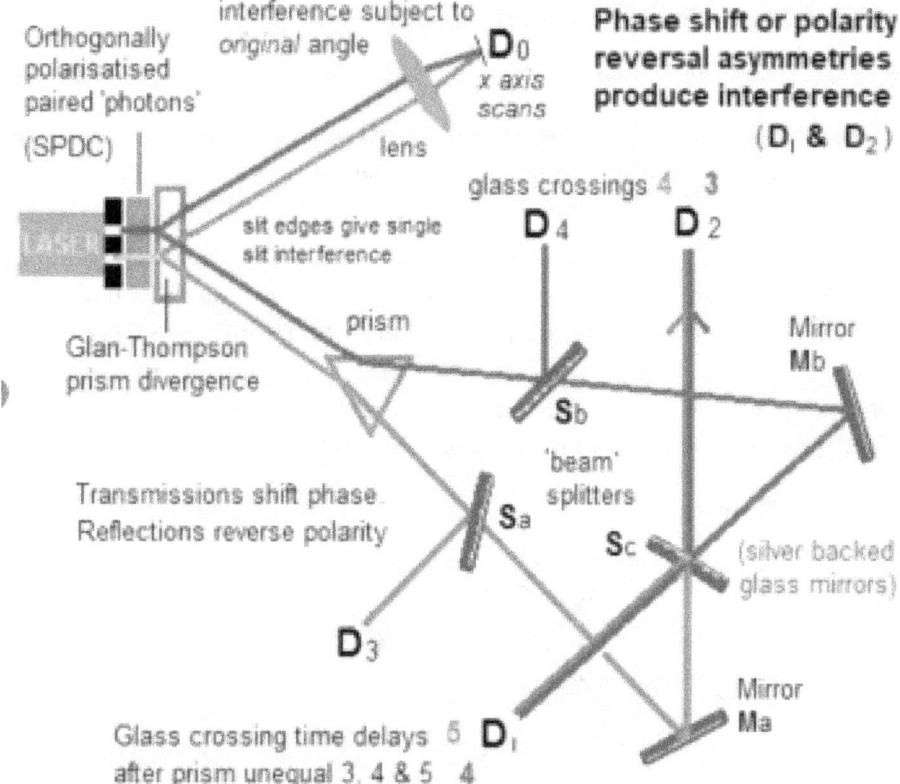

Where we are accustomed to thinking in terms of distance as being relevant, is in the fact that distance under conventional descriptions is limited to 1-causal path. That is, there is no description, yet, of distance from here to there when that distance is causally connected more than one time. There is a section later in this text regarding what it is Alice and Bob see of each other under the conditions that the distance between them is causally connected more than one time.

And it is agreeable that conventional $\Delta S$ entropy is defined and limited to 1-causal path and Shannon Entropy is described by >1-causal path. That is, the lower limit for Shannon Entropy would be >1-causal path. Again, this was:

## The Delayed Choice Quantum Eraser and Non-Preferential Frame of Reference

When $i = 1$, the entire term falls to zero.

$$H(X) = -\sum_{i=1}^{n=1} P(xi = 1) Log_2 P(x = 1) = 0$$

I think the confusion is that entropy is being treated the same in conventional descriptions, namely, associating conventional ΔS entropy as having any relevance to systems that are multiply connected. This cannot be the case. Merely sticking ΔS in front of a term and rendering math does not make it a property of nature.

I will reiterate once again:

- Conventional ΔS entropy is [upper and lower] limited to exactly 1-causal path.
- Shannon Entropy is lower limited to >1-causal path.

Being limited to 1-causal path, this is where the ontological and philosophical issues of time reversal come from. 1-causal path must be unidirectional, else if a time reversal is implied, then there are *2-causal paths*. The notion that time is reversable is not validated scientifically. Reversible time is the hypothesis that evolves out of the history of 2-slit observations; the hypothesis does not validate itself. Nor does the empirical evidence that evoked the hypothesis, validate the hypothesis.

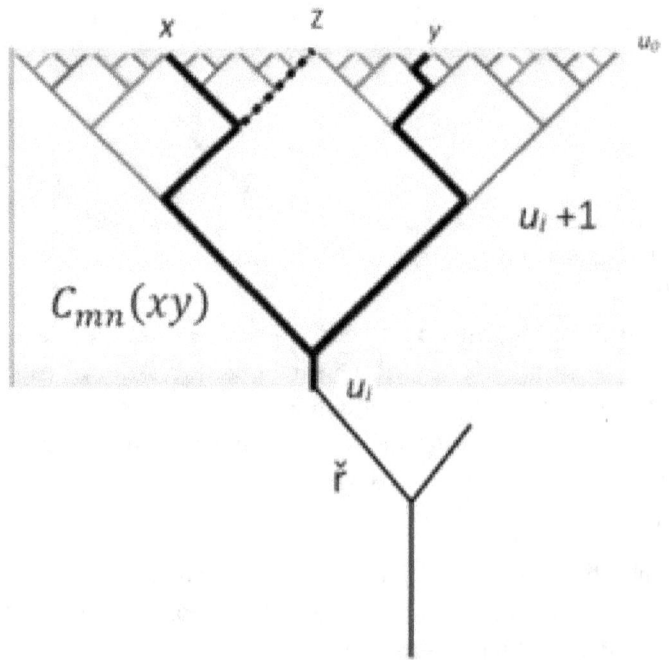

**The Delayed Choice Quantum Eraser and Non-Preferential Frame of Reference**

We can regard points $x$ and $Z$ as being causally connected. The path back to their common origin is unidirectional. Ultimately, they both originate at point $u_i$. However, we cannot regard points $x$ and $y$ as causally connected, because the path between them requires two changes in causal direction. Although they are both mutually connected, ultimately, at point $u_i$, to get from $x$ to $y$, we have to go back down the timeline to point $u_i$, then change temporal vector upward or forward, to get to point $y$. That is, where $x$ and $Z$ have a unidirectional path to a common origin, points $x$ and $y$ require changing causal direction from backward [or downward] to forward [or upward].

This is a violation of Causality Conservation, which is obviously a term I fashioned up for this text. However, the principle is clear, and the only evidence of time reversal is in fact the 2-slit. If we define and describe the 2-slit in a unidirectional manner that makes sense, perhaps we are on to something.

This was the same issue we had with the conventional Ryu-Takayanagi Path $l$. Information must travel figuratively 'inward' toward the larger Conformal Values epsilon, then at exactly the midpoint begin figuratively 'burrowing' back toward the present, in no case reaching the future, because the starting and endpoints are fixed at time-zero. Convention doesn't regard it in this way because I have never seen anyone give it any thought or even bring the matter up. I am not clear on why this is the case.

This may seem trivial, but it is not at all trivial. In the above image, scale is everything. For example, $x$ is Bob and $y$ is Alice. Point $u_i$ is the Big Bang. Now, some 14-billion years later Bob wants to send Alice a love letter. So, he sends the message back in time to the Big Bang. However, Alice is not there, she does not exist yet, it is 14-billion years ago. Time has to change direction and go forward another 14-billion years, so that Alice exists. That is perhaps over simplified, but it demonstrates the need for changing causal direction for Information taking some Path of type $l$, $l'$, $l_G$, and so on.

Preservation of Causality Conservation, among other things, prohibits changing temporal direction, *twice*.

That is *one of the problems* with the conventional Ryu-Takayanagi Path $l$: *[following page]*

## The Delayed Choice Quantum Eraser and Non-Preferential Frame of Reference

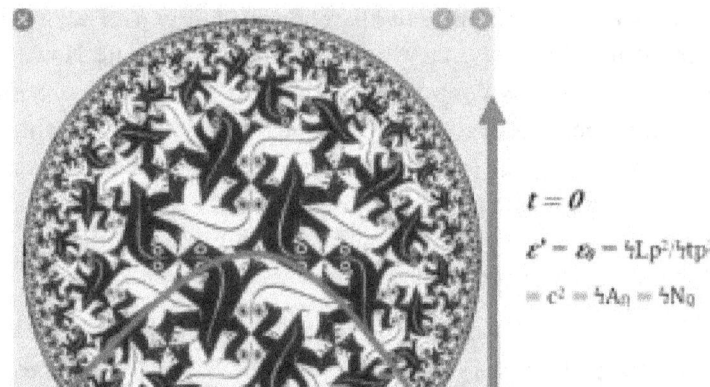

The diagram above is in fact the scenario I just described with Alice and Bob. Note again that the red arc actually passes through the very center of the diagram of 'lizards.' That center point is in fact the 'Big Bang.' In the above scenario, we will reverse the scenario so that Alice at time-zero $A\varepsilon_0$ sends a letter to Bob, who is on the exact other side of the cosmos, whatever that means. The Information travels inward toward the larger Conformal Values of epsilon until it reaches the center, where the Conformal Values epsilon are largest, the 'Big Bang,' whatever that is. In order for the information to travel inward toward the larger Conformal Values epsilon, the Information must be travelling causally *backward in time*.

That is the general problem. The larger Conformal Values of epsilon are past states of the AdS Horizon Surface. The smallest conformal Values of epsilon are always *on* the AdS Horizon Surface. Thus, if Information travels in such a way that it is passing through larger Conformal Values of epsilon of any scale, they are thusly travelling backward in time.

If Alice sends a message backward in time, then one has to consider that this path the Information is taking is going to *be a unique path*, because there is no causal path where the letter was sent forward in time from the 'Big Bang.' I think it is perhaps to consider the notion that a time reversal on a Planck scale [e.g., quantum scale] is not the same as sending a letter backward in time to the Big Bang, which is my point. However, again, there doesn't seem to be any time reversal on a Planck scale, either all of those notions evolve out of observations of the 2-slit.

However, at the exact midpoint, the center, the 'Big Bang,' the Information then changes direction back towards the AdS Horizon Surface, where the Conformal Values epsilon are at their smallest, *which is the present*. Information had to travel from Alice at time-zero $A\varepsilon_0$ causally backward in time to the Big Bang, change direction at the exact midpoint, and travel causally forward *to the present*.

## The Delayed Choice Quantum Eraser and Non-Preferential Frame of Reference

In the above diagram, Alice at $A\varepsilon_0$ and Bob at $B$ are not causally connected. Time cannot be dynamic in the system as described by convention, it must be static, 'frozen.' Causal connectivity suggests that there is a path from one point to another point and that the path is *traversable*. There is no 'traversing' in a temporally frozen, static system.

So, we fixed the problem:

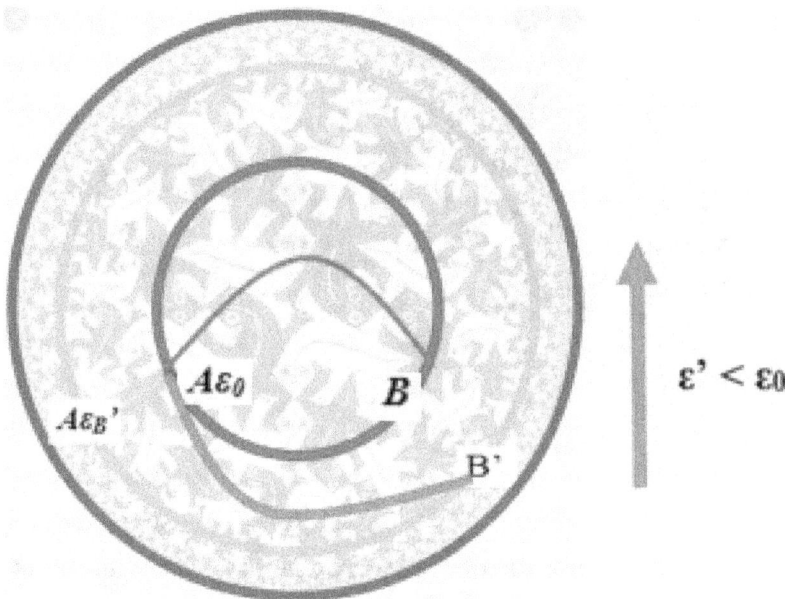

Now, things begin to make sense again. Alice at time-zero $A\varepsilon_0$ sends information to Bob at point $B$, who moves to point $B'$ in that time it takes for the Information to reach him, taking the green arc, the non-conventional hypothetical Path $l_B'$, which is the green arc.

Alice and Bob are now causally connected in the above diagram. The system is obviously dynamic and evolving as indicated by the 'new horizon' in yellow, where Bob at point $B'$ is. Alice has also pseudo-moved to her new location at $A\varepsilon_B$. The red arc in the above diagram is purely for reference's sake and has no meaning, unless Alice and Bob are sharing entangled Information.

Entanglement Swapping

## Entanglement Swapping

What does this all have to do with the Delayed Choice Quantum Eraser and Entanglement Swapping? Perez notion of Entanglement Swapping is already being coined, literally, 'Quantum Steering into the Past.'

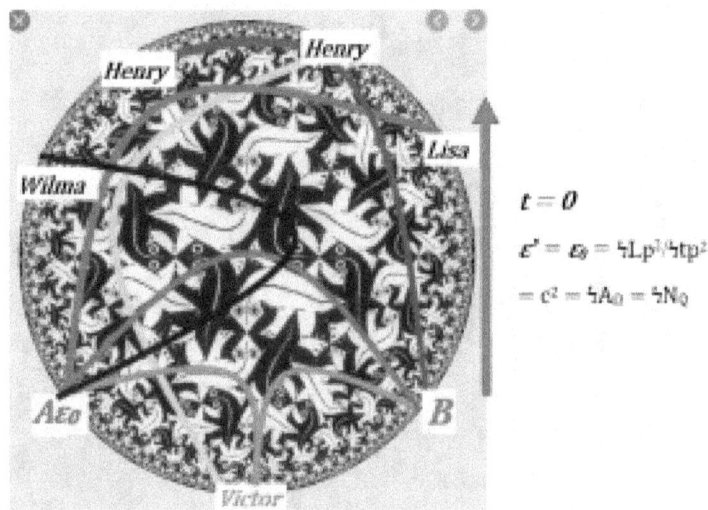

It really doesn't matter who is 'connected' to whom. We can connect any number of 'people' in the first diagram in any number of ways, provided such is limited to one causal path for each connection. Any one connection has no effect or affect on any other connection. In the diagram above, because the 'people' are sharing some entangled Information, the system is the conventional Ryu-Takayanagi Path $l$.

We discussed Causal Connectivity in an earlier section.

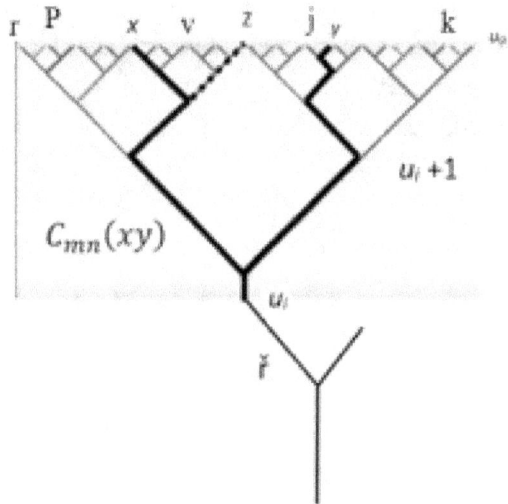

**Entanglement Swapping**

- $\{r, P, v, x, Z\}$ are all causally connected to each other.
- $\{j, y, k\}$ are causally connected to each other.
- $\{r, P, v, x, Z\}$ and $\{j, y, k\}$ are mutually exclusive, no element from one subset is causally connected to any element in another subset.

There is no 'steering into the past,' because the conventional Ryu-Takayanagi Paths are temporally frozen, there is no temporal evolution. The entire system is static. The exact same postulate put forth for Alice and Bob applies to everyone on the horizon surface.

Entangled systems do not observe Preferential Frame of Reference. This is because they uniquely traverse the conventional Ryu-Takayanagi Path $l$, which is a static system. There is no temporal evolution in the conventional Ryu-Takayanagi Paths.

The Big Dilemma in the DCQE is the notion that information must be travelling causally backwards to display the interference pattern at all detectors, *what is true in one Non-Preferential Frame of Reference is true in all Non-Preferential Frames of Reference; what is true in one system is true in all systems, because all systems are frozen at time-zero.*

How do I shake hands between entangled Alice and Bob without sending Information causally backward? How do I do it using conventional Ryu-Takayanagi Path $l$ ?

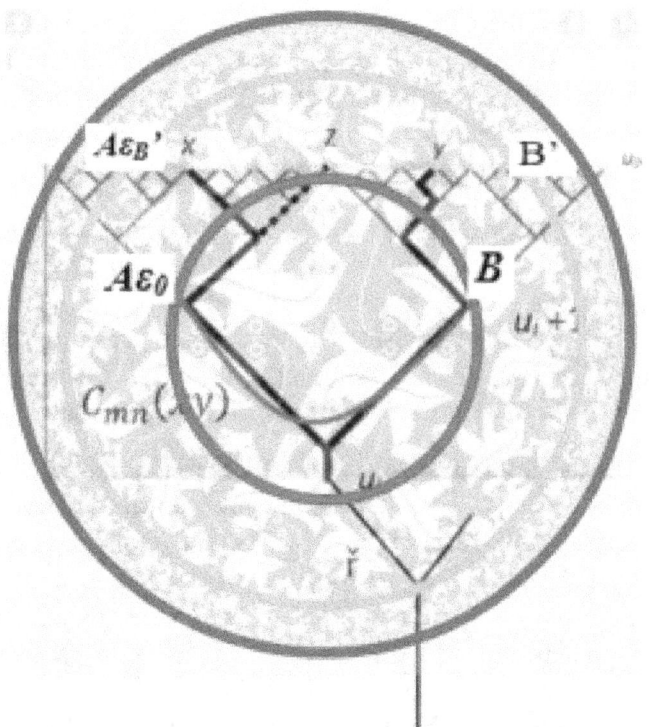

Here, I've flipped the Ryu-Takayanagi image so that the images would overlay nicely and perhaps even make some sense. Keep in mind that this image is that of the conventional Ryu-Takayanagi Path $l$, where I stated that Alice and Bob are not causally connected. That

## Entanglement Swapping

is the key. In order for them to communicate Information, the message, again, has to travel causally backward in time toward the inner portion of the 'lizards,' where they are larger. We cannot regard Alice and Bob as causally connected because this demands that time is evolving, progressing, it is not.

So, how can Alice and Bob display behavior described as, 'what is true in one system is true in all systems?' It is exactly as depicted, a frozen snapshot.

*They are connected, but not causally connected.*

Causality is a thing that demands a progression of time, whatever your philosophy of that means. The key term is *cause*. Then, what is a connection if not causal?

*Temporally frozen.*

Again, the problem with both conventions is the same, namely, changing causal direction, not once, but twice. I do not think even the ontology of the 2-slit considers this, that the descriptions they are rendering do not require time reversal, but *two-time reversals for every interaction*. This is why the systems cannot be causally connected.

We have the center of the Ryu-Takayanagi Path *l* going through the 'Big Bang' which in the Cayley tree image is point $u_i$.

If you take the image literally, it could appear that Alice and Bob are entangled at the origin, $u_i$. This has been a suggestion of many papers and lectures to date. However, I think the issue to date is a lack of a pure description of what entanglement actually is. There are so many thousands of descriptions but not a *single* description of what entanglement is.

So let us look at the conditions in the image. In the image, epsilon is not dynamic, *in the image epsilon is static*. In the image, Alice and Bob have only one common origin, the Big Bang. In the DCQE experimental setup, the common origin would be the BBO crystal, because that is how Alice and Bob begin as an entangled pair. In the image, Alice and Bob are not eigenstates, they are eigenvalues. I need to make this clear, if Alice and Bob are eigenstates then they are Boblice, not Bob and Alice. *They are sharing Information [perhaps spin] that is in an eigenstate, they are not, however, themselves eigenstates.*

That is, they are sharing information that is in some eigenstate, but they are not in themselves eigenstates. I think that gets cloudy in conventional thinking. Is the photon in an eigenstate or is it entangled with another photon and the information they share is in an eigenstate or states? If the photons themselves are eigenstates, then whatever we do to photon Alice happens also to Bob and there is no differentiating them. If we reflect Alice off of a mirror Bob spontaneously changes direction, would be the photons Alice and Bob in an eigenstate. What is in an eigenstate is Bob's spin and Alice's spin. All other factors of Alice and Bob are eigenvalues, not states. There is a very long list of values that can define the two photons Alice and Bob, too long to get into here. *There is a lot of information that defines a photon.*

*None of those other factors are in eigenstates, only spin, in this example.*

**Entanglement Swapping**

How do I truly know this? *Because that is the thing I am going to measure.*

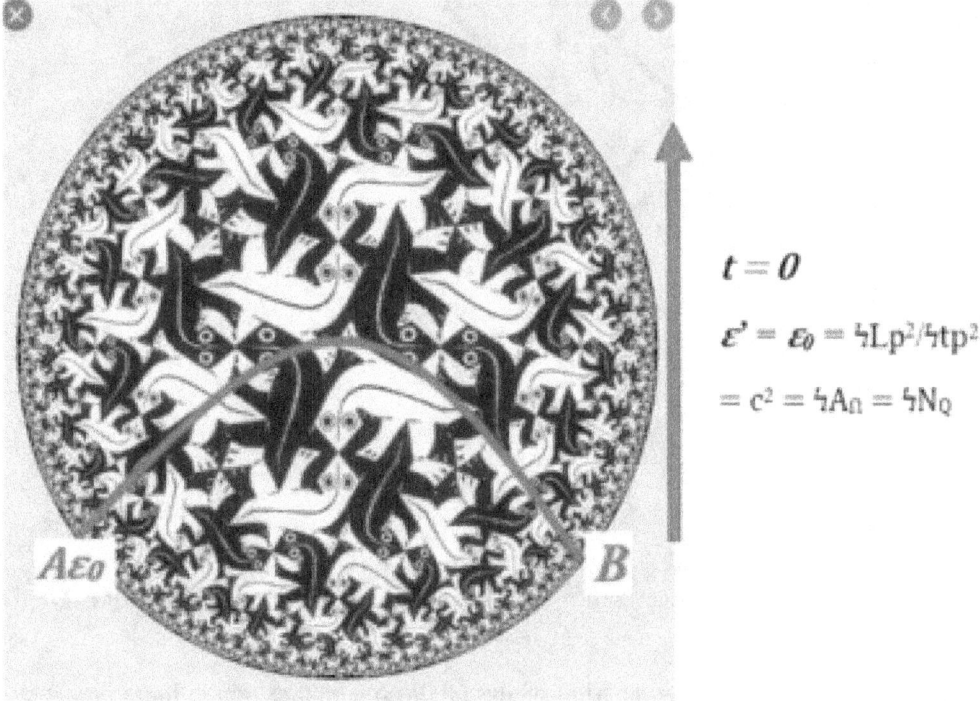

$t = 0$

$\varepsilon' = \varepsilon_0 = \hbar L_p^2/\hbar t_p^2$

$= c^2 = \hbar A_0 = \hbar N_Q$

The universe is static in this image, there is no 'Planck flow.' There is no time in any sequitur sense at all. We'll forego the fact that Alice and Bob are on opposite sides of the cosmos in this image because we cannot draw such infinitesimal resolution of them co-existing just one Planck interval 'below' the AdS Horizon Surface. Just imagine that they are causally connected at just one unitary Planck interval figuratively below the horizon surface. Everything in the image is static, Alice, Bob, and the entire universe.

This is why I proposed a Zeno Delayed Choice Quantum Eraser in an earlier paper and is summarized somewhere here in this text. The principle is a rather simple variation, we have detectors zero through four. We acquire scans from the PN junctions at some phenomenal rate so as to produce a Zeno Effect, then see if the systems behave the same or different. Qualitatively, there should be no Zeno Effect in the DCQE, because as stated, it is a temporally static system.

> *Note here that when I say Alice and Bob are entangled, I mean specifically that they are sharing some entangled feature, such as in these examples, spin.

Let's look again at the image:

**Entanglement Swapping**

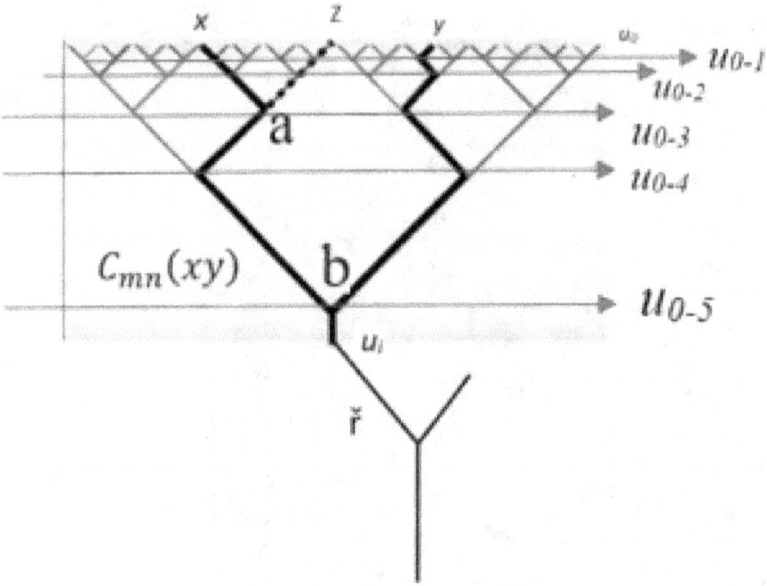

1. At $u_{0\text{-}1}$, the hypothesis is that Alice [x] and Bob [y] are entangled.
2. However, at $u_{0\text{-}2}$ as their common origin, Alice [x] and Bob [y] *are not entangled.*

At $u_{0\text{-}2}$, we are now at the lower limit of the HUP on a Planck scale, there is a sequitur 'direction' for causality, e.g., asymmetrically forward time evolution. There is as such a governing sense that time will evolve and what direction, for lack of a better term, it will evolve in. Time can now be regarded as extant. Alice and Bob will not reflect and bounce off of the walls of the lab in mysterious ways seemingly defying causality.

At $u_{0\text{-}2}$, as their common time index of origin, Alice and Bob *are not entangled, because the demand is that at point $u_{0\text{-}2}$ time has evolved to $u_0$, the present.*

Time cannot be resolved on a Planck scale of one unitary interval, that is by convention. The question is, what is the factor that defines Alice and Bob as being or existing at some value of one unitary Planck interval figuratively 'below' the AdS Horizon Surface?

In the BBO crystal, by convention, the two photons, Alice and Bob are co-created simultaneously, whatever simultaneous means. So, we will go with the hypothesis that 'time stands still' for Alice and Bob, at least momentarily. I put forth the argument in prior papers that it is Bob, by detection or observation, which can be any electromagnetic interaction, who figuratively 'forces' Information to the AdS Horizon Surface by, in effect, making them or it to precipitate from eigenstates to eigenvalues. That is somewhat by convention albeit I don't think it has ever been stated in that way.

Thus, going again with the hypothesis that Bob, or any Tom, Dick, or Harriet, who by electromagnetic interaction of observation is the causal component that makes an eigenstate or states precipitate to eigenvalue or values.

**Entanglement Swapping**

Question: Can an eigenstate bounce off of a mirror or interact with a beam splitter? Can a native wave function bounce off of a mirror or interact with a beam splitter? By native wave function I am referring to the treatment that the photon exists as a native wave function and that the electric and magnetic bosons associated with it arise via the HUP. Thus, we have the native wave function which by convention has no definitive locality in essence leaving a 'wake' of electric and magnetic boson(s). Thus, I treat the electric and magnetic bosons as discrete and massive. Because they are massive, arising from the HUP, even as a 'field effect,' would have discreteness and mass-energy, they cannot travel at the speed of light but are mass limited to $v \ll c$.

Part of the issue I have with the 2-slit variants is that we are chasing native wave functions around, trying to understand their behavior, by looking at the discrete, massive, electric and magnetic bosons travelling at $v \ll c$ in the lab.

I have to present some math here in order to describe the bosons in the way that I want to treat them.

$$\varepsilon = \frac{Lp^2}{tp^2} = c^2 = \frac{E}{m} = hN_Q = A_\Omega = 1HUP = Le^{-\frac{3l}{e}}$$

By one HUP I am referring to the notion that the HUP has a lower limit at the Planck scale of 2-unitary Planck intervals. That is by convention, regarded as the Bekenstein-Hawking limit.

Then,

$$\varepsilon = c^2 = E/m = Le^{-\frac{3l}{e}}$$

$$\varepsilon e^{\frac{3l}{e}} = L$$

And given:

$$\varepsilon = \frac{Lp^2}{tp^2} = c^2 = \frac{E}{m}$$

Then:

$$\frac{E}{m} e^{\frac{3l}{e}} = L$$

231

**Entanglement Swapping**

And

$$\natural\varepsilon' < \varepsilon_0$$

NOTE: $\varepsilon$ and $\varepsilon'$ are not the same value, $\varepsilon'$ represents the evolutionary state over time of epsilon. Thus, Bob's value **B'** at point $\varepsilon'$ is less in Conformal Scale than it was when he was at point **B** at point $\varepsilon$.

$$\natural B'_{\varepsilon'} < B_0\varepsilon_0$$

The term, $A'_\Omega$ defines the worldsheet, as is convention of the Bekenstein-Hawking relationship. The temporal distance is given by:

$$\natural tp'^2 = \frac{\natural Lp'^2}{\natural\varepsilon'} = \frac{e^{\ln(\frac{L}{\varepsilon})}}{L'} = \frac{c^2}{\natural A'_\Omega}$$

$$\natural\varepsilon' = \frac{\natural Lp'^2}{\natural tp'^2} = \frac{e^{\ln(\frac{L}{\varepsilon})}}{L'} = \frac{c^2}{\natural A'_\Omega}$$

$$\natural\varepsilon' = \frac{c^2}{\natural A'_\Omega}$$

What was the point to all of that? The relationship immediately above states that the horizon value epsilon is equated to the worldsheet, $A_\Omega$. Furthermore, it is a fractional value of the term $c^2$, which means, as the convention states, the worldsheet $A_\Omega$ is superluminal, *on the surface*.

*The surface of the worldsheet $A_\Omega$ is superluminal.*

That is not suggesting that Information travels faster than the speed of light on the worldsheet $A_\Omega$, it is a statement, again, that each Qubit of Information is superluminally isolated from each other Qubit that makes up the surface of $A_\Omega$.

As odd as it sounds, bouncing off of mirrors and splitters and such is non-sequitur, *there is no motion in the system. Time does not exist, and each Qubit of Information is superluminally isolated from each other Qubit of Information.*

## DERIVATION OF ½NQ HUP, The Orthogonal Components of The Electric ⊥ Magnetic EigenVectors

### DERIVATION OF ½NQ HUP, The Orthogonal Components of The Electric ⊥ Magnetic EigenVectors

I will get back to the Delayed Choice Quantum Eraser after this section. This aside is necessary in order to understand the relationships between native wave functions and their associated electric and magnetic boson content. Briefly, a photon is a boson, but a boson is not necessarily a photon. The adage that the electric and magnetic bosons are photons is entirely incorrect. That is not even remotely correct in Quantum field Theory. The electric and magnetic bosons are treated as discrete, massive particles, as they arise out of the HUP. Consequently, they are thusly mass limited to $v \ll c$. This is never accounted for and thus has to be corrected.

The HUP, described, as conforming to such Fractal Set:

$$\sigma x \sigma p \geq \frac{h}{4\pi}$$

Being a distribution, we regard $\sigma x$ as some value $\sigma Lp$ [x cannot be a non-integer of Lp], we make the first correction:

$$\sigma Lp \sigma p \geq \frac{h}{4\pi}$$

Which is then

$$\sigma Lp' \sigma p \geq \frac{h}{4\pi}$$

We then regard $\sigma p$, as consisting of the components, {t, m}: [which is the convention of milking mass-energy from the inequality], given that t is then tp' of ε'

$$\sigma Lp' \sigma tp' m \geq \frac{h}{4\pi}$$

Here, we regard <u>mass as a value N, in Qubits, which cannot be created, destroyed, nor borrowed.</u> The limit of such 'borrowing' based on the 'Negative Information' requirement, e.g. 'Anti-Information.' Whereas the instinct would be to regard $Lp^2$ as the root to such end, in this case, the arithmetic use of ± is non-sequitur; as the value in the denominator on the right of the inequality is described as $4\pi$, which describes *two circles*.

*This is where mass evolves.* Wherein we rewind to such terms, Chiral vs Helic. The HUP is not describing the evolution of mass-energy, whose consequential requirement is a

### DERIVATION OF ƕNQ HUP, The Orthogonal Components of The Electric ⊥ Magnetic EigenVectors

forbidden change in $N$, but whose obvious intent is that of *moving a set package of $N$ around*; as a distribution in the number of available superpositions. We look at the distribution as:

$$\sigma Lp' \sigma tp' vm \geq \frac{h}{4\pi}$$

Where $m$ described by a fixed value $N$, in Qubits

$$\sigma Lp' \sigma tp' v N_Q \geq \frac{h}{4\pi}$$

The function $\sigma$, rather than an alteration in Path length, be it L or type $l$, is non-sequitur, as this requires dynamics *below* the Horizon, else superluminal at the only dynamic of L, then, regard the distribution $\sigma$ as fixed, by the limits L being superluminal, and $l$ requiring a dynamic below the AdS Horizon; such that; given $\sigma$ is fixed, I will arbitrarily assign it some double-blind character, which is intended to represent $\sigma$ being a fixed value, described by $N_Q$: ƕ [which is literally a blindfolded selection from my character set]. The ƕ is the fixed quantity for $N$-Qubits, which was fixed at such time of whatever rendering of 'moment of creation,' e.g. Big Bang, either regarded as event or process. If regarded as process, then ƕ defines the fixed quantity at such time the process was complete.

$$ƕLp' ƕtp' v N_Q \geq \frac{h}{4\pi}$$

$$ƕ(N_Q) v\{Lp' tp'\} \geq \frac{h}{4\pi}$$

$$ƕ(N_Q) v\{Lp_\epsilon' tp_\epsilon'\} \geq \frac{h}{4\pi}$$

$$v \equiv nLp/xtp$$

$$v \equiv nLp'/xtp'$$

## DERIVATION OF ℏNQ HUP, The Orthogonal Components of The Electric ⊥ Magnetic EigenVectors

$$v \equiv nLp'_{\varepsilon'}/xtp'_{\varepsilon'}$$

$$\hbar(N_Q)(nLp''_{\varepsilon''}/xtp''_{\varepsilon''})\{Lp_{\varepsilon'}'tp_{\varepsilon'}'\} \geq \frac{h}{4\pi}$$

Here, my designation of Lp'' is to represent the orthonormal eigenvector against Lp'

Given that $h/4\pi$ is both fixed, and rational, and $v$ is a subluminal value as meeting the demand of taking Path $l$, as Path L is forbidden:

$$l_v = \frac{e}{3} Ln \frac{L}{\epsilon'}$$

$$l_v = \frac{e}{3} Ln \frac{L}{Lp'}$$

*Regarding again that 'e' is not an infinite perturbation, as per prior reference to the Mitrofanov limit $\Theta_P$. There is therefore no 'smoothing factor,' but some integer $h$.

$$l_{v''} = \frac{e}{3} Ln \frac{L}{Lp'_{\epsilon'}}$$

Where $v''$ is to represent the orthonormality:

$$l_{(nLp''_{\varepsilon''}/xtp''_{\varepsilon''})} = \frac{e}{3} Ln \frac{L}{Lp'_{\epsilon'}}$$

If not immediately obvious, $l_{(nLp''\varepsilon''/xtp''\varepsilon'')}$ is describing the orthonormal eigenvector against $Lp_\varepsilon'$ as taking on a different Conformally Scale Invariant (as per the Fractal Set) set of values. Meaning, the value $v$ is not an actual change in *velocity*, as the term 'velocity' requires dynamic activity below the AdS Horizon, which is forbidden. The transfer of Information is taking the form of a change in orthonormal eigenvector Conformal Scale Invariance; e.g. the description of convention for the E and M bosons, as following such underlying geometry; *without changing the geometry below the Horizon, which is non-Dynamic, fixed,* but altering $\varepsilon''$ orthonormal to $\varepsilon'$; $\varepsilon'$ is L, $\varepsilon''$ is $l$, which describes the

## DERIVATION OF ⊬NQ HUP, The Orthogonal Components of The Electric ⊥ Magnetic EigenVectors

'nipple' as seemingly dipping below the Horizon, L, into the region otherwise regarded as $l$.

This is not a description for just a Black Hole, but $G_{\mu\nu}$, e.g. 'geometry of space-time,' from the Trace Matrix $T_m$, as a function of $\varepsilon'/\varepsilon''$, as orthonormal eigenvectors, where $N$ cannot change [forbidden], but the values $\varepsilon'/\varepsilon''$ via the Fractal Set alter apparent Path length $l$, orthonormal to L, which is the fitting description for the 'nipple,' as apparently dipping below L into the region defined by $l$. The orthonormality, again is described by:

$$C_2 = \sum_{\check{r}} \sum_{IJ} e^{S_{\check{r}}} e^{\frac{S_m - S_{\check{r}}}{2}} I_m I_{\check{r}} \lambda_{IJ}{}^{(u_0 - u_i)}$$

$$e^{\frac{S_n - S_{\check{r}}}{2}} J_m J_{\check{r}} \lambda_{IJ}{}^{(u_0 - u_i)}$$

[Note that the scaffold of the equation immediately above was derived by Susskind in lecture and to the best of my knowledge I do not think he ever published that formally].

Where we isolated the term:

Because $\Delta I$ has the binary state: $\{0, 1\}$, when

$$\Delta I = 1; \ln(\Delta I) = 0$$

$$\Delta I = 0; \ln(\Delta I) = \pm\infty$$

$$\lim_{|X-Y| \to +\infty} \left(\frac{\delta_{IJ}}{|X-Y|^{2\Delta I}}\right) \doteq 0 \doteq \lim_{|X-Y| \to -\infty} \left(\frac{\delta_{IJ}}{|X-Y|^{2\Delta I}}\right)$$

For the singular state:

$$\Psi^*\{-\infty, 0, +\infty\}$$

As meeting the demand of the most fundamental Limits at Infinity, my prior references to the Bekenstein-Hawking limits yielded the Shannon and conventional $\Delta S$ Entropies diverging at 2 Lp above the Schwarzschild limits for the *logical, not ontological* relationships for the set:

{observer, observed, observable, unobservable}

## DERIVATION OF ⊬NQ HUP, The Orthogonal Components of The Electric ⊥ Magnetic EigenVectors

The elements is **bold** represent conditions that can only occur **at the Horizon**, but not *between* the Horizon and any non-dynamic *below* the Horizon [hence, logical relationships], as no 'observation'; which is a dynamic, has any sequitur meaning in a non-dynamic environment. The term, *unobservable* then represents everything below the Horizon, e.g. Holographic environment defines the AdS Surface as a Schwarzschild *Surface*.

Moving on:

$$\langle \theta_I(A)\theta_J(B)\theta_k(V) \rangle = \frac{C_{IJK}}{|A-B|^{\Delta I+\Delta J-\Delta K}|B-V|^{\Delta J+\Delta K-\Delta I}|A-V|^{\Delta I+\Delta K-\Delta J}}$$

We look again at, for example $|A - B|$, as either one or the other orthonormal eigenvector whose Conformation is described by either

$$|A - B| \perp |A - V|$$

Where $|B - V|$ was $v = c$ [for a massless photon] representing superposition across the entire Horizon L. My prior reference to Laws of Motion on a Planck Scale, albeit, amusing, describe the phenomenal aspect that $v \ll c$ is *not possible*. That is, again, only $v = c$ OR $v = 0$ is possible when Transfer of Information occurs across a Planck interval of any unitary step $(u_O + u_{O+1})$. That is, $u_{O+n}$, where $n$ is not an integer, is forbidden, as this requires a violation of $\{Lp, tp\}$.

As a result, $|B - V|$ is then, in terms of $v$

$$|B - V| \equiv \{0, c\}$$

The remaining terms become [arbitrary]

$$|A - B| \equiv l_{(nLp''\varepsilon'/\chi tp''\varepsilon'')}$$

$$|A - V| \equiv L_{(Lp'\varepsilon'/tp'\varepsilon')}$$

Dr. William Joseph Bray

### DERIVATION OF ℏNQ HUP, The Orthogonal Components of The Electric ⊥ Magnetic EigenVectors

$$|A - B| \equiv l_{(nLp''\varepsilon''/xtp''\varepsilon'')} \perp |A - V| \equiv L_{(Lp'\varepsilon'/tp'\varepsilon')}$$

$$l_{(nLp''\varepsilon''/xtp''\varepsilon'')} \perp L_{(Lp'\varepsilon'/tp'\varepsilon')}$$

The result is that $l$ is orthogonal, but not orthonormal, to $L$. The other result is that $l$ is equal to or less than the speed of light and that $L$ is some other value that is slightly greater in magnitude than $l$. The reason for this is that $(nLp''\varepsilon''/xtp''\varepsilon'')$ is evolved by *exactly one more step* than $\varepsilon'$. Thus, a mini-person with their local meter stick at some point $l$ would measure $L$ to be of relative slightly larger value. This is similar to, in the result only, convention, but not convention. Perhaps a bit bold of a hypothesis to put forth, but perhaps worth investigating.

The obvious question is; if $\Delta l = \{0, 1\}$ is the only set of possible values, then how does $(u_0 + u_{0+1})$ occur? The answer is that, you are on the Horizon, whose limit is L. There is no 'observation' of any event $\geq 2Lp$ via Path L, which is the coalescence of the Bekenstein-Hawking limit, which is defined by and defines 1 HUP. By *conserving* ℏ$(N_Q)$, as described by:

$$l_{(nLp''\varepsilon''/xtp''\varepsilon'')} \perp L_{(Lp'\varepsilon'/tp'\varepsilon')}$$

The meaning of the above equation is that the two eigenvectors are orthogonal, but not orthonormal to one another. They are of very slightly different magnitude, as the Path L portion takes a Fractal Iteration of one less iteration than the Path $l$ component. That is, the Path $l$ fractal is one greater iteration that the orthogonal Path L eigenvector. Thus, the Path $l$ eigenvector is of slightly different magnitude from that of Path L.

$$ℏ(N_Q)(nLp''_{\varepsilon'}/xtp''_{\varepsilon'})\{Lp_{\epsilon'}'tp_{\epsilon'}'\} \geq \frac{h}{4\pi}$$

And:

$$ℏ(N_Q)\left[\frac{nLp''_{\varepsilon''}}{xtp''_{\varepsilon''}}\right] \perp \{Lp_{\epsilon'}'tp_{\epsilon'}'\} \geq \frac{h}{4\pi}$$

### DERIVATION OF ℏNQ HUP, The Orthogonal Components of The Electric ⊥ Magnetic EigenVectors

Given $\{Lp'_{\varepsilon'} tp'_{\varepsilon'}\}$ are Path L, $[nLp''_{\varepsilon''} xtp''_{\varepsilon''}]$ are $l$, note that $\{Lp'_{\varepsilon'} tp'_{\varepsilon'}\} \equiv c$, whereas the set of values $[nLp''_{\varepsilon''} xtp''_{\varepsilon''}]$ described the arc, $l$. My prior reference to arc $l$ being 'fluid,' rather than static, is this dynamic relationship between:

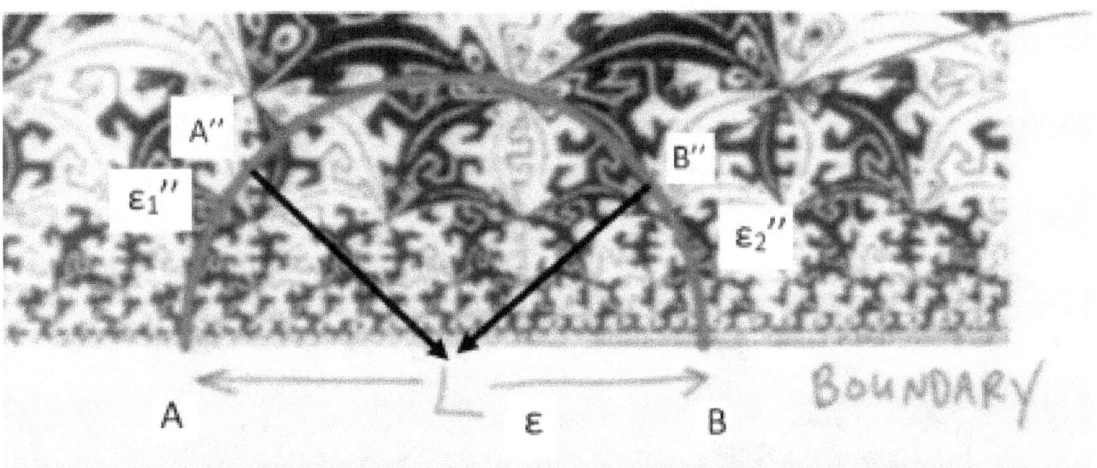

Where I've changed the interior conditions to $\varepsilon''$, $\varepsilon \rightarrow \varepsilon'$ is actually limited to the domain of 2 Lp.

Thus:

$$l_{(nLp''\varepsilon''/xtp''\varepsilon'')} \perp L_{(Lp'\varepsilon'/tp'\varepsilon')}$$

$$\hbar(N_Q)(nLp''_{\varepsilon'}/xtp''_{\varepsilon'})\{Lp'_{\varepsilon'} tp'_{\varepsilon'}\} \geq \frac{h}{4\pi}$$

$$\hbar(N_Q)\left[\frac{nLp''_{\varepsilon''}}{xtp''_{\varepsilon''}}\right] \perp \{Lp'_{\epsilon'} tp'_{\epsilon'}\} \geq \frac{h}{4\pi}$$

**DERIVATION OF ҟNQ HUP, The Orthogonal Components of The Electric ⊥ Magnetic EigenVectors**

$$\varepsilon = E/m = A_\Omega = ҟN_Q = Lp^2/tp^2 = c^2$$

$$(A_\Omega)\left[\frac{nLp''_{\varepsilon''}}{xtp''_{\varepsilon''}}\right] \perp \{Lp_{\epsilon'}' tp_{\epsilon'}'\} \geq \frac{h}{4\pi}$$

$$(A_\Omega) l \perp L \geq \frac{h}{4\pi} \leq ҟ(N_Q)\left[\frac{nLp''_{\varepsilon''}}{xtp''_{\varepsilon''}}\right] \perp \{Lp_{\epsilon'}' tp_{\epsilon'}'\}$$

The key factor is that the *magnitude* of the two orthogonal vectors is ever so slightly different, not some integer value of the Planck interval, but the system *fracks*, in a self-similar manner described in another section of this text.

Back to the DCQE

## Back to the DCQE:

The 2-slit variants are, as the hypothesis goes, chasing native wave functions, which have no sequitur locality, by detecting the discrete, massive, and orthogonal but not orthonormal electric and magnetic bosons. Keep in mind there is no science and no technology to *directly detect* anything other than the electric and magnetic bosons. Thus, we have the electric and magnetic orthogonal with the scale ever so slightly different in magnitude between the electric and magnetic. I do not think it is correct to say that universally the electric will be of slightly larger magnitude or visa versa. This really depends on the evolution of the Information as it exists *on* the AdS Horizon Surface with a vector figuratively 'downward' looking figuratively 'into' the AdS figurative interior or interiors [plural is possible]. The result would be exactly half of the time the electric would be of slightly larger magnitude and slightly off from a perfect $90°$ and exactly half of the time it is the other way around.

Rewinding back to Alice and Bob at 1-unitary interval 'below' the horizon surface, at point $u_{0-1}$, and I will show the image again:

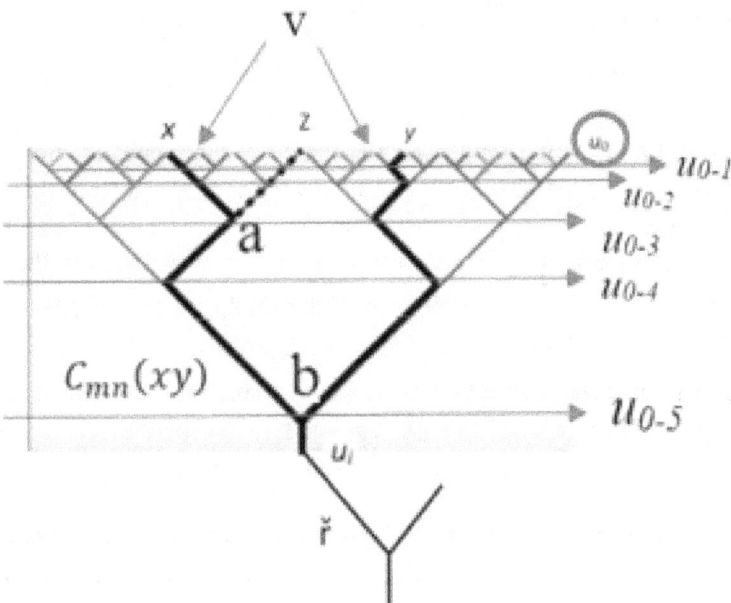

1. At $u_{0-1}$, the hypothesis is that Alice and Bob are entangled.
2. However, at $u_{0-2}$ as their common origin, Alice and Bob *are not entangled.*

That Cayley tree is a good visual metaphor for the orthogonal relationships between the vectors of the electric and magnetic bosons. The thing to know is that, at greater than 2-unitary Planck intervals the orthogonal relationships do not apply, because they are all derived as existing at the extant HUP lower [Bekenstein-Hawking] relationship of 2-unitary Planck intervals.

## Back to the DCQE

*The break occurs* at greater than 2-unitary Planck intervals, which in the above diagram is at $u_{\omega,2}$, because the vectors become *less orthogonal*. I don't know any other way to express it. If we look at the more conventional visual metaphor:

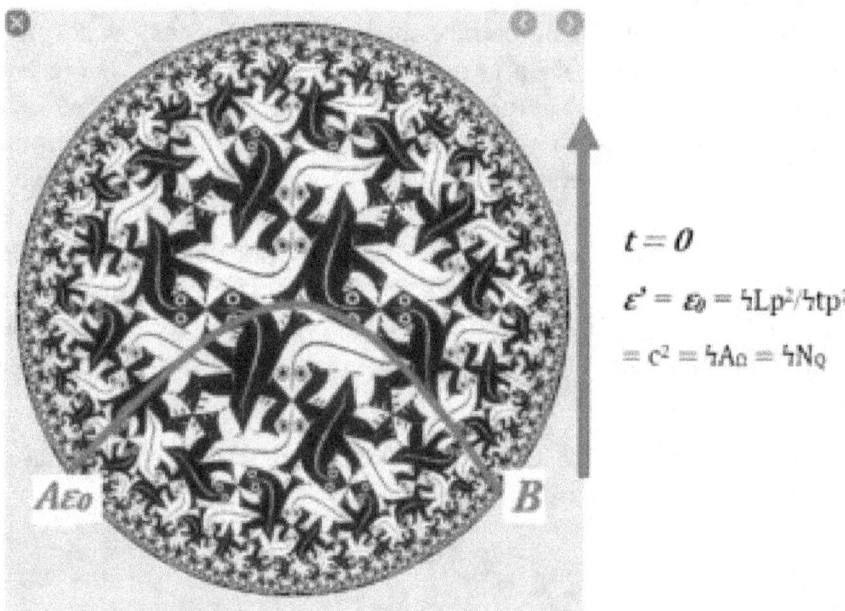

$t = 0$
$\varepsilon' = \varepsilon_0 = \hbar Lp^2/\hbar tp^2$
$= c^2 = \hbar A_0 = \hbar N_Q$

From time-zero at $A\varepsilon_0$, the first vector encountered is orthogonal to the AdS Horizon Surface. The same is true at point $B$. That extends to point $u_{\omega,1}$, and from there the vectors begin to 'turn' toward some other point on the horizon surface, rather than directly 'down into' the interior. Thus, the first unitary step figuratively 'downward' into the AdS interior is orthogonal [perfect $90^0$] and from there on deviates from $90^0$, not orthogonal, and never orthonormal.

Thus, at 1-unitary Planck interval, which is at $u_{\omega,1}$ we have the vector of Information and the evolution of temporal progression [e.g., Planck 'flow'] at the lower limit of the Bekenstein-Hawking bound and at greater than 2-unitary intervals, which begins at $u_{\omega,2}$ the Information is no longer orthogonal or becoming less orthogonal. That is important because the electric and magnetic bosons, by convention, are orthogonal to one another. Note that in Near Field Effects we can have the electric and magnetic 'fields' not at all orthogonal to one another. In Near Field systems we see a lot of use of this unique characteristic. Meaning, in Near Field systems, we make use of sending the electric field at odd angles to the magnetic field.

How do the Electric and Magnetic bosons move around the AdS system?

Given that they are by Quantum field Theory convention arise out of the HUP and thus are both discrete and massive, we have to assume that they move about the AdS system in the same manner as any wave function that possesses mass. Locality, on the other hand, *on the AdS Horizon Surface* is a sticky point. They cannot move about *on* the horizon surface

## Back to the DCQE

because the horizon surface is superluminal. Thus, the only other avenue is that region figuratively 'below' the horizon surface, which means they must be eigenstates as they move about and cannot be eigenvalues, else they would be forced to the horizon surface.

There is an episode of the TV show, *Lucifer,*' where an angel stops time. They are talking and moving about with everything around them 'frozen.' Then, the angle 'releases' time and everything just starts moving again. If Bob *never detects Alice's signal*, that system would in fact remain frozen in that inert, static state. *There is no nor shall there ever be any empirical evidence otherwise.* The instant Bob detects the signal, that inert, static system is put in motion again, because Bob, by electromagnetic interaction, with those orthogonal electric and magnetic bosons, figuratively 'forces' that Information from eigenstate or states to eigenvalue(s).

Bob 'forces' Information to the horizon surface as his observation and detection is any ElectroMagnetic phenomenon. Thus, the Information is an artifact of a native photon wave function, namely Electric and Magnetic bosons that are in eigenstates as they travel, and interaction via some other Electric and Magnetic bosons forces them to precipitate to eigenvalues, at which time they are now on the AdS Horizon Surface.

It is rather straight forward, in form, but a bit complex in the details. Keep in mind that there is no science nor technology that can *directly* detect any phenomenon other than by way of Electric and Magnetic bosons, e.g., electromagnetic. All of the fuss over detecting Higg's bosons is vastly misunderstood; there is no detection of a Higg's boson, the Higg's cannot even reach the detector as its life is about the diameter of a proton. 'Particles' of these sorts are pseudo-detected by a complex cascade of interactions, which must end in some electromagnetic phenomenon, else we cannot detect them.

## What do Alice and Bob see of each other when connected by >1-causal path?

### Alice and Bob on a Schwarzschild Horizon

### What do Alice and Bob see of each other when connected by >1-causal path?

**Preliminary discussion:**

This is a fascinating question that to the best of my knowledge has not been addressed. We have numerous descriptions of multiple connectivity but no physical description of what this might look like to our characters Alice and Bob, in the case where 1) they are multiply connected and 2) observing only each other. Invariably in any relativistic system, for instance, the arguments continually flip flop between frames of reference until such time the lecturer has no clue what is going on. We have numerous descriptions of what Alice and Bob may look like to some arbitrary third-party observer but no descriptions of what Alice and Bob see of themselves and each other. This is a critical point.

The reason this becomes important is that our Delayed Choice and Lunar Landing scenarios can be equated with respect to exactly what Preferential Frame of Reference as established by pure distance looks like, as well as what Non-Preferential Frame of Reference looks like. Then we have to establish what Preferential Frame of Reference looks like from the perspective of a Lorentzian system as well as a Schwarzschild system.

In earlier discussions, when PFR applies, we see that what is true in one system, Bob, Armstrong on the moon experiences a head on train crash with Alice, CAPCOM on Earth. However, Alice experiences no such set of events, no train crash, she wizzes by Bob without a scratch.

And [as the hypothesis was put forth], when PFR *does not apply*, what is true in one system, such as photons interfering to make a pattern is true for all the photons in the system, regardless of special geometry and time, *because time is the factor that does not exist under the conditions of Non-preferential Frame of Reference*. Everything is frozen in a non-dynamic, static state. The 'distances' between Alice and Bob are non-sequitur, they are connected by *the conventional Ryu-Takayanagi Path I*, which is both Alice and Bob, eternally at time-zero.

We find that in our Lunar Landing example what appears to be at first an anomaly, later in a high school level attempt at describing the results as simple signal delays does not work because we have *recorded* two distinct and true realities for Alice and a separate reality with a completely different set of causal events in a different order for Bob.

There is another caveat that was discussed; that entanglement does not occur in nature but is an observation in the lab that results from creating two systems simultaneously, which is *true simultaneity*.

**What do Alice and Bob see of each other when connected by >1-causal path?**

And for the record, entanglement does not occur in nature, ever. The notion that electrons are entangled in an orbital is simply nonsense, based merely on the notion that they have opposite spins. This scenario in the Pauli Exclusion Principle is defined as *reaction*, not entanglement. Electrons in an orbital are *not in eigenstates, they are well defined eigenvalues*. Things that are not in an eigenstate cannot be entangled. The proposal of such notions is a misunderstanding of the fundamental conditions of entanglement.

In all of these descriptions we have to avoid the human penchant to regard time as 'reversible' in any way on any scale, from the cosmological down to the Planck length. There can be no bidirectional symmetry to time on any scale, as this would then absolutely demand determinism. This is an artifact of human thinking architecture that is literally engrained into the physical human brain. There can be no temporal symmetry on any scale, again.

The reason this artifact in human consciousness appears in physics is because of the 'mystery' of quantum mechanical observations, in particular the history of the 2-slit experiment in all of its variations and iterations.

Entanglement of qualities such as force and velocity is also nonsense, applied to atom kinetics, a rocket's flame shooting out the back is not entangled to its forward motion. The same argument holds true for eigenstates, interactions and reactions are not eigenstates. The list of nonsense is quite long, never once validated by experiment, nor can it ever be, pure and poor supposition, and merely demonstrates a complete lack of knowledge of quantum systems. More so than any other topic in physics, entanglement is all alluring, because it cannot be explained, and therefore takes on a life of its own.

Ultimately the goal is to address the Delayed Choice experiments, where it is postulated that the Delayed Choice can be described as a system that is Non-Preferential Frame of Reference and applies to entangled systems. Where we see the Lunar Landing Anomaly is a system where the causal sequence of events seems to be out of order between Earth and moon as established purely by distance, yielding two distinct realities, e.g., a train crash in Bob's Preferential Frame of Reference but no train crash in Alice's Preferential Frame of Reference. The Delayed Choice, 2-slit variations in general, describes an entangled system wherein the train crash either is or is not true *in all frames of reference, e.g., is non-preferential*.

The term Preferential Frame of Reference will be used as a more specific definition such that these systems can be thusly differentiated. Again, PFR can result, or be associated with, distance, Lorentz transform, Schwarzschild transform. Preferential Frame of Reference is a term that is chosen because 'frame of reference' is non-specific and clouded with mysterious issues.

Understanding what entangled systems see of each other is important, and this must extend to multiple connectivity, for reasons that will become clearer as we move forward.

**What do Alice and Bob see of each other when connected by >1-causal path?**

We look at this as the non-ontological relationship {observable, unobservable}, purely as a function of proximity to a Schwarzschild Surface. I will avoid the term 'horizon' as it has a host of terms of penchant. I will use Schwarzschild Surface or Schwarzschild Horizon and they will be synonymous, and proper nouns.

In addition, there is no causal component to the relationships; we can render Shannon entropy as associated with {unobservable} as a system's condition. That is, Shannon Entropy describes a system that is unobservable, and a system that is unobservable is thusly described by Shannon Entropy. They are associated states, meaning that if a system is unobservable or *becomes unobservable*, it is then described by Shannon and not conventional $\Delta S$ entropy. This split from conventional $\Delta S$ as limited to 1-causal path to Shannon Entropy which is limited to >1-causal paths is described in greater detail in another section.

Likewise, if a system is {observable} as its state, it can *only be described* by conventional $\Delta S$ entropy, and only 1-causal path can describe the system.

The logical states {Observed, unobserved} will be differentiated out of this as a logical set of the sort: {observable, unobservable, observed, unobserved}; a non-ontological set of states completely dependent on proximity to a Schwarzschild Surface, where I want to derive that conventional $\Delta S$ and Shannon entropies diverge; as a direct association of:

$$\{\text{conventional } \Delta S\} \text{ 1-causal path} \rightarrow \text{>1-causal paths } \{\text{Shannon}\}$$

The progression from conventional $\Delta S$ entropy to Shannon entropy in this description is the result of falling into proximity to a Schwarzschild Surface such that multiple connectivity begins. For instance, our characters Alice and Bob are approaching a Schwarzschild Surface.

Ultimately, I might find that the arrow from conventional $\Delta S$ to Shannon Entropy is not unidirectional:

$$\{\text{conventional } \Delta S\} \text{ 1-causal path} \leftrightarrow \text{>1-causal paths } \{\text{Shannon}\}$$

However, I do not know the answer to this yet, nor do I know the conditions which might lead to, for instance, >1-causal path somehow going off to 1-causal path. Although this sounds unlikely, it cannot be ruled entirely out until it is fully probed. Black hole evaporation will not be a topic in this text.

The postulate is:

1. Conventional $\Delta S$ Entropy, exactly 1-causal path describes the system. {observable}
2. Shannon Entropy, >1-causal path describes the system; always. {unobservable}

The question becomes, what do Alice and Bob see of each other under such conditions of being connected by >1-causal path? There are thousands of papers regarding hypotheses to this end. However, I do not find them satisfying, because invariably they mix and match

**What do Alice and Bob see of each other when connected by >1-causal path?**

conventional ΔS and Shannon entropies as overlapping in the same domain, where I do not see it possible that this can be the case. This is because, and it should be rather obvious, being connected by greater than one causal path can have no definitive description in normal spacetime. As such, being connected by more than one causal path must be and can only be described by Shannon Entropy. Again, the rationale for Shannon Entropy as referring *only to* >1-causal path is another section.

*Very briefly*, if we look at the most fundamental description of Shannon Entropy:

$$H(X) = -\sum_{i=1}^{n} P(xi) Log_2 P(xi)$$

When $i = 1$, the entire term falls to zero.

$$H(X) = -\sum_{i=1}^{n=1} P(xi = 1) Log_2 P(x = 1) = 0$$

A single causal path is not being regarded as anything other than unidirectional. There will be no 'reverse causality' or backward time in any of these discussions. It does not happen on any scale. Which is the point, to describe entanglement without phenomenal or transcendent hypotheses.

Another postulate is that a system connected by more than one causal path cannot be described by conventional ΔS entropy. This should be self-evident. These are discussed in detail in another section.

## What do Alice and Bob see of each other when connected by >1-causal path?

### Alice and Bob, Multiply Connected

We connect Alice and Bob by a tether that is 2-Planck lengths long, as the agreeable lower limit of the Heisenberg length, which must be greater than one Planck length. We are lowering Bob onto the Schwarzschild Surface or Horizon.

At such time Bob reaches a 2-Planck length proximity to the Schwarzschild Surface, he coalesces with the Horizon, and he is no longer observable to Alice and Alice is no longer observable to Bob. One question is then, what does this look like to each Alice and to Bob.

At such point that Alice and Bob *both reach* that 2-Planck length proximity to the Schwarzschild Surface, they both coalesce with the black hole. They do not entropy via conventional $\Delta S$ entropy; they are dropping to infinite connectivity and can thus only be described by Shannon Entropy, as being >1-causal path. There is no description of conventional $\Delta S$ entropy for a system that is causally connected by >1-causal path. In fact, there is no description for what 2-temporal endpoints, regarded as Alice and Bob, see of each other at such time they are connected by >1-causal path.

There are some notes in the literature that Entanglement entropy is difficult to calculate, and I must agree. However, I think a simplification of the issue is that we can allow conventional $\Delta S$ and Shannon entropies to diverge, and then differentiate systems in this way. In order to understand this, it is good to start a dialogue of what Alice and Bob see of each other, rather than what we see from some arbitrary vantage point. Ultimately, we see nothing. What goes on inside such a system then is purely an issue for Alice and Bob to deal with. There is no arbitrary or arbitrarily distant frame of reference because they are {unobservable} as their condition.

A visual metaphor that helps takes the Bekenstein relationship:

$$N = \frac{A_\Omega}{4L_p^2}$$

Which became

$$ϞN_Q = Ϟ\frac{L_p^2}{t_p^2} = c^2 \equiv Ϟ1$$

And Koppa [Ϟ] is a police operator that forbids the penchant to violate Preservation of Information Conservation. This includes 'borrowing,' as there is no 'negative Information' sequitur to such violation. The HUP is defined later as Preserving Information Conservation, not 'borrowing' [negative] Information.

The worldsheet, $A_\Omega$ is defined by $4L_p^2/4L_p^2$ which is why I portray the *visual metaphor* as a trigonal bipyramid rather than a trigonal pyramid. It is just a visual metaphor for thinking around the arithmetic, not incredibly unlike a Feynman diagram, just more metaphoric and

### What do Alice and Bob see of each other when connected by >1-causal path?

less of a process. Because of its 9-causal paths, I refer to the tongue in cheek term, 'Yggdrasil.' We want to use this visual metaphor in some sense to come to an understanding of what Alice and Bob see of each other if and when they become multiply connected by >1-causal paths. Then back out some distance and try and visualize what we see of Alice and Bob at such time they become connected by >1-causal path. Although there is much hypothesizing regarding Tensor Networks, there is no consistency in describing what Alice and Bob see upon being multiply connected. What I find most unsatisfying is the propensity to describe multiply connected tensors with conventional S entropy, which simply cannot be.

First, we establish the definition for a causal connection as the unitary interval, with two temporal endpoints at either end:

$$\mathbf{Lp} = \mathbf{tp} \equiv (u_0 \pm u_{0+1})$$

In these discussions, I want to revert back to the classic Alice/Bob descriptions. As such I am going to assign Alice and Bob as the two temporal endpoints:

$$\mathbf{Lp} = \mathbf{tp} \equiv (u_0 Alice \pm u_{0+1} Bob)$$

Thus, Alice will be the temporal endpoint $u_0$ and Bob will be the *future temporal endpoint* $u_{0+1}$. I will refer to these as 'lone temporal endpoints.' This will represent the unitary interval. 'Information cannot be created or destroyed, *nor borrowed.*' There is no sqrt(Information) to balance such accounting. The two temporal endpoints above, $u_0$ and $u_{0+1}$ *each* represent a 'lone temporal endpoint.' Meaning, we can regard $u_0$ as {Alice} and $u_{0+1}$ as {Bob}. The thing that connects them is a causal path; the terms Lp and tp have no sequitur meaning until such time that {Alice} and {Bob} are causally connected. A lone temporal endpoint cannot actually exist, as we must stick with the conventions such as a system must be in or otherwise evolved from some eigenstate.

We want to look at what a system looks like in a metaphoric sense and what qualifies as sequitur Information:

## What do Alice and Bob see of each other when connected by >1-causal path?

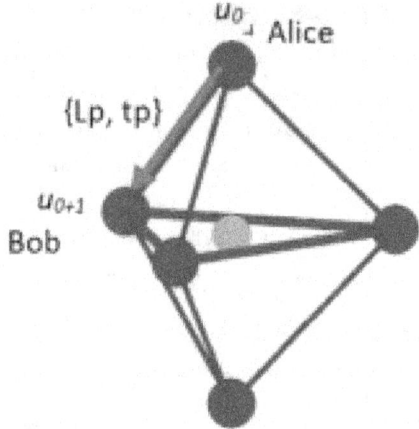

Note that the arrow is unidirectional. This trigonal bipyramid is purely a visual metaphor, nothing more. There is no sequitur shape on the Planck scale. A triangle has a height that is a non-integer value of Lp, a non-sequitur. The same issues arise in all Platonic notions of shape, obviously a circle cannot be smoothed by fractions of Lp; hence, π is quantized to a Planck length, and so on.

On the other hand, there can be no 'pixilation' of the Planck scale for this very reason. For example, the notion of a square to a cube:

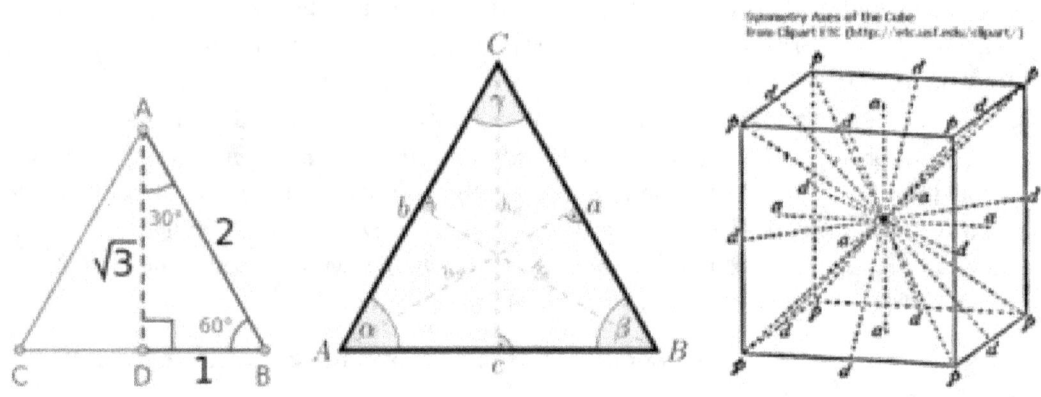

All of these shapes demand non-integer values of the Planck length. If we are going to resolve ourselves to describing a Planck scale, we have to stick with the rules, and the first rule is that we cannot regard non-integer values of the Planck length of the Planck scale, on a Planck scale. Thus, all Platonic shapes cannot exist on this scale, rendering the notion of 'pixelation' nonsensical.

Circles, spheres, arcs of any sort, require dividing spacetime in impermissible slices. The actual argument regarding the quantization vs infinite divisibility of spacetime, fields, and

## What do Alice and Bob see of each other when connected by >1-causal path?

so on is sectioned throughout this paper. It has to be discussed one principle at a time. This is by convention but needs to be outwardly expressed here in such use as visual metaphors.

I refer to the trigonal bipyramid visual metaphor as the tongue-in-cheek 'Yggdrasil,' because of its 9-edged features; the 9-edges are the causal paths in discussion. The grey dot in the center is merely a referent, without it the stick image has no illusion of volume.

Above, $\{u_0 \text{Alice}\}$ and $\{u_{0+1} \text{Bob}\}$ are connected by the causal path $\{Lp, tp\}$. In order to help determine what is and what is not Information, we can try adding an additional single temporal endpoint:

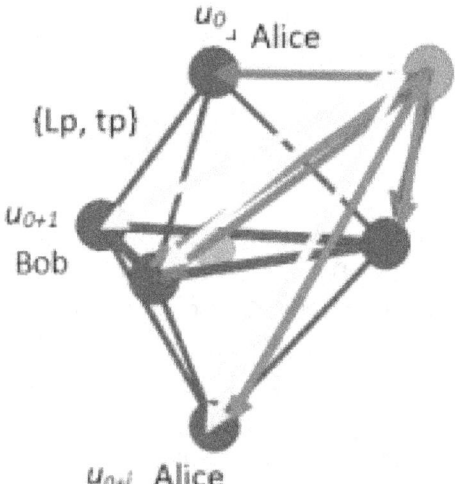

Just adding one temporal endpoint creates multiple additional systems. The red, yellow, and blue arrows are connected to different points of existing temporal endpoints. $k N_Q = k 1$ is not preserved. This is Preservation of Information Conservation Violation. Hence,

$$k N_Q = k 1 \text{ is defined by the quantity of temporal endpoints.}$$

The visual metaphor is described by $4Lp^2$, as one equilateral triangle. Hence, the trigonal bipyramid shares one face, which is poorly drawn in below.

We can now try causally connecting $\{\text{Alice}(u_0)\}$ and $\{\text{Bob}(u_{0+1})\}$ more >1-causal path:

**What do Alice and Bob see of each other when connected by >1-causal path?**

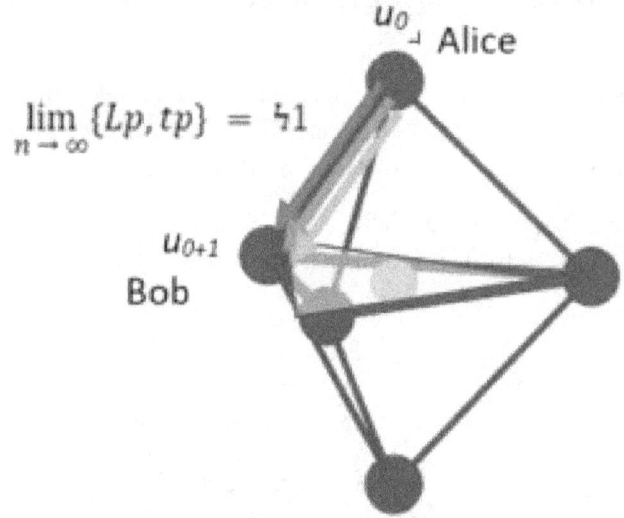

I cannot draw an infinite number of causal paths, however, from my perspective not in the Qubit, an infinite drop into any $n>1$-causal connections do not change $kN_Q = k1$. Note that regardless of the number of connections, the arrow still only points one direction. It does not point backward in time simply because it happens more than once. This places a great constraint on what Alice and Bob see of each other in being connected by >1-causal path. I think in general people go with the assumption that multiple connectivity allows for temporal bidirectionality, e.g., backward causality. This can never be the case for reasons discussed in detail in other sections, namely, the conditions of the Bob demand that the 'interior,' or that region figuratively 'below' the horizon surface does not exist, it is only a record of past states of the horizon, lacking any tangibility. The notion that a multiple connection somehow is bi-directional by some default has no foundation, it is aberrant thinking.

The unitary interval has been well defined:

$$\mathbf{Lp = tp} \equiv (u_0 \pm u_{0+1})$$

Here, the interval:

$$\mathbf{Lp = tp} \equiv (u_0 - u_{0+1})$$

*Does not mean, go backward.* It means:

$$(u_0 - u_{0+1}) = u_0$$

Go nowhere. Because, either the interval $u_{0+1}$ has or has not $\{0, 1\}$ occurred, there is no other option, it is a binary choice with no sliding value between $\{0, 1\}$; the temporal endpoint either does or does not exist:

**What do Alice and Bob see of each other when connected by >1-causal path?**

- If the temporal endpoint, $u_{0+1}$ does exist, then the result is $u_0$.
- If the temporal endpoint, $u_{0+1}$ does not [yet] exist, then the result is $u_0 - 0 = u_0$.

We can connect {Alice} and {Bob} an infinite number of times; as limited specifically to the way I have drawn them above, and there is no apparent change in the system with respect to *content*. $kN_Q = kI$ is preserved. Meaning, any two points in spacetime can be regarded as multiply connected in the manner above, because it is simply defining the path information will take from Alice to Bob and has no limit on how many times Alice can send information in this manner. One may be thinking that this is not multiple connectivity, however, connectivity is still defined as the passage of information, which can be a tensor or a love letter. Information in itself has no constraints, *everything is information*.

Nor is there any change in order of events in causality. What the other Bob are doing is not critical to this; {Alice} and {Bob} can be connected by any number of causal paths, provided at least once, and does not affect the *Information Content* with respect to quantity. Again, the Information Content will be the *quantity of Bob*, not the passage of information between them.

If $u_0$ is assigned as {Alice} and $u_{0+1}$ assigned as {Bob}, then the 3$^{rd}$ leg of that system cannot be an eigen*value* {Alice} OR {Bob}, it must be the eigen*state* {Alice | Bob}. Else, 2-systems that violate causality in the manner in which they are connected, {Alice, Alice} and {Bob, Bob} exist:

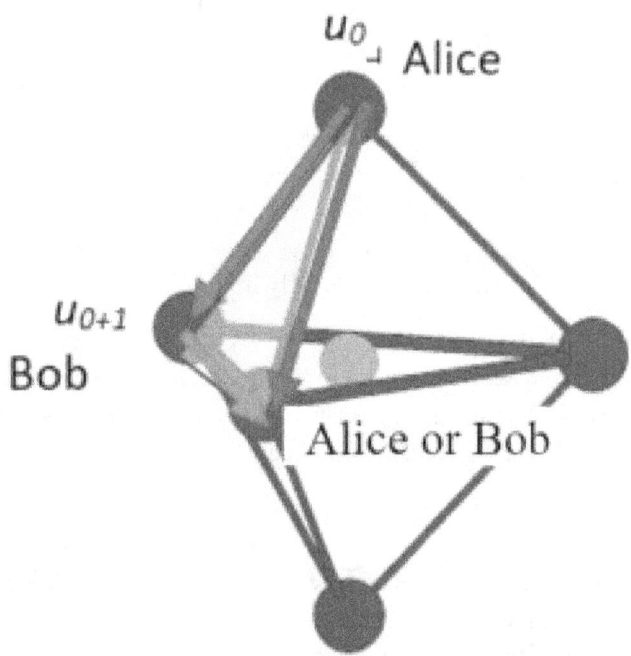

### What do Alice and Bob see of each other when connected by >1-causal path?

This is an important observation because Bekenstein's use of the visual metaphor of an equilateral triangle in representing $1Lp^2$. The yellow surface is the visual metaphor that represents the $1Lp^2$ with the entire upper half of the trigonal bipyramid representing $4Lp^2$, again, only as a visual metaphor. The upper and lower halves of the structure represent the $4Lp^2/4Lp^2$, where the structure is defined by:

$$N = \frac{A_\Omega}{4Lp^2}$$

The worldsheet, $A_\Omega$ defined as 1-Qubit, $4Lp^2$, which became

$$4N_Q = 4\frac{Lp^2}{tp^2} = c^2 \equiv 41$$

The system {Alice, Alice} is aberrant because this would imply that the lone temporal endpoint evolves to itself, which is rather illogical. Meaning specifically that Alice is at $u_0$ and evolves to $u_0$, is another non-sequitur. It seems clearly that the lone temporal endpoint would evolve to some other temporal endpoint. We are assigning each temporal endpoint a unique identity. Thus, {Alice, Alice} and {Bob, Bob} are non-sequitur. That is, the eigenstate is Alice | Bob, like a single coin flip, the coin's eigenstate either evolves to the eigenvalue Alice or the eigenvalue Bob. The coin cannot be made of Alice|Alice or Bob|Bob.

Above, I'm using the blue arrow to represent the {Alice$u_0$, Alice$u_0$}, Alice at $u_0$ is connected to Alice $u_0$. The green arrow to represents the system {Bob$u_{0+i}$, Bob$u_{0+i}$} who is also connected to himself. We're just going to assume for the time being that this is not possible, then derive something more fundamentally definitive later on in this text.

The assignment is some arbitrary $u_{0+i}$. One may be asking, who cares what the temporal endpoints are named? The answer is that Alice is defined as $u_0$ and Bob $u_{0+i}$. A {Bob, Bob} then is {$u_{0+i}$, $u_{0+i}$}, there is no causality. Likewise, the Alice systems would both be {$u_0$, $u_0$} and there is no causality. The goal is to understand causal paths. For example, we can label Victor as $u_{0+i}$.

At this time, I cannot say rather or not this is a problem; causally connecting some {Alice, Alice} and/or {Bob, Bob}. I am trying to understand what an {Alice, Alice} and/or {Bob, Bob} system looks like, from out here in macroscopic land as well as what {Alice, Alice} see of each other, and what Bob(s) might see of {Alice, Alice}. However, again, this is illogical, as each Alice and Bob are unique temporal endpoints.

**What do Alice and Bob see of each other when connected by >1-causal path?**

Ultimately, I might find that upon being {unobservable}, Shannon entropy prevails, and when Alice and Bob become connected by >1-causal path what they see of each other is the eigenstate Alice | Bob. Dropping into Shannon entropy forces a system causally back to its prior eigenstate. That will take some explaining, then derivations. First, we have to use this visual metaphor to understand life on the Planck scale, as viewed by a Planck unitary interval; to be Alice and be Bob.

The following is a happier seeming connectivity:

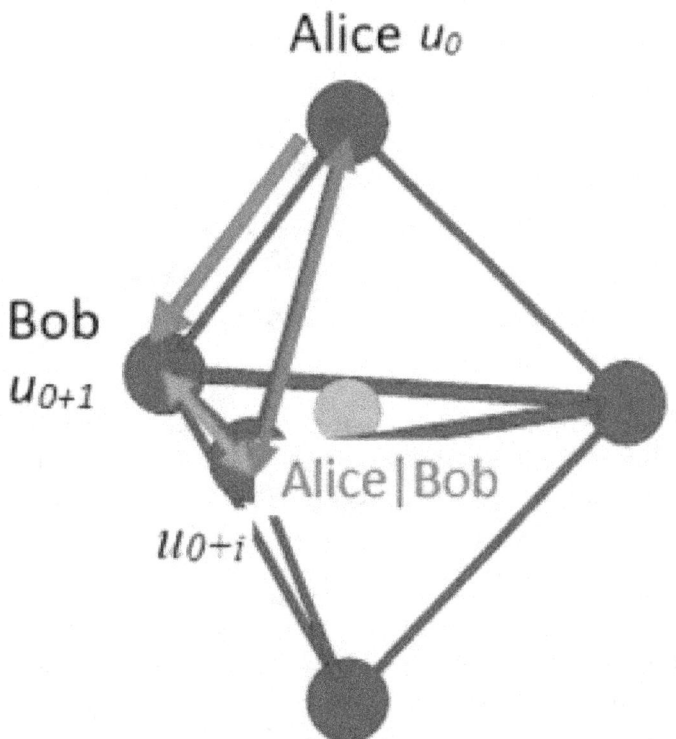

That 3$^{rd}$ leg is the eigenstate Alice | Bob and the directions of the arrows is now irrelevant, as connected to the eigenstate, but relevant to Alice$_{u0}$ and Bob$_{u0+1}$ as eigen*values*. To clarify, to Alice and Bob the directions *from them to* the eigenstate Alice | Bob is relevant, but to the eigen*state* Alice | Bob, directions to-from the eigenvalues Alice and Bob are not relevant. For the eigen*values* Alice and Bob time is a sequitur and has a defined arrow, points one direction and only one direction, but for the eigenstate there is nothing, just noise. In general, we will treat eigenvalues as obeying typical causality and eigenstates as unknown, but causality is likely not relevant.

**What do Alice and Bob see of each other when connected by >1-causal path?**

We'll explore around the loop:

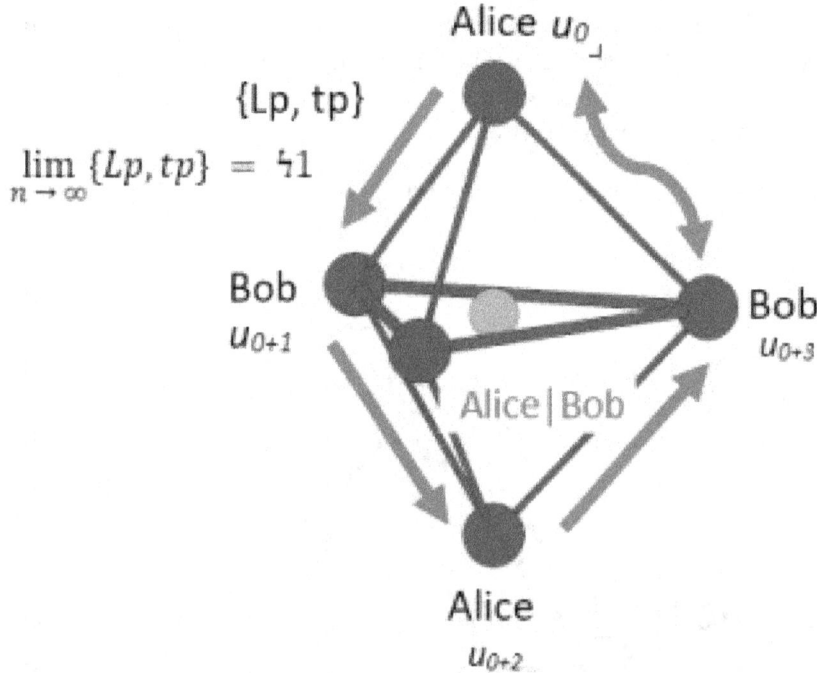

The curvy red arrow indicates a potential sci-fi temporal loop and seems unlikely. If the curvy arrow were pointing toward Bob$u_{0+3}$ that would seem OK, because that would be a proper {Alice, Bob} system and it is causally in the future of Alice at point $u_0$, merely skipping a beat. However, that particular Bob is at $u_{0+3}$, which makes it 3-unitary intervals, rather than 1-unitary interval, from Alice. This is however a non-issue for the moment.

Looking specifically at Alice$_{n0+2}$, this is completely legal in this scenario. Alice can change position or evolve temporally, and that would explain Alice at point $u_{0+2}$. The same is true for Bob$_{n0+3}$. What Bob$_{n0+3}$ cannot do, however, is connect to Alice at point $u_0$, because that is in Bob's causal past. Thus, the curvy fictitious red arrow.

Anyone and everyone at once can connect to Alice | Bob as the eigenstate and temporal direction [causality] is irrelevant. However, the temporal arrows *from* Alice | Bob can only be unidirectional, because Alice and Bob are eigen*values*. A precipitate eigenvalue is causally set in stone and cannot be violated in this fashion. Thus, we can regard eigenstates as unidirectional when pointing to eigenvalues and eigenstates as observing no sense of causality when connecting to other eigenstates.

The rules of the scenario include that the unitary interval defines itself as the unitary interval $\{Lp, tp\} \equiv (u_0 \pm u_{0+1})$, rather straight forward. So, if Alice$u_0$ is connected to Bob, he is required to be Bob$u_{0+1}$, wherein lies the problem. This is some sci-fi time travel, going around in a causal loop to back to the beginning. If the curvy red arrow points Bob to Alice, then it is going back in time by 3-unitary intervals to zero, the initial state, more sci-fi time travel. The Alice | Bob eigenstate in the middle has no choice but to be such, given the

**What do Alice and Bob see of each other when connected by >1-causal path?**

other eigenvalues; a Bob at the middle position would make 3 {Bob, Bob} systems, an Alice would make 2 {Alice, Alice} systems:

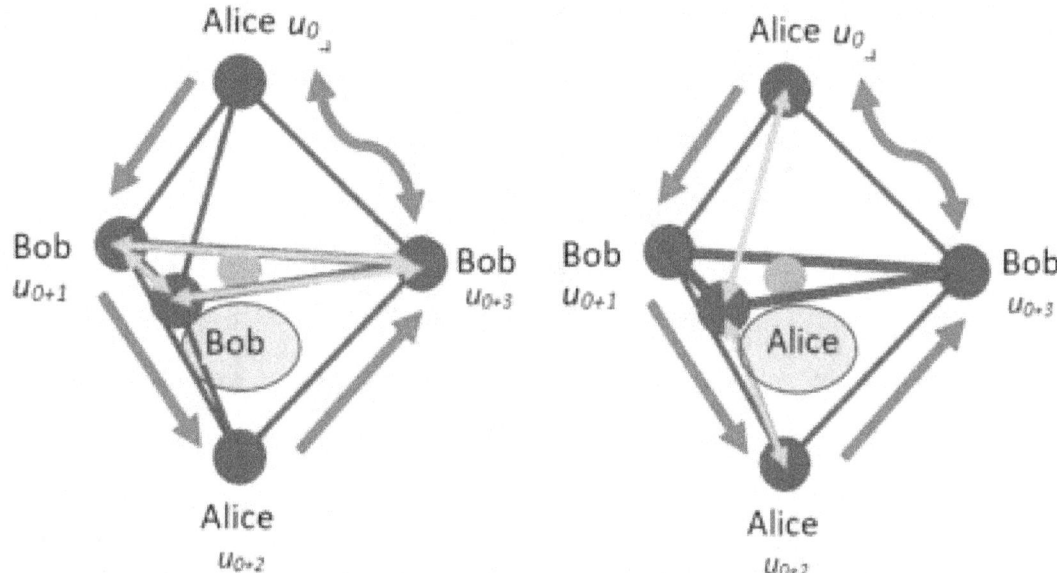

The obvious question is, why just Alice and Bob, why not introduce Victor, Wilma, etc? The answer is because we want to see what Alice and Bob see of each other when they become multiply connected. What we are doing is allowing two distinct temporal endpoints, Alice and Bob, each to evolve in a unitary stepwise fashion. Thus, Alice can take on values $u_{0+n}$, where n = {0, 2, 4, …} and Bob can take on values $u_{0+j}$, where j = {1, 3, 5, …}. If we were to introduce more characters, we would end up with a haze and that is not useful.

If the 12:00 position is assigned Alice and 9:00 position Bob. From these two positions and only these two positions:

- The middle position cannot be either eigenvalue Alice or Bob.
- The middle position must be the eigenstate Alice | Bob.
- The 3:00 position cannot be either eigenvalue Alice or Bob.
- The 3:00 position must be the eigenstate Alice | Bob.
- The 6:00 position can only be the eigenvalue Alice$_{u0+2}$, because this is logically connected to Bob at $u_{0+1}$.

**What do Alice and Bob see of each other when connected by >1-causal path?**

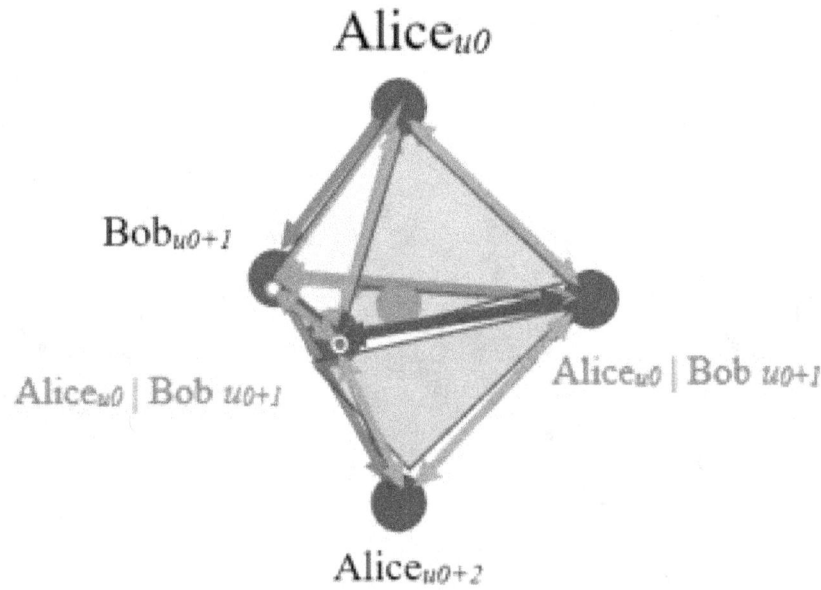

Alice is blue, Bob green, Alice | Bobs are purple.

Taking an inventory:

- There are three temporal endpoints that are eigenvalues. {Alice$_{u0}$, Bob$_{u0+1}$, Alice$_{u0+2}$}
- The three temporal endpoints {Alice$_{u0}$, Bob$_{u0+1}$, Alice$_{u0+2}$} are connected in such a way that there are two causal paths, e.g., 2Lp, which satisfies the lower boundary constraint for the Bekenstein-Hawking Heisenberg limit of two unitary Planck intervals.
- There are two temporal endpoints that are eigenstates. Alice | Bob
- The two temporal endpoints in eigenstate Alice | Bob are multiply connected to each other as well as the eigenvalues {Alice$_{u0}$, Bob$_{u0+1}$, Alice$_{u0+2}$}.
- All eigenvalues {Alice$_{u0}$, Bob$_{u0+1}$, Alice$_{u0+2}$} are *multiply connected* to the eigenstates Alice | Bob. That is, the eigenvalues {Alice$_{u0}$, Bob$_{u0+1}$, Alice$_{u0+2}$} are only connected once to other eigenvalues, but the eigenvalues are connected more than one time to various eigenstates Alice | Bob.
- There are exactly two surfaces [in blue shading] that have no edge connecting to a causal path established by an eigenvalue.

The two surfaces that are purely eigenstate are of interest. I will get back to that at a later time. Ultimately, we would find that eigenstates causally connected to other eigenstates gets rather mystical. I think one issue is thinking that Alice and Bob are entangled, but the fact is that only their spin states, for instance, are entangled. Two entangled photons are not entangled photons. A photon is energy, vector, spin, parity, charge, angular momentum,

## What do Alice and Bob see of each other when connected by >1-causal path?

and so on. Only the spins are entangled, and we do not know what other properties, if any, are entangled. The vectors, for instance, are not entangled. If they were, when we deflect one photon a certain way with perhaps a mirror, the other photon would spontaneously deflect. That has never been observed to the best of my knowledge. We thus have to clarify what property or properties of a wave function are entangled, which we know, because it is going to be the one that we measure. This seems a bit deterministic, but it goes back to the entangled properties of wave functions travelling the conventional Ryu-Takayanagi Path $l$.

Interestingly, the system must remain frozen in this set of eigenstates. For example, if the temporal endpoint in the middle precipitates to Bob, then again, we have a {Bob, Bob} connection, similarly if the eigenstate precipitates to Alice, we would have Alice at the top connected to an Alice in the middle temporal endpoint position.

The only possible out for evolution is to make an entirely new 'Yggdrasil' describing 1-Qubit of Information. That too will be discussed in a later section. We find the AdS Horizon Surface consisting of a foam [not Wheeler foam] of these Qubits, the interior of the AdS domain literally ceasing to exist in exactly 2-Planck intervals of time to, in a manner of speaking, recreate itself as a new AdS Horizon Surface.

Note that the orientation of the trigonal bipyramid visual metaphor is not entirely arbitrary: the three planar temporal endpoints {$Bob_{u0+1}$, [$Alice_{u0}$ | $Bob_{u0+1}$]} are each causally connected 4-times:

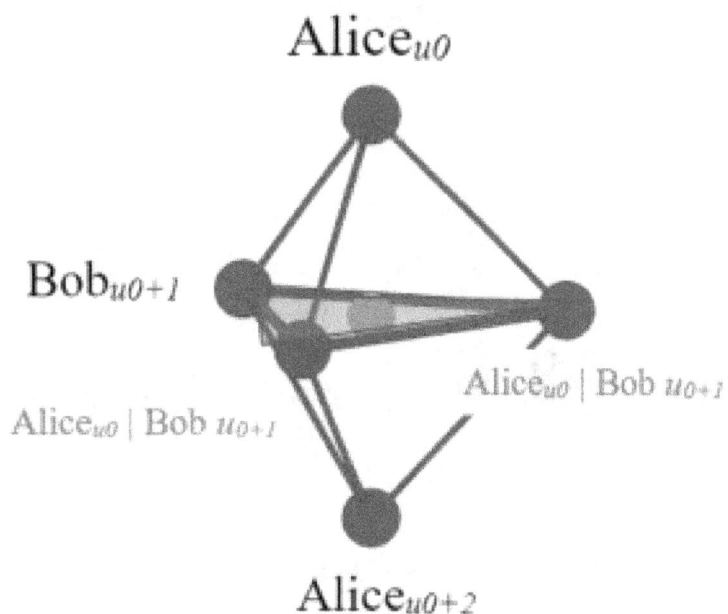

### What do Alice and Bob see of each other when connected by >1-causal path?

At the top and bottom, only Alice's$_{u0}$ are each connected 3-times. This {Bob$_{u0+1}$, [Alice$_{u0}$ | Bob$_{u0+1}$]} in fact establishes a plane and will eventually map out to causal symmetry. The surfaces are in fact surfaces. If I select one of the planar temporal endpoints to be Alice$_{u0}$

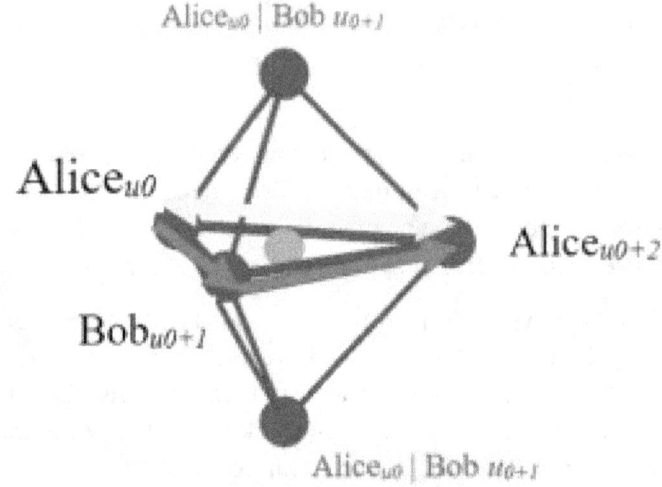

Alice$_{u0}$ | Bob $u_{0+1}$

Alice$_{u0}$

Alice$_{u0+2}$

Bob$_{u0+1}$

Alice$_{u0}$ | Bob $u_{0+1}$

Here, the problem demonstrates how Alice$_{u0}$ must be placed in such a way that she cannot be planar, else she ends up causally connected to herself from Alice$_{u0}$ to Alice$_{u0+2}$, which is a thing we are trying to avoid, as this represents backward [in time] multiple connectivity from Alice$_{u0+2}$ back to her past state at Alice$_{u0}$.

Thus, the 3:00 position must be an eigenstate:

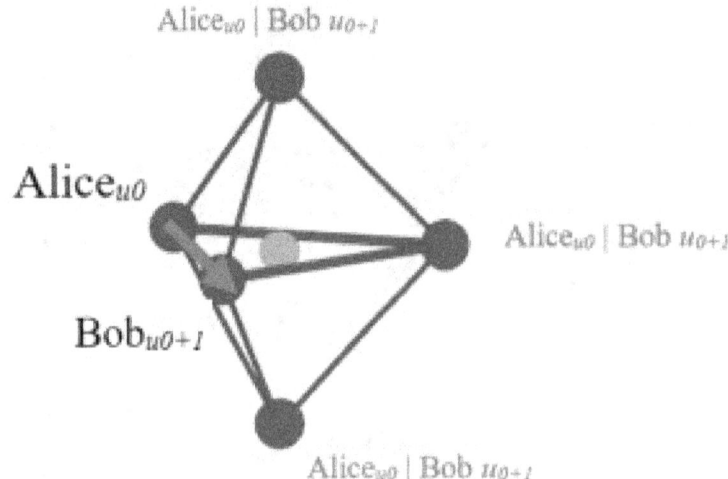

Alice$_{u0}$ | Bob $u_{0+1}$

Alice$_{u0}$

Alice$_{u0}$ | Bob $u_{0+1}$

Bob$_{u0+1}$

Alice$_{u0}$ | Bob $u_{0+1}$

**What do Alice and Bob see of each other when connected by >1-causal path?**

Here, the problem is that there is only 1-causal path. The 'Yggdrasil' must consist of 2-causal paths, else violate the quantized lower limit of the HUP. The requirement of 2 Lp thus 2{Lp, tp} is to satisfy the Bekenstein-Hawking Lower Limit:

$$S = \frac{A_\Omega}{4Lp^2}$$

Where the worldsheet $A_\Omega$ is defined here as 1-Qubit:

$$S = \frac{4Lp^2}{4Lp^2}$$

And again, $A_\Omega$ is the 'worldsheet' that defines the system. The entropy is defined as the number of Planck areas that make up the surface, this visual metaphor was actually rendered by Bekenstein:

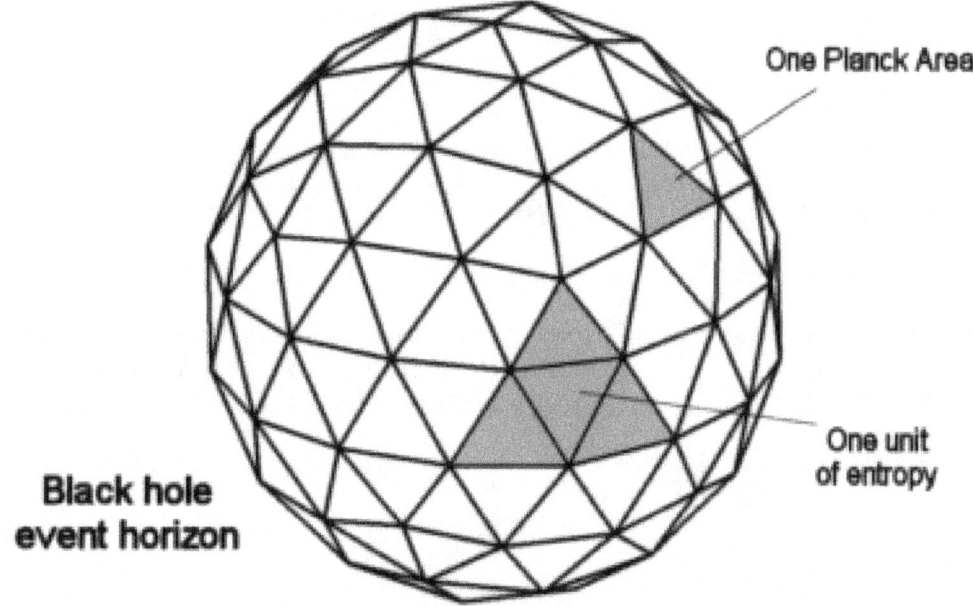

Up and to the right, Bekenstein has depicted 'One Planck area' as an equilateral triangle. Over to the right, where Bekenstein has stated, 'One unit of entropy,' is the $4Lp^2$ in question. What I have done here is to fold those sides in to form a trigonal pyramid, and the $4Lp^2/4Lp^2$ represented by taking two of those 'One unit of entropy' equilateral triangles and forming a trigonal bipyramid. The trigonal bipyramid is useful as a visual aid and has its line of causal symmetry on the planar edge between the two trigonal pyramids, indicated in blue:

**What do Alice and Bob see of each other when connected by >1-causal path?**

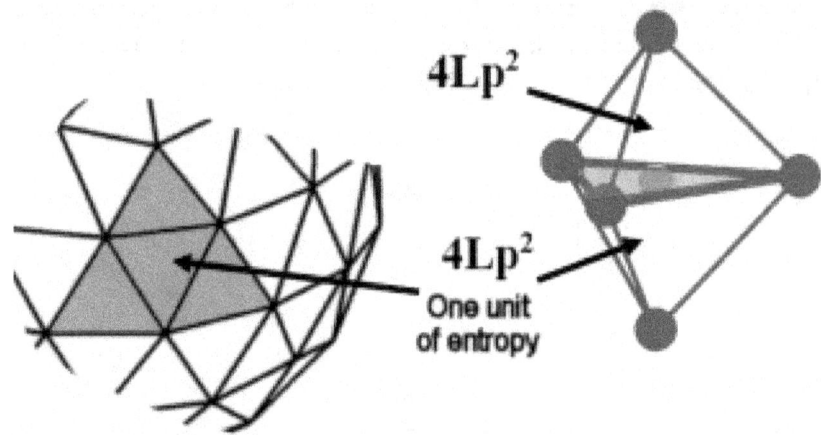

This can be regarded as one half of the bipyramid as representing the worldsheet, $A_\Omega$, and the other half as $S$ entropy, which is equated with $N$-Qubits, where $N \equiv \hbar 1$. Which, again is simply manipulated:

$$N = S = \frac{A_\Omega}{4Lp^2} = \frac{4Lp^2}{4Lp^2} = 1$$

$$\hbar N_Q = \hbar \frac{Lp^2}{tp^2} = c^2 \equiv \hbar 1$$

The term, $N = S$ is rather straight forward, the entropy is in units of Information, which is defined as the Qubit. Here, we are just setting the worldsheet, $A_\Omega$ to its least possible value, which is $4Lp^2$. As a visual metaphor, $1Lp^2$ was rendered by Bekenstein as:

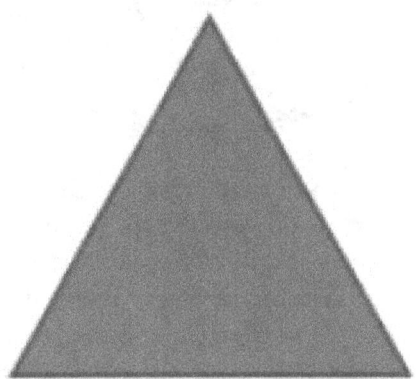

## What do Alice and Bob see of each other when connected by >1-causal path?

And $4L_p^2$ as:

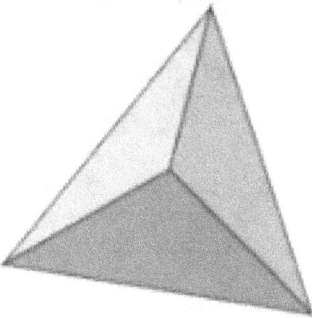

As the area is defined by $4L_p^2$, the linear component is sqrt($4L_p^2$), $2L_p$. This is also defined as the point of coalescence for that reason. In order to be unity, the term $4L_p^2/4L_p^2$ is being rendered as a trigonal bipyramid. Again, the nine sides of the structure is why I use the term, 'Yggdrasil.' Oddly, the nine-sided image has a real temporal meaning that we will get to later.

Simply, a Qubit is more complex than just a linear path. Furthermore, it is very touchy about how its internal structure of causal paths are connected. We do not have the option to just leave everything as eigenstates, for the obvious reasons that there would be no temporal order of the sort ($u_0 \pm u_{0+1}$), which are eigenvalues. Also, because a Qubit must give way to some eigenvalue or values, else has no substance nor tangibility.

We go back to:

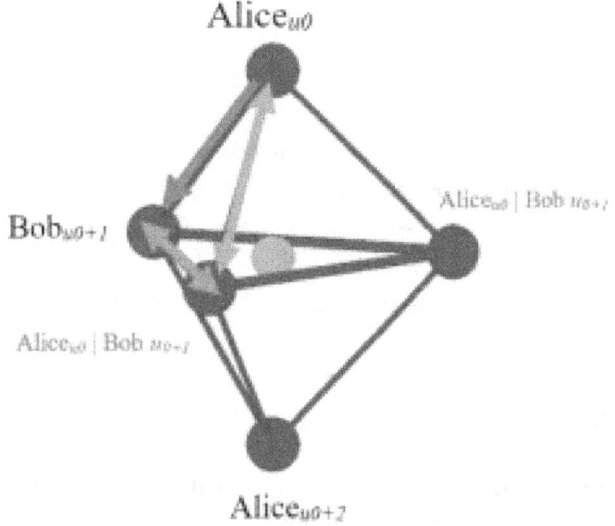

Alice is blue, Bob green, Alice | Bobs are purple.

**What do Alice and Bob see of each other when connected by >1-causal path?**

Note that Alice$_{u0}$ has the double headed arrow connecting to the Alice | Bob eigenstate. This, again, is because the eigenstate has no sequitur nor quantifiable temporal progression until such time it evolves to an eigenvalue. Also, Bob$_{u0+1}$ is connected via the double headed green arrow to the Alice | Bob eigenstate for the same reason. I left out the other connections for a bit of clarity.

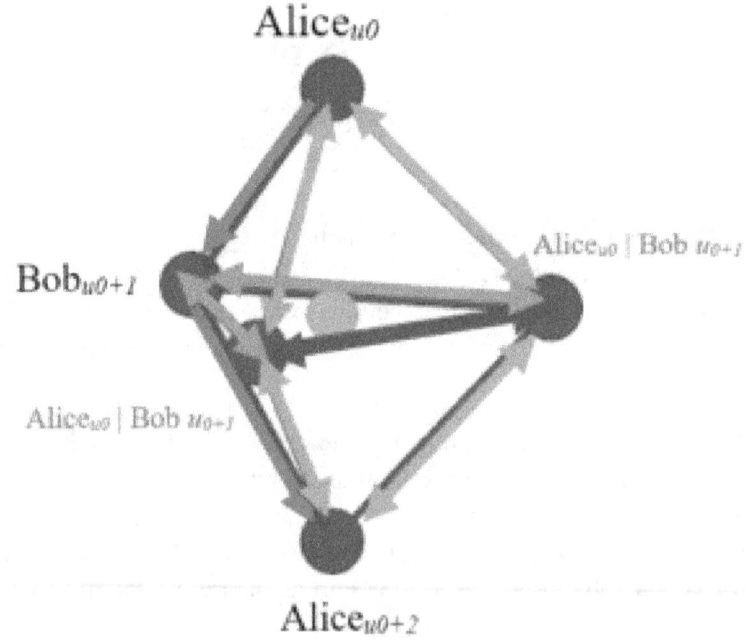

Alice is blue, Bob green, Alice | Bobs are purple.

- All of the Alice eigenvalues, which are Alice$_{u0}$ and Alice$_{u0+2}$ and the Bob eigenvalue Bob$_{u0+1}$ are connected in red and are unidirectional. *There are exactly two causal paths connecting eigenvalues*. Eigenvalues cannot be bidirectional as this violates causality. There are no causal violations in this.
- All of the [Alice$_{u0}$ | Bob$_{u0+1}$] eigenstates are connected to the Alice eigenvalues [top and bottom] in the light blue double headed arrows.
- The [Alice$_{u0}$ | Bob$_{u0+1}$] eigenstates are connected to the Bob eigenvalues in the green double headed arrows.
- The [Alice$_{u0}$ | Bob$_{u0+1}$] eigenstates at 3:00 and center are connected to each other by the purple double headed arrows.

Again, temporality is not regarded as bidirectional with respect to the [Alice$_{u0}$ | Bob$_{u0+1}$] eigenstates in this, merely that temporal progression has no sequitur description and as such no logical 'direction' with respect to temporal direction. Meaning, there is no sense of causality as time is not a thing that has emerged or evolved at this point, until such time something precipitates from an eigenstate or states to an eigenvalue or values.

## What do Alice and Bob see of each other when connected by >1-causal path?

One can regard time as 'frozen,' in some sense with respect to the $[Alice_{u0} | Bob_{u0+1}]$ eigenstates. This is the same description I rendered for the properties of the AdS Horizon Surface with respect to the conventional Ryu-Takayanagi Path $l$.

Prior to the evolution of Alice at $u_0$ to $u_{0+1}$ Bob:

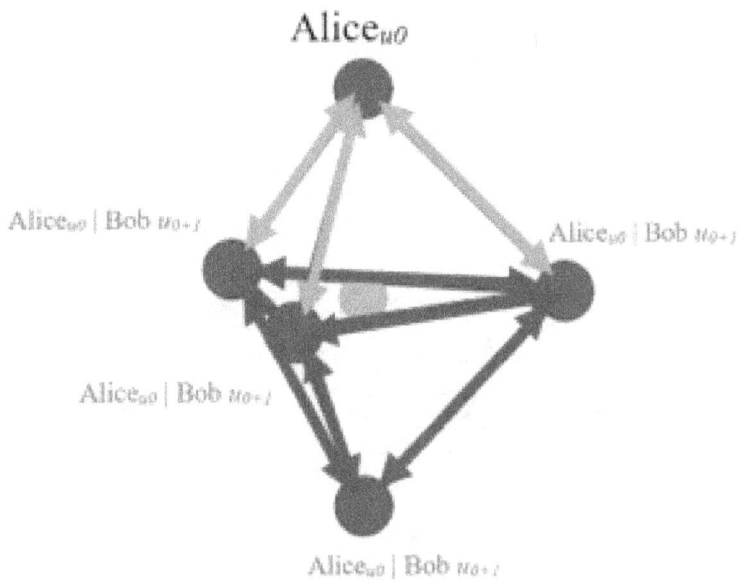

Alice is blue, Bob green, Alice | Bobs are purple.

- $Alice_{u0}$ is multiply connected to the eigenstates $[Alice_{u0} | Bob_{u0+1}]$ with double headed [light blue] arrows, time or temporal progression [causality] is non-sequitur from $u_0$ to all available eigenstate points.
- The eigenstates $[Alice_{u0} | Bob_{u0+1}]$ are all multiply connected *to each other* via the purple double headed arrows, again, temporal progression and causality are non-sequitur.
- The plane of symmetry in the middle is non-sequitur.
- The eigenstates form eigen-surfaces that are not drawn in for lack of clarity. There are three eigen-surfaces at the top connected to Alice, and three on the bottom half connected only to each other. The upper eigen-surface is relevant with respect to causal direction from Alice's perspective, but not from that of the eigenstates attached to Alice. The reason is that Alice can and thus will evolve temporally to one of these localities. It is noteworthy that the three planar points are arbitrary with respect to which evolves to a 'Bob' temporal endpoint. Meaning, vertical rotation of the bipyramid at this point is arbitrary. Horizontal rotation will be discussed as relevant later.
- There is an eigen-surface dividing the two halves of the bipyramid. At this point this eigen-surface cannot be regarded as relevant.

**What do Alice and Bob see of each other when connected by >1-causal path?**

Of all of the systems that can represent such, it appears that the Alice | Bob eigenstate systems are multiply connected to one another. The Alice eigenvalue is connected to more than one thing, non-selectively a 2-way temporal direction. However, again, temporal direction is non-sequitur as these are eigenstates, not eigenvalues.

Thus, Alice is only multiply connected in any of these scenarios to eigenstates, as is Bob, as below: If we add just one eigenvalue $Bob_{a0+1}$:

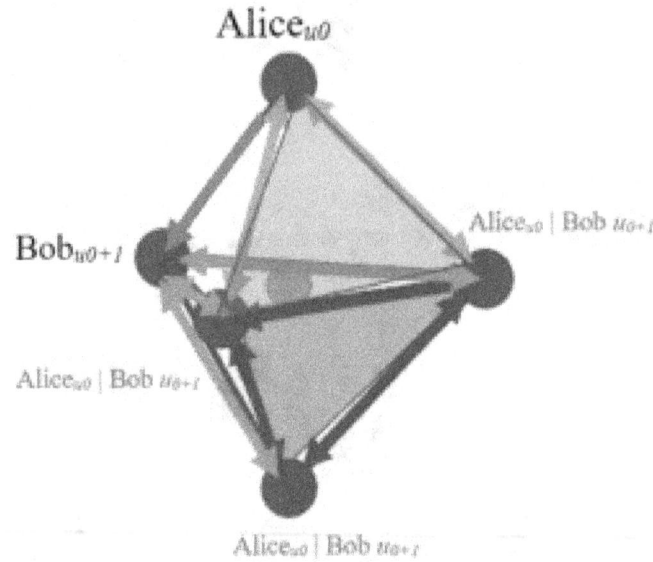

Alice is blue, Bob green, Alice | Bobs are purple.

Only $Alice_0$ and $Bob_{a0+1}$ are connected via exactly one unidirectional arrow [in red].

- $Alice_{a0}$ is multiply connected to 2-[$Alice_{a0}$ | $Bob_{a0+1}$] eigenstates and causal direction is non-sequitur.
- $Alice_{a0}$ is not planar and connected 3-times, once to the eigenvalue Bob and twice to eigenstates.
- $Bob_{a0+1}$ is causally connected once to $Alice_{a0}$.
- $Bob_{a0+1}$ is acausally connected [in green] to 3-[$Alice_{a0}$ | $Bob_{a0+1}$] eigenstates.
- $Bob_{a0+1}$ is planar and thus connected 4-times. Once to the eigenvalue Alice and three times to eigenstates.
- The 3-[$Alice_{a0}$ | $Bob_{a0+1}$] eigenstates are acausally connected to each other [in purple].
- The 2- planar [$Alice_{a0}$ | $Bob_{a0+1}$] are each connected 4-times.
- The 1-non-planar [$Alice_{a0}$ | $Bob_{a0+1}$] at the bottom is connected 3-times.
- There are effectively two surfaces of $ILp^2$ indicated in blue and purple that make up the eigenstate content. The other potential surfaces are mixed eigenstates. These surfaces are important because it is not just the causal paths but the causal surfaces that define the worldsheet, $A_\Omega$.

## What do Alice and Bob see of each other when connected by >1-causal path?

But neither Alice nor Bob can be said to be multiply connected to an identical system, e.g., an Alice or Bob eigenvalue. In fact, *it is the unidirectional temporal arrows that limit the feature of multiple connectivity*. In fact, everything that is multiply connected, is multiply connected to an Alice | Bob system, and all multiple connectivity is lacking a preferential sequence of events, e.g., causality.

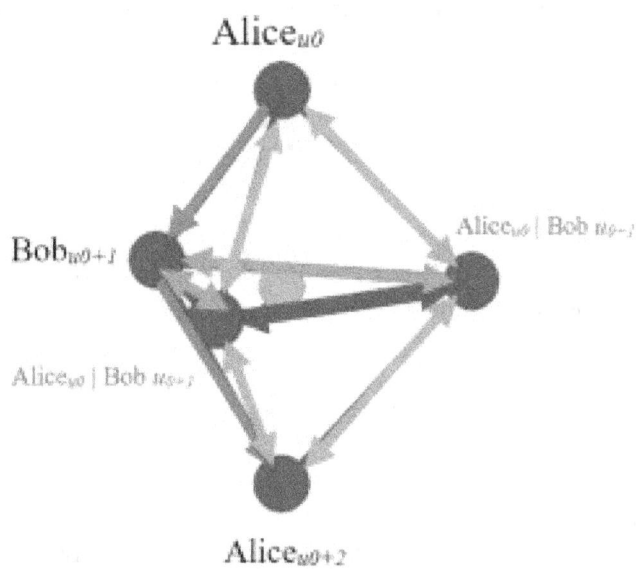

Alice is blue, Bob green, Alice | Bobs are purple.

- The temporal order, {Alice$_{u0}$, Bob$_{u0+1}$, Alice$_{u0+i}$} is 2{Lp, tp}. This satisfies the lower Bekenstein-Hawking Heisenberg boundary of 2 unitary Planck intervals.
- Alice-$u_0$ is connected once to Bob $u_{0+1}$.
- Bob $u_{0+1}$ is connected once to Alice-$u_0$ and once to Alice-$u_{0+2}$; they differ in identity. They are thus not connected simultaneously, as Alice$_{u0+2}$ is in the future from Alice$_{u0}$.
- Alice-$u_{0+2}$ is connected once to Bob $u_{0+1}$.
- All temporal endpoints are connected to both Alice | Bob eigenstates and lack temporal preference in direction.
- The eigenvalues Alice and Bob are multiply connected to Alice | Bobs eigenstates and lack temporal direction.
- The eigenvalues Alice and Bob are not thusly multiply connected in any other way, to an eigenvalue Alice nor Bob, *if and only if*, we preserve causality in the sense that time is unidirectional. *If* we connected Alice at $u_{0+2}$, for example, to the eigenstate Alice | Bob at the 3:00 position, the 3:00 position would have to be a Bob$_{n+3}$, and eigenvalue, not an eigenstate.

### What do Alice and Bob see of each other when connected by >1-causal path?

However, this would be one causal path of unitary Planck interval and this interval, $Bob_{a0+J}$ cannot connect to $Alice_{a0}$ as this, again, would be inverse causality, thus, the interval $Bob_{a0+J}$ cannot occur.

- Alice | Bob are connected to one another and lack any identifying feature of temporality to set them apart.
- Given the Alice | Bob systems are indifferentiable, each is connected any number of times to the other. That is, there is no differentiating Alice | Bob in the middle from being connected to Alice | Bob on the right, or to itself, and visa-versa. There is also no causal preference for a system in an eigenstate.

Leads to the conclusions:

- When an eigenvalue Alice or Bob is multiply connected it sees an *eigenstate*.
- When an eigenstate is multiply connected it sees an *eigenstate*.

*Multiple connectivity looks like some prior eigenstate.*

This seems intuitively correct as well as satisfying. At such time Alice and Bob become connected by greater than one causal path what they see of each other is their *prior eigenstate*. What *we see* of them is intuitively nothing and not relevant to this discussion. Meaning, Alice sees Alice | Bob systems when she is multiply connected to Bob, and vice versa.

The Lunar Landing Anomaly

## The Lunar Landing Anomaly

### Introduction, The Lunar Landing Anomaly

Here, we use the Lunar Landing example [Feynman, 1969]. The 1969 Lunar Landing was recorded, both sets events as they occurred on the moon aboard the Lunar Module and the recordings of events as they occurred on Earth at CAPCOM.

- In the CAPCOM recording [as events occurred on Earth] there is a pause and stutter in Armstrong's words, 'Houston…. Ahhh….. tranquility base here, the Eagle has landed.' There is no apparent cause for Armstrong's pause and stutter.
- In the lunar recording [as events occurred on the LEM], however, the reason for Armstrong's pause and stutter is clear, because CAPCOM cuts Armstrong off just as Armstrong begins to speak, with [unknown CAPCOM operator] 'Ok everybody. T-1, stand by for T-1.' Armstrong is cut off mid-word and pauses.

In most cases people at a high school level will try and explain the anomaly off as a simple issue of time delays of signaling between Earth and Moon. However, as we shall see momentarily this is quite impossible. There is a clear causality problem [again, Feynman, 1969] here that cannot be explained as a result of time delays and signaling issues. That is to say, convention would regard the details as otherwise an indication of causality violation, and summarily dismiss it. If this were the case then conventional $\Delta S$ could be applied regarding unipolar time and yield a $\Delta S$ result, is not the case. We could in fact explain the anomaly off with the two-trains scenario, however, in one frame of reference a 2-train wreck occurs, and in another frame of reference no train wreck occurs. There is no conventional $\Delta S$ resolution for systems where events occur from one frame of reference but do not occur in another.

### The Two Train Scenario

If the 'Lunar Landing Anomaly' [Feynman, 1969] were just an issue of signaling delays, the high school teacher approach at an argument, then a two-train scenario, which is the classic argument for two systems in motion, could resolve this problem. The two-train solution is in fact a simple conventional $\Delta S$ solution, as all systems must obey simple conservation laws. Every series of causal events in one system, that of Alice, must be equal to the sum of events in the second system, Bob.

It is noteworthy that Feynman pointed this 'anomaly' out just a month after the Lunar Landing took place, as he had opportunity to listen to the recordings and noted the issue.

## The Lunar Landing Anomaly

He, however, never managed to resolve it, only to point out that this problem was in his own words, 'not a paradox, but something we haven't considered as of yet.'

Meaning, in a classic 2-train scenario, trains Alice and Bob start off directly toward one another at arbitrary speed. For Bob [Armstrong, moon] *he crashes into train Alice* [CAPCOM, Earth], a train wreck occurs. This is the unknown CAPCOM operator cutting Armstrong off as he begins to speak, causing him to stutter and pause, as events were recorded on the LEM. Thus, in Bob's Preferential Frame of Reference there is a clear causal component for Armstrong's pause and stutter, namely he gets cut off mid-word, e.g., train crash into Alice.

However, for train Alice, *she does not crash into train Bob*, no train wreck occurs at all in Alice 'reality.' This is the pause and stutter in the Earth [CAPCOM] recording, 'Houston, …. Ahhh….. tranquility base here, the Eagle has landed. In Alice's Preferential Frame of Reference there is no causal component for Armstrong's pause and stutter. There is no train crash in Alice's reality.

Furthermore, both trains are recorded in each Preferential Frame of Reference as being two differing sets of events, one [Bob's lunar reality] in which there is a train wreck and another [Alice, CAPCOM] where no train wreck occurs at all.

This is not a paradox, merely Preferential Frame of Reference establishing two distinct and real realisms, one for Bob [does crash into Alice] and a different one for Alice [does not crash into Bob].

The second key factor in this text is that in the train wreck scenario, Alice and Bob are not entangled, so Preferential Frame of Reference applies and *what is true in one system may not be true in another system*, e.g., Bob crashes into Alice is true for Bob, but Alice does not crash into Bob; the crash is not true for Alice.

However, if Alice and Bob were entangled, *Preferential Frame of Reference does not apply*, and what is true in one system is true in all systems. Meaning, if Alice and Bob are photons with entangled spins in a Delayed Choice Quantum Eraser experiment, either a train wreck [interference pattern] occurs or no train wreck [particle-like behavior] occurs, is true in all systems.

What differentiates the entire entanglement scenario and descriptions is that Alice and Bob, because they are photons with entangled spins, traverse the conventional Ryu-Takayanagi Path $l$, which is a static, non-dynamic system. If they are not entangled, then they traverse information between them on a non-conventional Path $l'$ [prime], which is the result of the *dynamic vanishing rate of epsilon* [$\varepsilon$] and defined by the Zeno Effect, as precipitating eigenstates to eigenvalues according to the rate of observation. This will be discussed in great detail.

And for the record, eigenvalues, such as electrons in an orbital, cannot be entangled. Only systems in eigenstates can be entangled. Properties such as force cannot be entangled, along with a host of other things. I say this because I have seen a lot of aberrant statements

### The Lunar Landing Anomaly

suggesting, for example, that electrons in an orbital are entangled, merely because they have opposite spins. I have also seen aberrant statements to the effect that the momentum of the electron is entangled with its impinging photon of given energy, these are all incorrect notions.

There is no conventional ΔS solution to this 2-train scenario, as conventional ΔS can only deal with one causal path. Shannon Entropy *may possibly define* this condition because Shannon entropy is not limited to one causal path but applies only to systems where the number of causal paths are greater than one. That will be discussed at length in later sections. [see hyperlink]

- Conventional ΔS Entropy, which is limited to 1-causal path: {observable}
- Shannon Entropy, which is defined as >1-causal path, *always*: {unobservable}

Any *observable* system must be limited to 1-causal path, this is by convention. Shannon Entropy is defined as >1-causal path, always, and is thus *unobservable*. Again, that was explained in derived in prior [Bray] papers and texts and will be discussed and derived at length in later sections of this text.

When we look at the Delayed-Choice Quantum Eraser and the Delayed Choice Entanglement Swapping, we will look at the *dynamic conditions* of the AdS Horizon Surface value epsilon and see that the apparent causality violation of the Lunar Landing and the DCQE and DCES are the same phenomenon. Namely, that entangled systems traverse the conventional Ryu-Takayanagi Path $l$, which is non-dynamic, and that non-entangled systems traverse the non-conventional [derived later] Path $l'$ as the result of the dynamic vanishing rate of epsilon.

Again, this takes a lot of verbiage, so I have moved as much of the math as possible to the appendixes to facilitate readability.

This will be our 'two train scenario.' Simply, if the Lunar Landing Anomaly were a case of simple signaling delays, every set of events and the order in which they would occur in both *frames of reference* would be identical. I think this goes without saying. In typical causality we expect that what is true in one system is true in all systems. However, this is not the case, 'what is true in one system is true in all systems' is only the case for entangled systems, which is also obvious.

Interestingly, in 2-slit variant experiments, such as the Delayed Choice Quantum Eraser, it is the fact that what is true in one system is true in all systems that seems to be the dilemma. There is an expectation that events would differ between detectors, and it does not happen that way.

Obviously, if Alice and Bob are both systems in reference to one another and they are sending signals back and forth, conventional ΔS entropy could and would describe the magnitude of the content in terms of N-Qubits and the ordering of events as temporal markers. We can definitively say that if the Bob train crashes into the Alice train it is

## The Lunar Landing Anomaly

sufficient to say that the two-train collision is true in both frames of reference; Alice train would be expected to crash into Bob train. The Lunar Landing does not occur in this fashion, it is an example of Preferential Frame of Reference as established purely by distance and no other factor.

Meaning, if Bob train crashes into Alice, then Alice train crashes into Bob. This is definitively would be true in each frame of reference and definitively true *in both frames of reference*. However, this is not what happened.

The Earth recording clearly indicates that there is no apparent cause for Armstrong's pause and stutter because CAPCOM does not cut Armstrong off as he begins to speak. There is nearly a second of dead air that has no explanation. In fact, there have been some inquiries regarding this pause and stutter, if only the CAPCOM recording is available. There is actually a documentary where the interviewer asks Armstrong what the pause and stutter was about, but Armstrong cannot even remember in that much detail. Meaning, without the lunar recording, history has no record of such events even taking place as they occurred, and in that causal series of events.

- There is no train wreck in the CAPCOM [Earth] frame of reference. There is no causal source for Armstrong's pause and stutter.
- However, there is a train wreck in the Lunar frame of reference. CAPCOM clearly cuts Armstrong off just as he begins speaking, resulting in his pause and stutter.

Also, keep in mind, again, that I am using the term, Preferential Frame of Reference over that of just frame of reference. The reason is that there are simply too many conflicting descriptions and lack of clarity regarding what 'frame of reference' actually means.

Keep in mind that scenarios in relativistic systems invariably are different between frames of reference. The Twin Paradox is a simple Lorentzian example of this. There is a description of systems in Lorentzian conditions that differ in *the ordering of events*. However, this Preferential Frame of Reference issue is a result of Lorentzian transform, this paper will focus on Schwarzschild metrics.

Preferential Frame of Reference can be the result of:

- A Lorentzian system
- A Schwarzschild system
- Distance and no other factor.

Non-preferential Frame of Reference only occurs in entangled systems. This will be discussed in detail in later sections of this text.

All of these things were described in detail with respect to the conditions and demands of the AdS Horizon Surface in:

**The Lunar Landing Anomaly**

1. Bray. Artificial Alteration of Spatial Geometry via the Quantum Zeno Effect. December 2020 DOI: 10.13140/RG.2.2.13527.29601, DOI: 10.13140/RG.2.2.13527.29601/1 [updated file]
2. Bray. Faster Than Light Propulsion via the Zeno Effect. March 2021 DOI: 10.13140/RG.2.2.28575.48809
3. Bray. A Zeno Dynamic Delayed Choice Quantum Eraser. April 2021 DOI: 10.13140/RG.2.2.19726.08003
4. Bray. Altering the Rate of Time Evolution on the AdS Horizon Surface. April 2021 DOI: 10.13140/RG.2.2.35703.75682

The order and timing of events differ, between the moon and Earth systems, although the quantity of N-Qubits is the same. The quantity of N-Qubits in the two frames of reference are the same regarding the Information they contain, although the total duration of the sequence of events differs in temporal length by 1.2 seconds, the time distance to the moon. That is, the two frames of reference differ by 1.2 seconds but the quantity of Information in the form of N-Qubits is exactly the same.

Information cannot be created or destroyed, *nor borrowed*.

That is, the Heisenberg Uncertainty Principle [HUP] cannot 'borrow' Information, as there is no sequitur 'negative information' to balance such a thing. Mass-energy and so on can be 'borrowed,' provided of course that conventional $\Delta S$ entropy conservation is obeyed. However, Information is too rudimentary and cannot be 'borrowed.'

We will look at why this occurs with respect to the AdS Horizon Surface.

There are numerous scenarios where Preferential Frame of reference applies in general relativistic [Schwarzschild metrics] type systems. When a system is coupled to its source of mass*energy in a General Relativistic system the conventional rules apply. However, when a system *is not coupled to its source of mass*energy*, such as in the case of Gravitational Waves, Preferential Frame of Reference does not apply. This has been described in prior [Bray] papers as the scaffold of the Zeno Effect.

The description requires a lot of verbiage in describing, the subsequent derivations are relatively straight forward. The seeming excessive verbiage is dealing with *pure distance as Preferential Frame of Reference*, which is not regarded as such by convention, and a host of several issues that are not discussed by convention or otherwise regarded in some other hypothetical fashion. A perfect example of pure distance as Preferential Frame of Reference is in fact the Delayed Choice Quantum Eraser. It is the light distance to the secondary detectors that establishes pure distance as Preferential Frame of Reference. This Preferential Frame of Reference would not apply if the systems are entangled, which is the dilemma. To date there has been no satisfying description why 'what is true in one system is true in all systems,' regarded as the history of the 2-slit. We expect that what is true at detector-1 [e.g., interference pattern] *would not be true* at detector-zero, because detector-

## The Lunar Landing Anomaly

zero is a different [lesser] distance from the source [BBO crystal]. Thus, we get into reverse causality to the extent that it becomes ontological.

When Preferential Frame of Reference applies to a system we see temporal effects, among other things. This can seem like causality violation. However, as we look at this it becomes clear that there is no difference between the Delayed Choice Quantum Eraser and the Lunar Landing Anomaly. The notion that such issues are limited to quantum [Planck] scales is erroneous. The experiments are defined as the observed system coupled to the observing system, which is typically on a benchtop scale and larger. For example, a nuclear explosion does not occur on an atomic scale, but requires a kilogram of fissionable mass, which is a macroscopic scale.

I will use the term, causality violation, flamboyantly and not entirely correctly. In nature, a system cannot actually be said to bear causality violation *if it is a natural system*. That would infer nature defying itself. The term, causality violation is simply a choice of words that describes our pondery of the natural system's behavior. There are no actual violations in natural systems, just lack of clarity in describing their behaviors.

The Preferential Frame of Reference as a function of pure distance will be related to Paths L vs $l$ [Ryu-Takayanagi] of the AdS Horizon, however, it is clear that Path $l$ cannot be rigid and fixed as regarded by convention but must be fluid, because the Horizon value epsilon [ε] is changing in real time. The scope and rate of the change in epsilon, typically only regarded in cosmological time scales, is quite immediate as the definition of $c \equiv 1Lp/1tp$ [Planck length Lp, Planck time, tp]; the scope being 1Lp at a rate of 1tp, the Horizon is changing in scope and rate at the speed of light.

As a result of the Horizon Surface value ε changing at the speed of light, a 'Gap' is left in the Path $l$/Path L system that is equal in scope to the distance, regarded as figuratively 'below' the Horizon Surface for lack of a better term, to the normal light distance travelled, Path $l$. [I use the term 'Gap' so as not to confuse with the gravitational constant, G] Path $l$ in the visual metaphor below, is the normal light distance between points, regarded as the rigid red arc, conventional Ryu-Takayanagi Path $l$. While Information is traveling the normal light distance between points, the value epsilon is diminishing in scope and rate in real time at the speed of light. As a result, the rigid red arc no longer suffices to connect the endpoints of Path $l$, but appears as visualized below as the yellow arc, extended off to greater distance, as measured by ⱨn{Lp, tp} which have decreased in size, but not in number. The *number*, quantity of N-Qubits of a system cannot change. Thus, I have assigned a police pseudo-operator, ⱨ, to identify those quantities that must remain fixed.

Thus, ⱨn{Lp, tp} remains fixed with respect to quantity, but the *magnitude of* both the Planck length [Lp] and Planck interval of time [tp] have decreased, as a result of the real-time evolution of the Horizon Surface [Ryu-Takayanagi] value ε. In prior papers and texts, I derived epsilon as equated to the unitary Planck intervals of time [tp] and length [Lp]. The summary derivation is in one of the appendixes of this text.

**The Lunar Landing Anomaly**

Thus, the red arc describes the rigid Path $l$ the Information travels, which is the normal light distance between points A and B, as describing the 'Normal' distance of Information transfer. However, the yellow arc is visualizing the Information transfer in such case that the Horizon value $\varepsilon$ were fixed, unchanging. This may seem bass-ackwards, however there is a rationale for visualizing it in this way that will be clearer as we proceed.

This Lunar Landing analysis may all seem like detail overkill, however, to the best of my knowledge no one has attempted a detailed analysis of this set of recordings. Keep in mind that the notion that this is not a signal timing delay issue but one of questionable causality is not novel. In fact, I think it was Feynman who first brought the subject up in lecture not months later in 1969. And if I recall I think there may be a note or footnote on this in one of the very early Feynman grad texts, since they were based on his lectures.

As we get into this series of events, hopefully it will become clear that there is no way to reverse engineer this argument back to signal delays, there are different events in two preferential frames of reference in one universe that, if they were entangled systems, one would shrug your shoulders and think, well, that is not atypical.

The key feature here is to determine that entangled systems actually are the only systems whose information pseudo-traverses the conventional Ryu-Takayanagi Path $l$, and non-entangled systems traverse a different pseudo-path, just regarded as Path $l'$ *[prime]*.

There is quite a bit of information in these descriptions that has to be compare point-by-point in a rather Permutation type of way. Thus, it may seem repetitive in one sense, but each go around has a completely different endpoint as the description and the goal.

Note that because of the human penchant to read papers on pads and even cell phones, sometimes I spell out $l'$ [l-prime] in that manner. Also, because of the human penchant to read papers on pads and even cell phones, indexing can be difficult, if at all possible, thus I tend to repeat images and equations rather than refer or reference back to them, as pads and phones have no such indexing. This is all out of the necessity to get with the times, where most people are reding scientific papers in the most passive modes as possible. The idea that a paper is formatted and such as with the hard print journals is a thing of the past with a distributorship that is dwindling to very nearly zero at this time.

The summary of math derivations that describe all of these are again, moved to the Appendixes of this text. I chose to do this because of the modern human penchant to read papers on pads and even phones. The math is nearly impossible to read under those conditions and makes the readability of the entire text worse than it already is.

## The Lunar Landing Anomaly

### LIGO and GRACE, Coupling to Mass-energy

The key concept in all of these discussions is Preferential Frame of Reference as established *purely by distance and no other factor*. Note that I use the term *preferential frame of reference*. This is because the latent term, frame of reference, has become so clouded with so many hypotheses that it is impossible to discuss the topic in any clarity, as no two descriptions agree completely.

This discussion is in several parts that are interdependent and a bit lengthy. The three portions of this text need to be read and then connected by the reader at the end. I have tried to move as much of the math derivations to the appendixes as possible to improve readability. This is primarily the modern human penchant to read papers on pads and even cell phones, the math is not readable on those scales. There are internal links to the appendixes that go directly to the bookmarked derivations being discussed. Thus, the first major section of this text is verbal descriptions, and the appendixes carry the math and derivations. Most of the derivations are at length in prior papers and summarized here in the appendixes.

Preferential Frame of Reference *always applies* in Special Relativistic [Lorentzian] systems. In prior papers I described how Preferential Frame of Reference only applies to General Relativistic [Schwarzschild] systems when the observed system is *coupled to* its local source of mass-energy. Gravity Waves are an example of the observed system of spatial geometry not being coupled to its local source of mass-energy. Preferential Frame of Reference *does not apply* for Gravity Waves.

Exactly what Preferential Frame of Reference is and what it has to do with causality is the key concept of this text and requires a lot of explaining because it is novel science, not novel technology.

For example, LIGO is a classic, albeit large interferometer that can detect its own Schwarzschild change in the geometry of spacetime. This is because Preferential Frame of Reference does not apply for gravity waves. To clarify that unambiguously, LIGO cannot detect its motion around the sun, e.g., a classic Michelson-Morley experiment, via the same pulsed metering of photon wavelength between its mirrors. This Lorentzian change does in fact bring about a change in path length of the interferometer. However, this change in path length cannot be detected by the interferometer itself [LIGO].

For example, the notion that, as with the Gravitational Recovery And Climate Experiment [GRACE], is based on physical changes in the distance between the satellites that result from gravimetric changes on or near Earth's surface and oceans, does not imply that the changes in gravimetrics does not result from a Schwarzschild transform. There is no change in spatial geometry, that results from mass or mass-energy, that is not a Schwarzschild transform.

### The Lunar Landing Anomaly

There is no such thing as Newtonian physics. GRACE, like LIGO, cannot detect any change of distance between the satellites as a result of its [2-satellite system] motion around the sun.

We then describe GRACE as detecting a secondary effect, much like a ball does rolling downhill. The ball is following spatial geometry. The notion that Gravitation is a 'Force' is obsolete thinking.

However, again, GRACE cannot detect any change in distance between the satellites that results from a Lorentzian change in path length, as the satellites, too, are in motion around the sun with the Earth. This key factor should bring to light the notion that as system detecting *its own* Schwarzschild transform does in fact occur under certain conditions. That set of conditions is when the spatial geometry in question is not coupled to its source of mass-energy, such as gravity waves.

Neither GRACE nor LIGO can detect any physical change that results from motion, a Lorentzian transformation. So, we put GRACE in the 'ball rolling downhill category;' it can detect the changing spatial geometry as it sinks deeper into Earth's gravity well but cannot detect its change of state as a result of its Lorentzian motion down the hill.

The obvious question is, 'is the spatial geometry that GRACE is detecting, and quantifying coupled to its local source of mass-energy?' That is a lengthy description that appears throughout later sections of the text. Again, when the local spatial geometry is not coupled to its local source of mass-energy, Preferential Frame of Reference does not apply, and the local spatial geometry can be detected, e.g., its own Schwarzschild transform. A gravity wave is the most obvious example of this.

LIGO, however, is a large interferometer that sends one laser pulse down an arm length followed by a second pulse and measures the difference between the two. This is not a static quantity, such as GRACE or a ball, but a dynamic change that only results purely as a consequence of a change in spatial geometry between pulse one and pulse two. There is no mass-energy associated with this change in spatial geometry. The source can be a billion lightyears away and as such a billion years ago. There can be no 'coupling' to such a distant phenomenon, as this demands a 2-way transfer of Information across vast distances and times.

LIGO, again, cannot detect its change of state using this exact same procedure as a result of its motion around the sun, a classic Michelson-Morley result. Thus, LIGO *can detect* its own Schwarzschild transform by whatever means but not its own Lorentzian transform under any conditions. That is the key factor.

- When a system is not coupled to its [local] source of mass-energy, Preferential Frame of Reference *does not apply* and the system can detect its own change of state under a Schwarzschild transformation, be it directly or indirectly.
    - This can only apply to General Relativistic systems, because,
    - there is no condition in which a Lorentzian system is not coupled to its local source of mass-energy, that description is non-sequitur.

## The Lunar Landing Anomaly

Furthermore, there is no such thing as an actual Newtonian change in spacetime or the things in it. If two objects change apparent time and/or distance as a result of spatial geometry, that is a gravimetric change in spatial geometry that results from a Schwarzschild transform. If the change were the result of Lorentzian transform, LIGO could not detect it, such as our inability for LIGO to detect its own motion around the sun, albeit we agree that a sufficient change in state does occur under Lorentzian transform. If we regard all of these systems as artifact of observation, then no real change in clock time would occur. If the argument is that the time transformation is real, but the length transform is artifact of observation, then that has to be revisited because that is a classic example of violation of Preservation of Information Conservation. Convention seems to regard the *quantity of* unitary intervals of time as changing, which is a violation of Preservation of Information Conservation. If, however, this quantity was changing, and the length transform regarded as artifact, this would be a second violation of Preservation of Information Conservation, in just one observation.

Nonetheless, if the length transformation is an artifact of observation, then such artifact would be detectable, because by definition it is an observation, provided the system is a Schwarzschild, not Lorentzian metric.

## The Lunar Landing Anomaly

### Overview of Lunar Landing Anomaly

*The same series of events in the two Preferential Frames of Reference differ in magnitude, as defined by the scope of Path L [Ryu-Takayanagi] and Path l:*

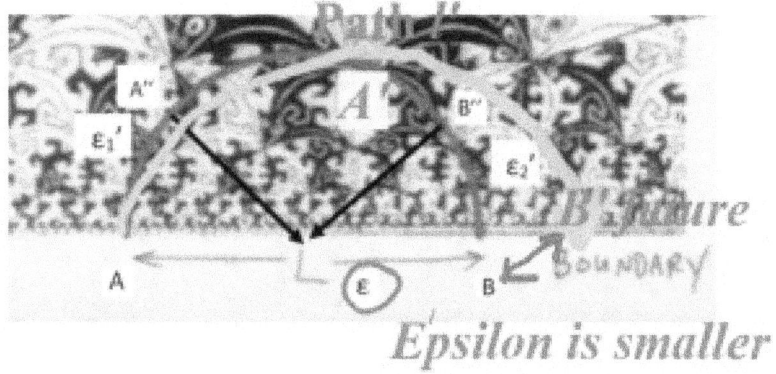

$$l = \frac{e}{3} Ln \frac{L}{\epsilon}$$

$$G_p = \frac{e}{3} Ln \frac{L}{\epsilon}$$

$$G_p = l$$

Again, I use the term 'Gap' so as not to confuse with the gravitational constant, $G$. The 'Gap' is designated above as $G_p$. Path $l'$ [prime, in red] is defined by the Information vector at such time it reaches the Horizon Surface at point B', rather than B. B, figuratively, does not exist; it is the locality of B if and only if the value epsilon were static, rather than dynamic. This is the conventional Ryu-Takayanagi Path $l$, it is static, non-dynamic.

Because epsilon is changing in real time as the Information is en-route this results in the double-headed red arrow, the 'Gap.' [designated as Gp]. And again, the magnitude of the Gap is equal to the normal light distance between points A and B as indicated by the red arc. In our Lunar Landing example, it is this Gap, from B to B' that is equal in magnitude, always, to the normal light distance between points, which is Path $l$. $G_p$ is always equal in magnitude because the vanishing rate at c is equal to the normal light distance between points.

## The Lunar Landing Anomaly

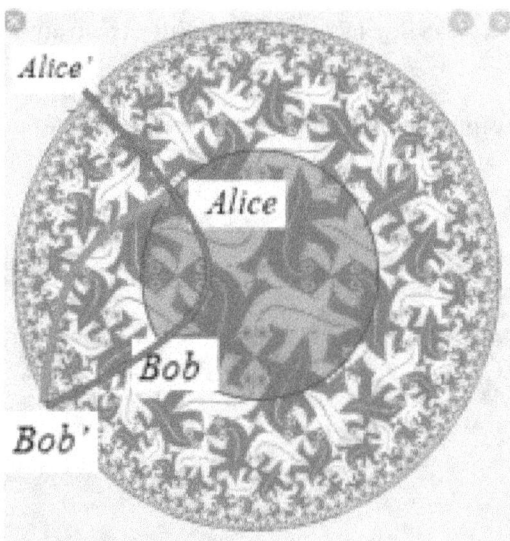

This is discussed in greater detail in another section of this text. At time zero Alice sends a signal to Bob. The path that the Information takes cannot be the conventional description, which is the red arrow connecting two *current points, both at time-zero*, on the Horizon Surface. Because the horizon evolves, regardless of scope and rate, the above visualization of the green arrow represents the true path of Information from Alice to Bob.

It is this difference between B and B' that is the two-reality scenario of a trains wreck of words occurring in Armstrong's frame of reference but not in the CAPCOM frame of reference. This will be described shortly.

Keeping in mind that the red arc is passing through Conformal Values that *are always larger in scope* than the Horizon Value, $\varepsilon$. The Horizon Value $\varepsilon$ is always the smallest Conformation of $c \equiv 1 Lp/ltp$ in the AdS System because the value epsilon is vanishing, regardless of scope and rate. However, that value is changing in real time at a scope and rate certainly more tangible than cosmological time scales.

And as we shall see, causal series of events in the CAPCOM Preferential Frame of Reference lack a causal component altogether; where the Moon Preferential Frame of Reference has and obeys proper causality. In terms of our 2-train scenario, we keep it such that a train wreck occurs in the Lunar frame of reference *but does not occur in the CAPCOM frame of reference*. The 'train wreck' is figuratively Armstrong's pause and stutter as CAPCOM cuts him off mid-word. This event [CAPCOM cutting in as he speaks] does not occur in the CAPCOM frame of reference on Earth.

## The Lunar Landing Anomaly

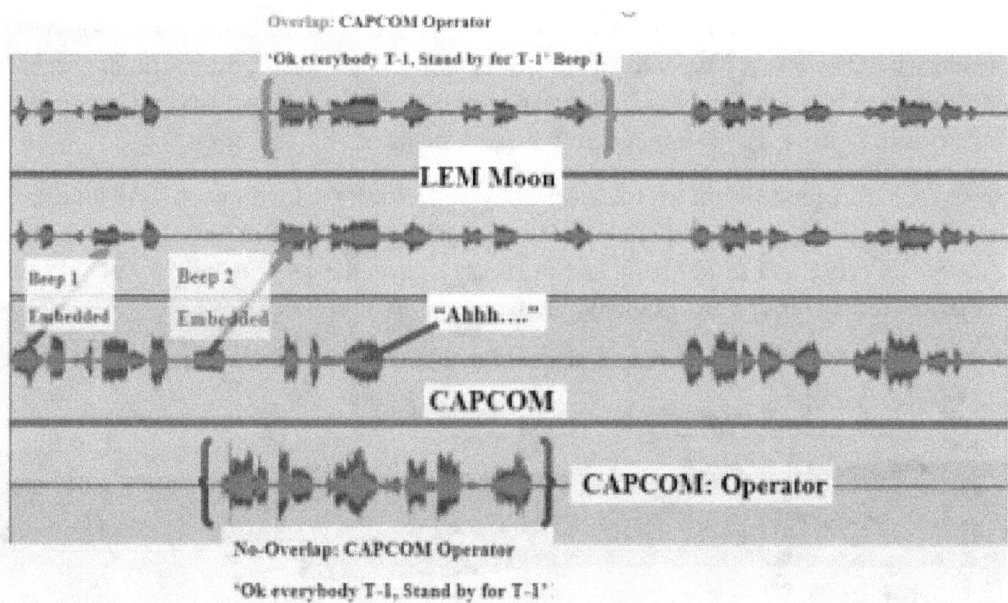

Focusing in on:

Transcribed from NASA archives: transcript of lunar landing at NASA.org: http://www.hq.nasa.gov/alsj/a11/a11.landing.html#1024540

**As recorded on Earth at CAPCOM: [upper portion of above image]**

CAPCOM: We copy you down Eagle….(beep$_1$)..

Armstrong: Houston……..ahhh……Tranquility base here, the Eagle has landed…

Total length **8.7 seconds** [amount of Information]

The upper waveform is the Earth [Alice, CAPCOM] frame of reference and recording of events. There is a large gap of about 1.2 seconds that has no explanation. In this recording, as recorded on Earth, there is no causal component for Armstrong's pause and stutter, there

### The Lunar Landing Anomaly

is a 'beep$_2$' embedded in his 'ahhhh' that is not in the written transcription, discussed momentarily. There is 8.7 seconds of Information in the recording. Keep the 8.7 seconds of Information in mind. If the lunar recording did not exist, we would never have any idea *why* Armstrong paused and stuttered at that exact moment.

However, as recorded aboard the LEM on the Moon [Bob], the lower waveform image, the series of events occur in a different order, to such extent that the causal component of Armstrong's pause and stutter is in fact made clear; it is the second 'beep$_2$' as CAPCOM cuts in just as he is beginning to speak, interrupting him mid-word:

[Figure: Waveform comparison showing Overlap: CAPCOM Operator. Top waveform labeled Charles Duke CAPCOM and LEM Moon with 'Ok everybody T-1, Stand by for T-1'. Bottom waveform showing Beep 1, Beep 2, "Ahhh....", Armstrong Pauses, CAPCOM.]

Transcribed from NASA archives: transcript of lunar landing at NASA.org: http://www.hq.nasa.gov/alsj/a11/a11.landing.html#1023820

**As recorded on the LEM on the Moon: [lower portion of the above image]**

CAPCOM: We copy you do.. [Armstrong then begins speaking, cutting CAPCOM off]
Armstrong: Houston.. [CAPCOM now cuts in, cutting Armstrong off]
CAPCOM:..wn Eagle...
Armstrong: ..ah......
CAPCOM:... (beep$_2$)....
Armstrong: ....Tranquility base here, the Eagle has landed....

The total length of this broadcast is **7.5 seconds** [amount of Information]

Armstrong's pause and stutter has a clear causal source, CAPCOM cutting in as he begins to speak, in the Lunar recording. This is our train wreck. The train wreck occurs in Bob's Preferential Frame of Reference, CAPCOM cutting Armstrong off as he begins to speak. However, in CAPCOM's [Alice] Preferential Frame of Reference, no train wreck occurs, there is no overlap of signal at all, but are in fact separated by about a second of dead air.

- Lunar Preferential Frame of Reference: Bob, Armstrong. CAPCOM cuts Armstrong off as he begins to speak, train Bob crashes into train Alice.

**The Lunar Landing Anomaly**

- <u>CAPCOM-Earth Preferential Frame of Reference:</u> Alice, CAPCOM. CAPCOM *does not* cut Armstrong off as he begins to speak, train Alice *does not crash into train Bob*.

Again, in an argument based on simple signal delays, conventional $\Delta S$ entropy would describe the difference between the two systems. However, conventional $\Delta S$ only allows for 1-causal path and the sum of events between systems must be in the exact same causal ordering and be the exact same set of events. Keep note that there are Lorentzian as well as Schwarzschild systems where the same seeming causality violations occur, discussed in detail later.

Without the Lunar recording, we would never know that the systems differ in the ordering of events. Here, however, we have separate recordings in each frame of reference and the ordering of events differs.

For a conventional [not Shannon] $\Delta S$ entropy, the two systems have to be identical. This is in fact the dilemma with the Delayed Choice Quantum Eraser and various 2-slit results. Again, in the DCQE *we expect* a different result between detector-zero and detectors one through four. However, the systems *are identical*, because, as the postulate stands, Preferential Frame of Reference *does not apply to entangled systems*. This will be described at length later.

- Preferential Frame of Reference does not apply to entangled systems.

This will evolve to:

- When Preferential Frame of Reference applies, what is true in one system [train wreck] may not be true in another [no train wreck].
- When Preferential Frame of Reference *does not apply*, what is true in one system is true in all systems. [e.g., an interference pattern appears at all detectors in the DCQE].

Preferential Frame of Reference may be established by:

- a Schwarzschild metric,
- a Lorentzian metric,
- *or purely by distance.*

When a Preferential Frame of Reference is established purely by distance, this is the result of the conditions that differ between epsilon [ε] at the time a signal [Information] is sent [Alice] via Path *l* [Ryu-Takayanagi] and the condition of epsilon at the time the signal reaches the AdS Horizon Surface [Bob].

Keep in mind that by ordering of events we are describing that a distinct set of <u>N-Qubits occurs in a different order in any one frame of reference but differs *between* frames of reference.</u> This is true to the extent that Armstrong has no reason for his pause and stutter

### The Lunar Landing Anomaly

in the Earth based recording [no train wreck occurs] but does have a reason for his pause and stutter in the Lunar frame of reference [train wreck does occur]. Metaphorically, again, train Bob crashes into train Alice, but train Alice does not crash into train Bob. Bob will be Armstrong on the moon and Alice is at CAPCOM on Earth.

The Earth based recording has 8.7 seconds of Information in it, the Moon based recording has 7.5 seconds of Information in it. Information Conservation must be obeyed. Because:

$$ℵ7.5s \neq ℵ8.7s$$

What has happened here is that the AdS Horizon Surface has expanded at a scope and rate defined by $c \equiv 1L_p/1t_p$ [one unit of Planck length divided by one unit of Planck time]:

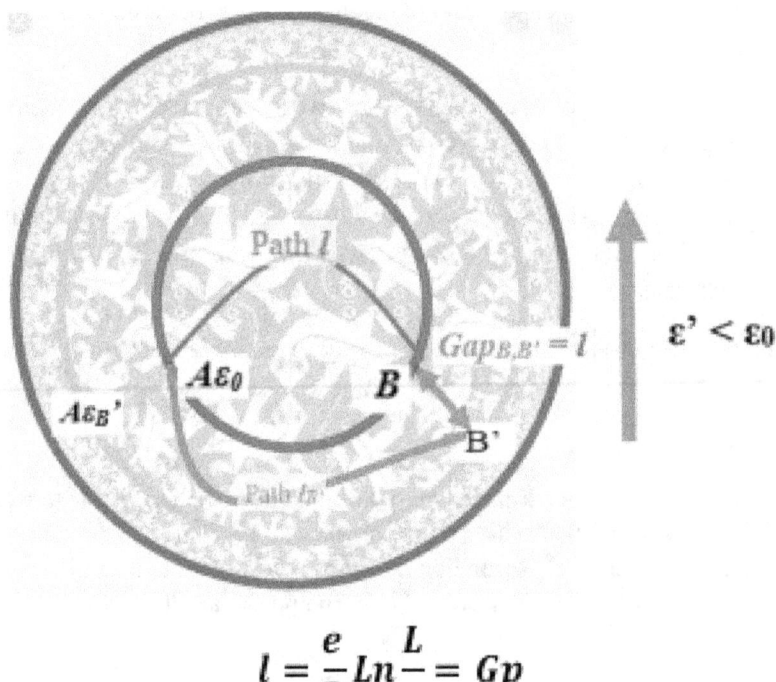

$$l = \frac{e}{3} Ln \frac{L}{\epsilon} = Gp$$

The term, Gp is the double-headed red arrow and is always equal in magnitude to the normal light distance between points, Path $l$, because the epsilon vanishing scope and rate is set by $c = 1L_p/1t_p$. Epsilon at point $B'$ is lesser in magnitude than epsilon at Alice point $A\epsilon_0$ [$\epsilon_{B'} < \epsilon_{A0}$]. The conventional Ryu-Takayanagi Path $l$ stretches from point $A\epsilon_0$ to $B$ and thus must be regarded as static, because both Alice at $A\epsilon_0$ and Bob at $B$ are at time-zero.

As will be discussed, only entangled systems traverse the conventional Ryu-Takayanagi Path $l$.

Alice is at point A [CAPCOM], which figuratively does not exist, because it is in the past, provided the system obeys Preferential Frame of Reference.

## The Lunar Landing Anomaly

Point A exists at the time of sending the signal but has diminished in scope while the Information transfer takes place across the normal light distance between points. Again, conventionally epsilon is regarded on cosmological time scales, however, the definition $c \equiv L_p/t_p$ demands that the value epsilon is vanishing at a scope and rate equal to the speed of light. The actual derivations for this are lengthy, appeared in prior papers, and merely summarized in the appendixes of this text. Briefly, the dynamic nature of epsilon is equal in magnitude, always, to the normal light distance between points, regarded as the conventional Ryu-Takayanagi Path $l$. More on that later.

Bob was at point B [Armstrong, moon]. What does exist is Bob, who is now at point B'. The double red arrow is exactly 1.2 seconds, equal to the normal light distance between points. It looks smaller because the Conformal Scale of epsilon [$\varepsilon$] has diminished in magnitude and rate according to $c \equiv 1L_p/1t_p$. The double red arrow and the normal light distance from A to B' are both 1.2 seconds. Ryu-Takayanagi Path $L$ would represent the normal light distance between points [according to the AdS transform $l=e/3 \ln(L/\varepsilon)$] if and only if the AdS Horizon Surface was static. However, the AdS Horizon Surface by convention is dynamic, forever vanishing toward zero. The only novel notion here is that the scope and rate of the vanishing value $\varepsilon$ is a real-time dynamic defined by the quantized speed of light. [Again, derivation in appendix]. In order for this to not be the case quantization would have to be violated. Path $L$ could be equal to $l$ if the system were both flat and static, but by convention, Path $L$ is superluminal.

The demand of the geometry is that the rigid red arc of convention *is fluid*, taking the form of the green arrow.

The Gap, the double red arrow, is that 1.2 light-second difference that causes the [by convention] causal violation of the same quantity of N-Qubits in both systems [preserving Information Conservation] but ordered different. This is the apparent point $B$; if the Horizon was static there would be no difference in the recordings between Earth and Moon because Armstrong at point $B$ and CAPCOM at point $A\varepsilon_0$ would be the normal light distance between points. What is true in one system would be true in both systems. This would require the conventional Ryu-Takayanagi Path $l$, regarded as essentially static, vanishing only on cosmological scales.

That would be the intuitively correct way of thinking. However, as the recording clearly indicate, the intuitive response is not what happened. In addition, our two-train scenario clearly shows that the series of events cannot be explained by simple signal delays, as the two-causal series of events would then have to be the same resulting in the recordings both indicating CAPCOM cutting Armstrong off as he begins to speak.

- Again, proper causality exists on the moon, as Armstrong's pause and stutter have an obvious rationale of being cut off by CAPCOM as he begins to speak.
- Proper causality is not demonstrated in the CAPCOM frame of reference, there is no cause for Armstrong's pause and stutter.

### The Lunar Landing Anomaly

However, what qualifies as a causal violation by convention is not a true causal violation. For example, we could say that the Speeding Twin's less-advanced age in comparison to his [older] twin on Earth is a causal violation, as one has experienced an entire lifetime of events in a timeframe impossible for the other, which is the dilemma. However, this is just nature. *All systems are in motion* and thus are subject to such seeming paradoxes and violations.

*However*, if we were to fix Alice and Bob at opposite ends of a rigid rod, such as the Earth and moon, for the most part, velocity is not the issue, when we get to macroscopic scales we can begin to detect and measure seeming anomalies, such as Bob crashing into Alice who does not crash into Bob. This is also a key, and it describes the Delayed Choice Quantum Eraser, it is on *macroscopic scales, which are not entangled* that we begin seeing these two systems demonstrate seeming causal violations.

- When Preferential Frame of Reference as purely a result of distance exists, we can see the two systems such as Alice and Bob have two distinct and true 'realities.' What is true in one system may not be true in another system. [e.g., train crash vs no train crash].
- When Preferential Frame of Reference *does not apply*, such as postulated for entangled systems, the two realities for Alice and Bob are then anomalously identical. What is true for Alice [detector-zero DCQE] is also true for Bob [detectors 1, 2, 3, 4].
- Again, Preferential Frame of Reference can be established by a Schwarzschild metric, a Lorentzian metric, or purely by distance.

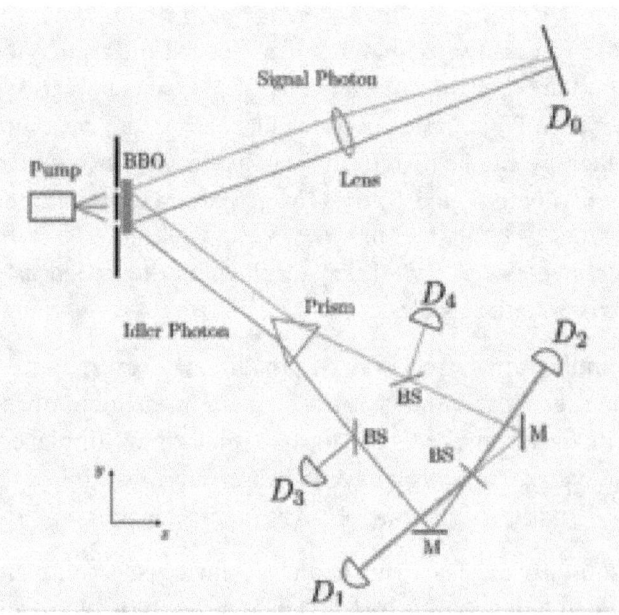

I like that diagram of the DCQE because it has good symmetry. When we add the Delayed Choice Quantum Eraser into the description this will become clearer. That is for later. In brief, the light distance between points in the DCQE experiment is treated the same as the

### The Lunar Landing Anomaly

normal light distance between points in this description of the AdS Horizon Surface, which is Ryu-Takayanagi Path $l$. However, as noted in the diagram on the prior page, non-entangled systems do not traverse conventional Ryu-Takayanagi Path $l$, but take the form of the large green arrow, which is thus designated Path $l'$.

- In the Delayed Choice Quantum Eraser, [and Delayed Choice Entanglement Swapping] because the system is entangled, Preferential Frame of Reference does not apply and what is true in one system must be true in all systems.
- Thus, in the DCQE, Information is traversing the conventional Ryu-Takayanagi Path $l$
  - Ryu-Takayanagi Path $l$ is a non-dynamic, static system. Information starts with Alice at time-zero and reaches Bob, who is also at time-zero, via the conventional Ryu-Takayanagi Path $l$.
  - Ryu-Takayanagi Path $l$ is equated with the horizon value epsilon, which is only conventionally regarded as dynamic on cosmological time scales. However, this cannot be. The local effects of epsilon's dynamic nature is constantly observed, but not understood as a locally dynamic epsilon.

The same casual sequence of events of the Lunar Landing, because Preferential Frame of Reference applies:

- Occurred in a causal sequence that differs in each Preferential Frame of Reference.
- The causal sequence as they occur on the Moon is coherent, there is a clear reason for Armstrong's pause and stutter.
- The causal sequence as they occur on Earth, *recorded on Earth*, has no coherence, there is no cause for Armstrong's pause and stutter.
- The amount of Information was of the same causal sequence of events.
- The amount [$\aleph N_0$] of Information *appears to differ* between the two preferential Frames of reference. However, this is artifact of observation, where Bob's [lunar] Conformal value B'$\epsilon$' at the moment of detection is *smaller than* his value Bo$\epsilon_0$ at time zero, when Alice [CAPCOM] sent it. Meaning, CAPCOM [Alice] sends a message but recedes, for lack of a better term, from Alice as the value epsilon diminishes.

If you take the bullet list above and apply it to a DCQE experiment, they are identical, except opposite. In the above Lunar Landing example, Preferential Frame of Reference applies, and is the direct result of pure distance. There are two distinct realities with different outcomes and series of events that occurs in one does not occur in the other.

In the DCQE, Preferential Frame of Reference *does not apply*, because it describes an entangled system. All outcomes are identical.

We will look at that in detail later. What has to be equated is why the entangled system does not obey the basic principles of distance on and figuratively 'below' the AdS Horizon Surface.

## Details of the Lunar Landing Scenario

### Details of the Lunar Landing Scenario

As noted, the voice and events as recorded by the LM-P60 (Lunar Module model P60 descent control) were recorded on the Eagle, these were also [communications and events] recorded on Earth, e.g. CAPCOM. The CAPCOM recordings were of two sorts, the voice recordings as an immediate local [e.g. not transmitted from the Eagle, but recorded on the ground in real time] and the LM-P60 data as transmitted from the Eagle to Earth. In effect, there are now more than two things to compare:

- The voice recordings as recorded on the tapes aboard the Lunar Module, Eagle.
- The voice recordings as recorded on the tapes at CAPCOM.
- The LM-P60 data as recorded on the Eagle
- The LM-P60 data as transmitted in real time from the Eagle to CAPCOM [and hence recorded as on Earth].

Keep in mind there are the P60, which is the actual computer, the P66, which is the program running on the computer, P63 and P64 which monitor the landing controls.

In all four cases, *the order of events is different*. For example, the coded error for the information overflow of the LEM's all of 1K computer is also out of sequence with the CAPCOM recorder's LM-P60 interface. Even the fuel indicator as recorded on Earth is out of sequence with the LEM LM-P60's warning messages; the CAPCOM recorder having about a second [1.2s] more fuel than the LEM recorder. If you watch some documentary on the landing, you find that the number of seconds of fuel left at the time of touch down is a critical issue, simply because historically the LEM was only 14 seconds from crashing, rather than landing on the Moon. However, it was 12.8 seconds aboard the LEM, 14 seconds on Earth. This is not a 'time signaling delay' issue. There are two distinct realities. The LEM touched down according to CAPCOM with 1.2 seconds more fuel than as recorded aboard the LEM, which is a critical piece of information. Even the scalar quantities of events differ. This is an example of quantitative information differing between Preferential Frames of Reference, more later.

I had this P60 transcript at one point on the NASA archive, but NASA has since taken it down, no message indicating why. My best guess is that it was just superfluous information, and it appears they have simply cleaned the site up in terms of keeping navigation simple. They had both the LEM audio transcripts as well as the CAPCOM transcripts and have since removed the LEM transcripts. However, I will provide a link to the actual audio recordings of the LEM and CAPCOM.

- In the CAPPCOM [Alice] Preferential Frame of Reference the LEM touched down on the moon with 14 seconds of fuel remaining in its tanks.
- In the lunar [Bob] Preferential Frame of Reference the LEM touched down on the moon with 12.8 seconds of fuel remaining in its tanks.

**Details of the Lunar Landing Scenario**

This brings up the obvious question, if the LEM crashed on the moon because it ran out of fuel there would be no LEM P60 recording of how much fuel was in the tank. Meaning, according to CAPCOM it would not have crashed as a result of running out of fuel, as there was 1.2 seconds of fuel remaining at the time it crashed. History would then have recorded that the LEM crashed on the moon for some other reason, not because of running out of fuel. Possibly it would be attributed to pilot error.

There would be:

- In the lunar [Bob's] Preferential Frame of Reference the LEM crashes into the moon because it ran out of fuel.
- In the CAPCOM [Earth, Alice] Preferential Frame of Reference the LEM did not crash as a result of running out of fuel but because of some unknown reason.

These are two distinct and true realities. This isn't a lack of information in the common sense, the information that CAPCOM has establishes that the LEM had fuel remaining in its tanks when it crashed. If we did not record that information or otherwise destroyed it, then we would have a 'paradox.'

- Because Preferential Frame of Reference applies, what is true in one system *may not be true* in another system. This can be quantitative information, states, events.

This will make a bit more sense when we get to the more technical descriptions and derivations of the [Ryu-Takayanagi] Paths of Information under the conditions of a dynamic AdS Horizon Surface. The same principal both does and does not apply for entangled systems, which describes the Delayed Choice Quantum Eraser and Delayed Choice Entanglement Swapping.

Thus, in Alice [CAPCOM] Preferential Frame of Reference we have the Eagle touchdown with 14 seconds worth of fuel and in Bob's [lunar] Preferential Frame of Reference we have the Eagle touchdown with 12.8 seconds of fuel. This is a difference of scalar quantity of a measurable substance between frames of reference at the same moment. In our 2-train scenario, the Eagle would have crashed on the moon [Bob, lunar frame of reference] because of zero fuel, and in Alice [CAPCOM] frame of reference the Eagle would have crashed for no reason, with sufficient fuel remaining. If one is thinking, the information regarding fuel content would have eventually reached Earth, that is not correct because the LEM would be destroyed before sending the information.

The scalar difference in quantities is an example of scalar difference as established by Preferential Frame of Reference as a function of distance. This is not a violation of Information Conservation; it is an example of; 'what is true in one system may not be true in another system.' Again, this can also be the result of Lorentzian or Schwarzschild systems.

## Details of the Lunar Landing Scenario

We're just going to use the voice recordings for simplicity's sake. The LM-P60 data is considerably more difficult to describe, primarily because the information is in antiquated computer format and language, more like *Assembly* hexadecimal than any modern recognizable language by today's standards. Thus, the first step is in translation. There is only one person, the MIT grad student who wrote the code and was present at CAPCOM translating the error messages, who was qualified to do the translation in any case.

In the 1969 Lunar Landing event there are two recordings of the landing procedure, again, one as recorded on the LEM on the Moon and the other as recorded at CAPCOM on Earth. In that classic recording as heard on Earth, there is a pause and stutter as Armstrong's announcement.

The NASA transcripts have been updated and they have deleted the *transcripts* of the lunar landing as recorded on the moon [LEM]. However, there are two Youtubes you can listen to in order to affirm that these original transcripts are in fact correct. The audio according to the CAPCOM recordings is *currently* at:

https://www.youtube.com/watch?v=xc1SzgGhMKc

The portion of the actual landing begins at 18:00 minutes:

> Transcribed from NASA archives: transcript of lunar landing at NASA.org:
> http://www.hq.nasa.gov/alsj/a11/a11.landing.html#1024540
>
> **As recorded on Earth at CAPCOM:**
>
> CAPCOM: We copy you down Eagle….(beep$_1$)..
>
> Armstrong: Houston……..ahhh……Tranquility base here, the Eagle has landed…
>
> Total length 8.7 seconds [amount of Information]

In this recording, as recorded on Earth, there is no causal component for Armstrong's pause and stutter, there is a 'beep$_2$' embedded in his 'ahhhh' that is not in the written transcription, discussed momentarily. There is 8.7 seconds of Information in the recording. Keep the 8.7 seconds of Information in mind.

This Preferential Frame of Reference is CAPCOM, Earth, Alice. There is no clash of words between Armstrong and the unknown CAPCOM operator. There is merely an unidentified pause as Armstrong begins to say, 'Houston…..' In our 2-train scenario, this is train Alice, and no collision occurs with train Bob. [Alice, no train crash]

However, as recorded aboard the LEM on the Moon, the series of events occur in a different order, to such extent that the causal component of Armstrong's pause and stutter is in fact made clear; it is the second 'beep$_2$' as CAPCOM cuts in just as he is beginning to speak, interrupting him mid-word.

The lunar landing as recorded on the LEM are *currently* at a Youtube site:

https://www.youtube.com/watch?v=ae6VJ6YU8uo

## Details of the Lunar Landing Scenario

The portion of the actual landing begins at 13:00. In it, you can clearly hear CAPCOM cutting Armstrong off as he begins to speak. The primary speaker at CAPCOM is Duke. However, the voice that cuts Armstrong off is an unknown CAPCOM operator.

> [Originally] Transcribed from NASA archives: transcript of lunar landing [originally] at NASA.org [but now overwritten with the CAPCOM transcript]:
> http://www.hq.nasa.gov/alsj/a11/a11.landing.html#1023820

> **As recorded on the LEM on the Moon:**

> CAPCOM: We copy you do.. [Armstrong begins speaking, cutting CAPCOM off. Specifically, Charles duke is saying, 'We copy you..' and the unknown CAPCOM operator cuts everybody off with his T-1 statement].
> Armstrong: Houston.. [CAPCOM operator cuts in, cutting Armstrong off]
> CAPCOM:...wn Eagle…
> Armstrong: ..ah……
> CAPCOM:… (beep$_2$)….
> Armstrong: ….Tranquility base here, the Eagle has landed….

The total length of this broadcast is 7.5 seconds [amount of Information]

Armstrong's pause and stutter has a clear causal source, CAPCOM cutting in as he begins to speak. Again, the voice that cuts in is not Charles Duke, but an unknown CAPCOM operator. It sounds like he is saying,

> **'Ok everybody. T-1, stand by for T-1.'**

The original lunar recording transcript had the content but the current one does not identify the person speaking or what he says.

Also keep in mind that Aldrin's voice is in the recordings in both frames of reference. However, the key content here is what Armstrong is saying as he is cut off by CAPCOM. Aldrin's voice can be heard in the background and prior to the roughly 8-second portion in question.

The CAPCOM [unknown] speaker is our train wreck in our 2-train scenario. The Preferential Frame of Reference is the moon, Armstrong, Bob. The clash of words as the CAPCOM operator cuts in as Armstrong begins to say,

'Houston…[CAPCOM cuts in; 'Ok everybody. T-1, stand by for T-1,']…ah…' The Bob train crashes into the Alice train.

So, we have the lunar recording, Armstrong, moon, Bob, whose train crashes into train Alice [CAPCOM]. The train crash is CAPCOM cutting Armstrong off as he begins to speak. The lunar Preferential Frame of Reference exhibits typical causality.

Then, we have the CAPCOM recording, Earth, Alice, whose train *does not crash into train Bob*. There is no explanation for Armstrong's pause and stutter in the CAPCOM recording.

## Details of the Lunar Landing Scenario

There is a clear distance between Armstrong's pause and stutter that is simply an empty gap. The CAPCOM Preferential Frame of Reference does not exhibit typical causality.

I very strongly encourage the reader to go to these two Youtube sites and listen carefully to these two different recording in these two frames of reference. The reason these recordings are so important is, oddly, this is the furthest distance man has ever been from Earth. I know one has to think for a moment on that, but oddly, it is true. The lunar distance of 1.2 light seconds is the furthest man has ever made it beyond Earth. There are many such examples in these audio-video recordings, but we can only focus on the one exemplary historic note of the actual landing itself.

Preferential Frame of Reference as established purely by distance and no other factor has rendered our 2-train crash scenario such that:

I:

- Lunar, train Bob crashes into train Alice.
- Earth, *train Alice does not crash into train Bob.*

II.

- The lunar Preferential Frame of Reference establishes typical causality, there is a reason for Armstrong's pause and stutter.
- The Earth Preferential Frame of Reference does not demonstrate typical causality, there is no reason for Armstrong's pause and stutter.

III.

- Because Preferential Frame of Reference applies to the 2-party system, what is true in one frame of reference is not necessarily true in the other frame of reference.
- Preferential Frame of Reference, however, does not apply to entangled systems. In this case, what is true in one system is true in all entangled systems.

IV: Later, this will expand out to:

- Preferential Frame of Reference does not apply for entangled systems; thus, they traverse or exchange information via the conventional Ryu-Takayanagi Path $l$.
    - What is true in one system is true in all systems.
- Preferential Frame of Reference applies for non-entangled systems; thus, they do not traverse or exchange information via the conventional Ryu-Takayanagi Path $l$ but traverse or exchange Information via the fluid Path $l'$.
    - What is true in one system may not be true in all systems.

We will visit this PFR not applying to entangled systems in a later section of this text when we discuss the Delayed Choice Quantum Eraser and Delayed Choice Entanglement

### Details of the Lunar Landing Scenario

Swapping. The DCQE and DCES are chosen because they best represent PFR not applying, with the result, which is the key concept:

*What is true in one system is true in all systems when Preferential Frame of Reference does not apply.*

The Earth based recording has 8.7 seconds of Information in it, the Moon based recording has 7.5 seconds of Information in it. Information Conservation must be obeyed. Because:

$$\nu 7.5s \neq \nu 8.7s$$

It isn't the *quantity of seconds*, because in the lunar [Bob] Preferential Frame of Reference the CAPCOM and Armstrong events overlap, indicating the proper causality as CAPCOM cuts him off as he begins speaking.

In the Earth Preferential Frame of Reference [Alice] there is a 1.2 second gap between events that does not represent proper causality, as there is no rationale for Armstrong's pause and stutter. Thus, the lengths are different, but the amount of Information is the same.

A waveform of the two recordings again, looks like:

In the Earth [CAPCOM, Alice] recording you can see that Armstrong's pause and stutter, noted by '…ahhh…' is 1.2 seconds removed from the beep, which is embedded with the unknown CAPCOM operator's words 'Ok everybody. T-1, stand by for T-1.' And this is clear in the lunar audio recording at https://www.youtube.com/watch?v=xc1SzgGhMKc The portion of the actual landing begins at 18:00 minutes.

In the lunar [Bob] recording the beep, the unknown CAPCOM operator's words, and Armstrong all overlap, which is why Armstrong pauses and stutters. https://www.youtube.com/watch?v=ae6VJ6YU8uo The portion of the actual landing begins at 13:00.

The 1.2 second overlay is not a simple signal delay, it is the Preferential Frame of Reference establishing two distinct and different 'realities,' for lack of a better word. And again, if this were an entangled system, what is true for Bob [train crash] would be true for Alice

## Details of the Lunar Landing Scenario

[would be a train crash]. However, upon reviewing the audio files you can clearly see that there are two distinct 'realities' for Bob [train crash] and Alice [no train crash].

And again, I use the 2-train scenario because it demonstrates conventional $\Delta S$ has no typical solution when one 'timeline' is not the same as another 'timeline.' Suffice it to say that if this were a mere signal delay issue than either both trains would crash or both trains would not crash, e.g., what is true in a 2-train system is true for both trains. Your high school teacher may have tried to explain this off as a simple signal delay. Your high school teacher did not have a knowledge of quantum mechanics.

I think it is sufficient to say that, in any classic scenario we would expect the outcomes and series of events to be the same in every frame of reference. But this is more than outcomes and series of events, there is a clear causal component of an event, the train crash with the CAPCOM operator, that happens in one frame of reference but not in the other frame of reference. And again, if you are thinking this can be true of Lorentzian and Schwarzschild systems, that is correct. Preferential Frame of Reference can result from Lorentzian, Schwarzschild systems, or purely as the result of distance.

Ultimately later on we will look at the Schrodinger Double-Dead Cat Paradox, which is fascinating. Very briefly, we send Schrodinger off at the speed of light and in his Preferential Frame of Reference he opens the cat box and finds a dead cat. In his Preferential Frame of Reference, the Alive | Dead cat eigenstate has precipitated to the eigenvalue, dead cat. However, from our Preferential Frame of Reference, because Schrodinger is speeding away at the speed of light, the cat remains forever in the eigenstate Alive | Dead. This is a simple example of a Lorentzian system where Preferential Frame of Reference applies and what is true in one system, Schrodinger, is not true in all systems, ours. Then we will drop Schrodinger on to a Schwarzschild Surface and play the same game with the same result of separate outcomes, dead cat for Schrodinger and forever eigenstate for us.

Lunar Landing: Preliminary Summary

## Lunar Landing: Preliminary Summary

As a demonstration of how a conventional $\Delta S$ solution cannot be rendered, as is the typical case, to resolve the Preferential Frame of Reference issues I will continue to use the 2-train scenario. We have the Alice line [Earth] and the Bob line [Moon]. A simple argument that the anomaly is the result of conventional timing signal delays [associative property of addition] will be dispatched by looking at the real, different, realities that emerge from the conditions.

Thus, some rules:

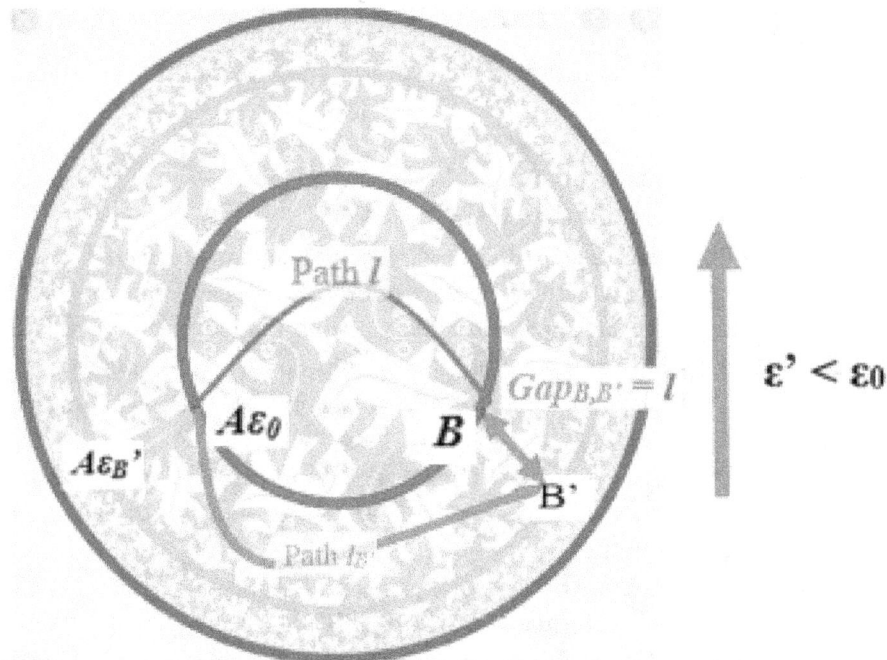

- The 'Gap,' $Gp$, is the result of epsilon diminishing in magnitude on immediate time scales, as envisioned above, figuratively moves Bob at point $B$ to the 'new horizon' at point $B'$. As a result, there is not only a temporal distance from Alice to Bob, which is the normal light distance between points, Ryu-Takayanagi Path $l$, but the additional temporal relocation from points $B$ to $B'$ as well as $A\varepsilon_0$ to $A\varepsilon_B'$.
- The conventional Ryu-Takayanagi Path $l$ is the red arc extending from $A\varepsilon_0$ to $B$. This is essentially a static system, as regarded on cosmological time scales.
- The magnitude of the Gap is equal, always, to the normal light distance as indicated by the red arc. Where epsilon is typically regarded in cosmological time scales, the vanishing rate is actually defined by $c \equiv 1Lp/1tp$, and as the AdS Horizon Surface value $c^2 \equiv Lp^2/tp^2$.
- Because the Gap is equal in magnitude to the red arc, Path $l$, the normal light distance between points, *seeming* causality violations arise. In this example we look at the 1969 Lunar Landing of Apollo 11 transcripts, recordings, and digitized visuals of the recordings, plural.

Lunar Landing: Preliminary Summary

With the Lunar Landing:

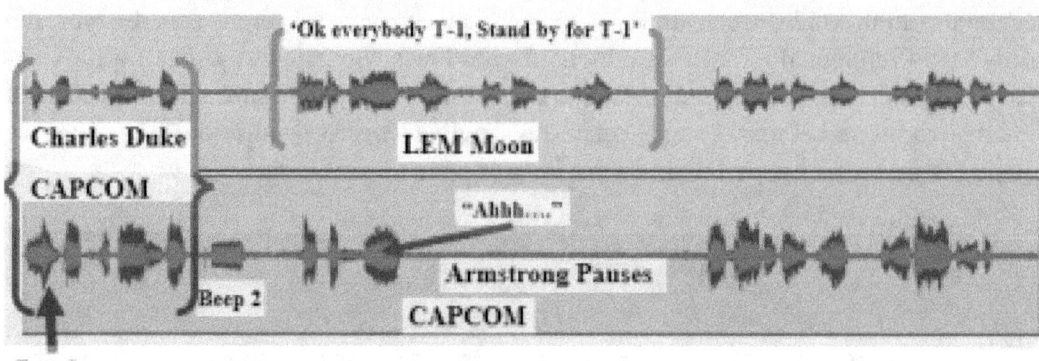

- There is a recording of events as recorded at CAPCOM on Earth.
- There is a recording as recorded on the Moon.
- The two recordings have events *out of causal sequence*, and in the case of the CAPCOM recording on Earth, an event occurs *without a causal source*. Systems that are 1) out of causal sequence and 2) lacking a causal source in one but not both frames of reference is sufficient to dispatch signaling delays. However, we will break things down to a finer level.
    - B, Moon, Bob: A normal causal sequence of events, complete with an event [train-crash of words] takes place.
    - A, Earth, Alice: The causal sequence of events occurs *in a different order of events*, and a causal event [stumble of words] occurs *with no causal source for that event*, namely, Armstrong's pause and stutter.
    - A, Earth, Alice: No train-crash of words takes place. The message from CAPCOM is well separated from the stutter and pause of Armstrong [Bob].
- An event [train crash] of trains A and B occurs in one Preferential Frame of Reference that does not occur in another Preferential Frame of Reference, as recorded in each frame of reference for the same sequence of events.
    - B, Moon, Bob: trains A *and B, crash into one another.*
    - A, Earth, Alice: No train crash occurs. In fact, the trains A and B in this Preferential Frame of reference never meet.

There are thus two real, tangible, and *recorded* realities of the same set of events between Alice [Rath] and Bob [Armstrong, Moon]. The events are *recorded* as 1) being in a different causal sequence and 2) no overlap of CAPOM-Armstrong occurs in the Earth frame of reference, e.g., no train crash 3) Armstrong's pause and stutter has a causal source in the Moon frame of reference [CAPCOM cutting him off as he begins to speak] but no causal source in the Earth frame of reference [those events are separated in time, by the exact magnitude of the 'Gap,' which is equal to Path $l$, the normal light distance between Earth and Moon. This is not a paradox; distance establishes a Preferential Frame of Reference. Preferential Frames of Reference can and do have distinct, isolated realities.

Lunar Landing: Preliminary Summary

This type of phenomenon is unique to Preferential Frame of Reference, in this case established purely by distance and no other factor, and relates specifically to the AdS Horizon Surface description as being superluminal, with normal light distance being that Conformally Scale Invariant [fractal] set regarded for lack of a better term as 'below' the Horizon Surface; all coupled with the phenomenon of $\varepsilon$ diminishing at a considerable scope and rate that is equal to the normal light distance descriptions of the 'red arc.'

Again, if signal delays could explain this then there would be some conventional $\Delta S$ solution. However, there is no $\Delta S$ solution for an event that includes both Alice and Bob, where Bob crashes into Alice, in Bob's frame of reference but Alice does not crash into Bob in Alice's frame of reference. A conventional $\Delta S$ solution would conserve Information. Information cannot be conserved in such a condition where, for example, the Information that is the causal source of an event [Capcom doesn't cut in] is *zero* in the Earth frame of reference, but *not zero* [CAPCOM cuts in] in the Moon frame of reference.

This is not a paradox, merely an anomaly that arises purely as a result of the real-time evolution of the Horizon value epsilon. Again, convention only treats the evolution of epsilon in cosmological time scales. However, the vanishing rate of epsilon is defined by $c \equiv L_p/t_p$.

Lunar Landing: Destruction of Information

## Destruction of Information

We are going to look at the scenario, what would happen if we could destroy the Information regarding the events of the Lunar Landing as recorded on the moon. Later, we will look at LIGO and compare this to the cessation of causal path that leads back to the operand black holes as they existed before they coalesced. Then we will look at the Delayed Choice Quantum Eraser in these same contexts.

A digitized visual of the recordings as they were taken on Earth at CAPCOM and the Moon is as follows:

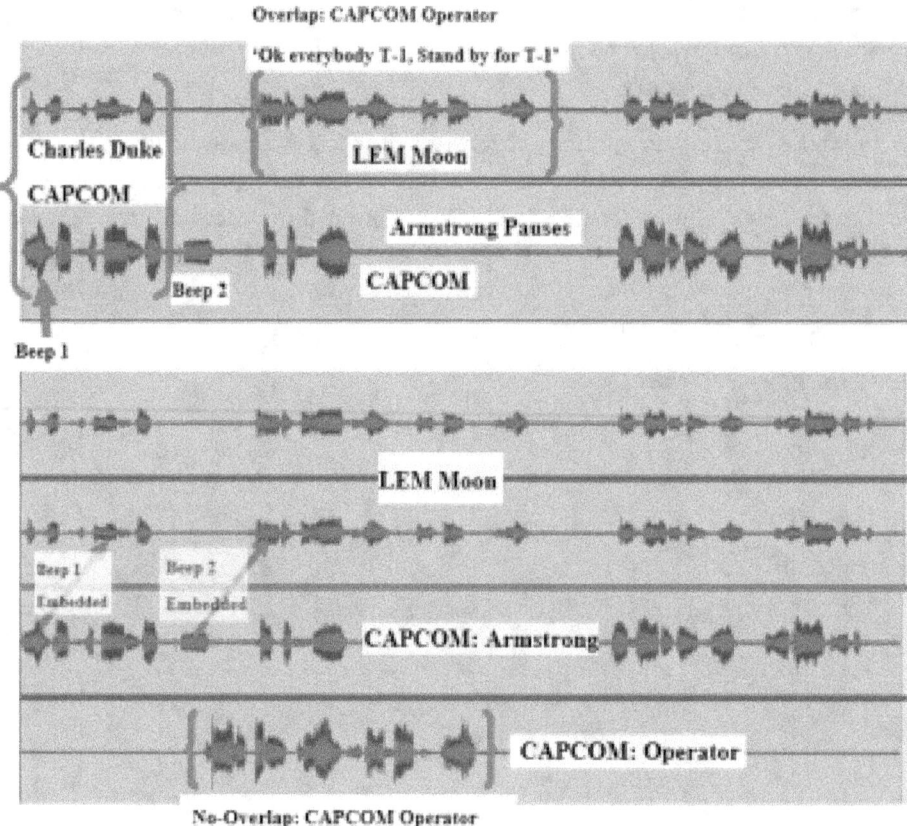

The same casual sequence of events; again: [The 2-trains scenario]

> 2 identical trains with the *numbered cars out of sequence* depending on frame of reference. Meaning, in Bob's 'reality,' [moon, Armstrong], the numbered railroad cars are in a certain order as compared to Alice's railroad car numbering, and in Alice's [Earth, CAPCOM] reality, her numbered cars are in a given order as compared to Bob's. However, the two numberings do not agree. As shown in the graphic depiction of the conversation above, Alice's CAPCOM reality, there is a large distance between the cars of the Armstrong stutter, and in Bob's lunar

Lunar Landing: Destruction of Information

reality, those cars have collided in a heap. There is no space between the railroad cars in the red circled blotch in Bob's reality, thus, the numbering of the collided cars is non-sequitur.

- The numbered events Occurred in a causal sequence that differs in each Preferential Frame of Reference.
- The causal sequence as they occur on the Moon is coherent, there is a clear reason for Armstrong's pause and stutter. [Train wreck occurs in that frame of reference as CAPCOM cuts in].
- The causal sequence as they occur on Earth, *recorded on Earth*, has no coherence, there is no cause for Armstrong's pause and stutter. [No train wreck occurs in that frame of reference].
- The amount of Information was of the same causal sequence of events. [Two identical trains].
- The amount [ϟnQ] of Information *appear to* differ between the two preferential Frames of reference. Two identical trains *appear to differ in the number of cars*, depending on frame of reference. However, this is not the case. The *magnitude of* Alice and Bob's metersticks differ according to Preferential Frame of Reference. Any change in the number of Qubits that define the system would be a violation of Preservation of Information Conservation. The koppa [ϟ] symbol is a police pseudo-operator that identifies a value that must be held constant.
- 2- identical trains *do differ* in the *numbering of* [e.g., sequence] cars and the numbered cars are out of sequence dependent on frame of reference, where Preferential Frame of Reference is established purely by distance and no other factor.

To clarify that point, destruction vs preservation of Information, we go back in time sci-fi style and destroy the lunar recording as it was made:

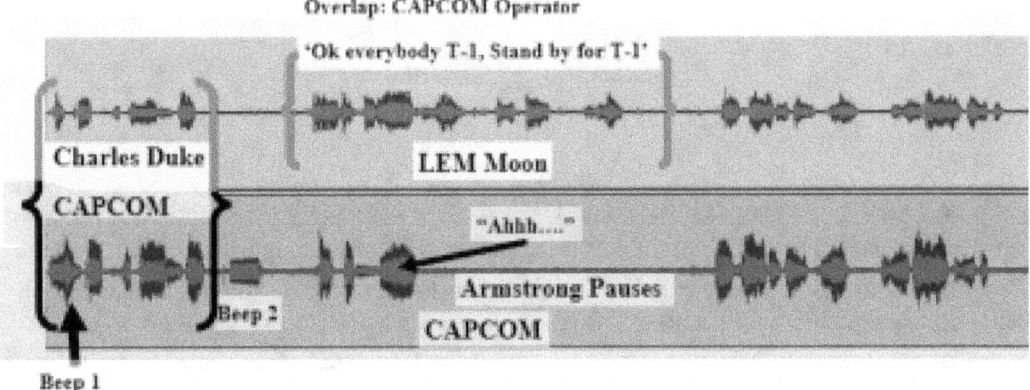

The double headed yellow arrow is intended to represent the double headed red arrow in the AdS image a couple of pages back, the 1.2 missing seconds, e.g., the 'Gap,' $Gp$. The double-headed yellow arrow is in fact 1.2 seconds [always equal to the conventional Ryu-Takayanagi Path $l$], as recorded on Earth at CAPCOM, the 'train wreck' that occurs in the Lunar Preferential Frame of Reference does not happen because it is separated by the 1.2 seconds distance between $B$ and $B'$:

299

Lunar Landing: Destruction of Information

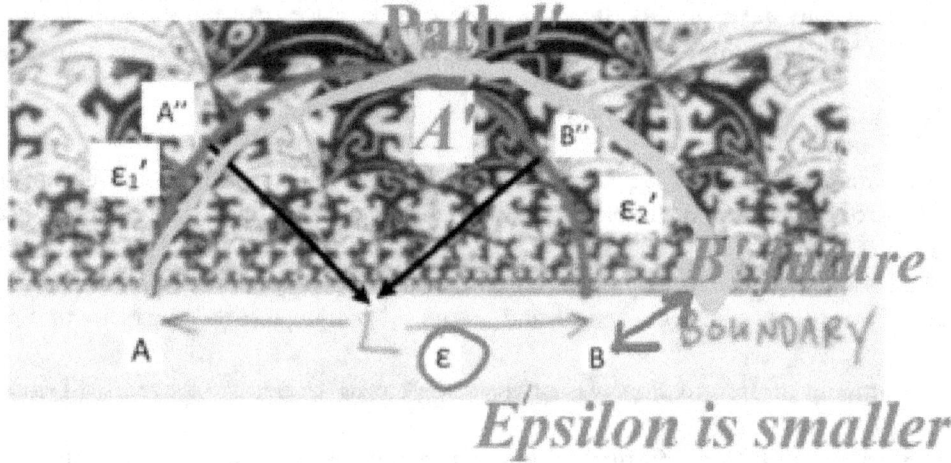

Again, the magnitude of the double-red arrow is always equal to the length of the red arc Path *l*, which is the normal light distance between points, in this case, between Earth and Moon. The double headed red arrow is referred to again, as the 'Gap.' [*Gp*] Again, because epsilon's vanishing scope and rate is defined by c ≡ 1Lp/1tp:

$$\frac{Lp^2}{tp^2} = c^2 = A_\Omega = \varepsilon$$

Where the term $A_\Omega$ is conventionally the 'worldsheet.' The worldsheet simply defines the scope of a system, for a single Qubit is the term above and immediately below. For the Lunar Landing Anomaly is that cross section of cosmos in 2D [Holographic] of the Earth-Moon system. It is treated purely as a Surface phenomenon.

$$\varepsilon \equiv \frac{Lp^2}{tp^2} = A_\Omega = \daleth N = c^2$$

The term *N* is number of standard N-Qubits. The actual quantity is irrelevant but must be preserved with respect to Information Conservation. The purpose of koppa is, again, a pseudo-operator that disallows the human penchant to beg, 'borrow,' or steal to change that number, typically by misuse of the HUP. In the simplest sense, as Information, there is no *negative information* to balance such an aberration in thinking. Borrowing mass-energy and such is another matter, Information cannot be 'borrowed.'

Everything is equated to epsilon, the exhaustive derivation in prior papers, summarized here. Epsilon is by convention being used as a unitary interval, here equated as above. For a single N-Qubit of Information; is dimensionality $Lp^2$, temporally $tp^2$, is the Worldsheet for that Qubit, is the AdS Horizon Surface value ε, epsilon. Each epsilon is superluminally isolated from each other epsilon, Qubit.

Lunar Landing: Destruction of Information

$$\varepsilon \equiv \text{ItoPath}L \equiv L_p^2/t_p^2 \equiv A_0 \equiv c^2 \equiv \hbar N_Q \equiv 1 \text{ HUP} \equiv 1 \equiv 2\,(L_p/t_p)$$

$$\text{length } l = \frac{e}{3} Ln \frac{L}{\epsilon}$$

$$\varepsilon \equiv L_p^2/t_p^2 \equiv c^2 \equiv \text{HUP} \equiv \hbar N_Q \equiv 1 = 2\,(L_p/t_p) \equiv \text{ItoPath}L$$

$$L_p^2/t_p^2 \equiv c^2$$

*The magnitude of the double-red arrow is always equal to the length, Path l, of the red arc; as defined by the vanishing scope and rate of epsilon, which is given by* $c \equiv 1L_p/1t_p$.

In this sci-fi scenario of destruction of Information, the Moon recording never occurred. It may have existed at one time, but the causal path back to that Information has ceased to exist. This does not infer that the event did not take place, or that the train wreck did not occur, only that the causal path back to that sequence of events has ceased to exist. This is true destruction of Information.

There is a stutter with no causal component. Without the lunar recording, there is no causal path back to the causal event, if one regards $\Delta S$ in terms of 'loss of Information regarding the microstates of a system.' That is infinite $\Delta S$, which is impossible in normal spacetime. A black hole is not normal spacetime.

Meaning, under conditions of typical causality, we would expect the Information in Alice's [CAPCOM] frame of reference to be identical in every aspect to Bob's [lunar] frame of reference. If we never made the lunar recording, there would, in that case, be no issue. However, *the systems are not identical in causal components*. There is a causal event in the lunar Preferential Frame of Reference that does not exist in the CAPCOM Preferential Frame of Reference. Thus, if we destroyed that Information as it occurred on the moon, there would be a causal 'paradox.' In terms of our 2-train scenario, the Bob train returns from the moon looking as though it had smashed into another train, however, no such event occurred, as Alice's recoding shows, there was a huge *Gap* between events such that the two trains never collided. This is lending the scenario to science fiction. Alice's recording at CAPCOM clearly shows no train wreck occurred. [This would be a great episode of JJ Abram's '*Fringe.*'].

We will then argue the definition, 'loss of Information regarding the microstates of a system,' as purely a technological issue that becomes less true every day. 'Loss of Information' is quite impossible, is purely cognitive and technological. Heat Death of the universe is conventional $\Delta S$, but *not loss of Information; $\hbar nl$ remains fixed*, merely spread out.

However, breaking a causal path back to a system occurs in nature, when two black holes coalesce. Again, this is demonstrated by LIGO and is essentially unassailable. In breaking a causal path back to a system, we are not destroying Information. I don't know if that is clear. In the LIGO example we have two black holes, Alice and Bob. Upon coalescing they entropy by whatever means infinitely. As a consequence, there can be no causal path back to Alice and no causal path back to

Lunar Landing: Destruction of Information

Bob. That would require infinite causal paths to connect back to Alice and back to Bob. Thus, the causal paths back to Alice and the causal path back to Bob has been, *deleted*. This goes back to Shannon entropy, as defined here, the lower limit to greater than 1-causal path, always. And conventional ΔS entropy has the limit as both upper and lower of exactly 1-causal path, always.

If you want to revert back to the conventional ΔS: 'Entropy is the loss of Information regarding the microstates of a system,' this example is irreversible entropy. If you do not have the Moon recording, there is no causal component. Going back to the visuals, you in fact have the wrong series of events [on the Earth recording] as they occurred on the Moon. There is no correcting the series of events as they occurred on the Moon without the recording by any conventional approach to date.

In the sci-fi scenario, there is no ΔS correction for this spacetime anomaly, now only an anomaly because the lunar recording does not exist. As such, there is no explaining the anomaly via time delays. We have Alice's train looking 'normal,' and Bob's train returns from the moon smashed, with zero knowledge of indication that any such event took place. In fact, the two trains crashed into *one another*. Thus, Bob's smashed train is truly paradoxical.

We then stitch the scenario back together, both the Earth CAPCOM and LEM Moon recordings are intact, and again; *there is still no ΔS correction, there is no 'time signaling delays' argument*. Again, because the two realities of Alice and that of Bob differ in the sequence, series, and even types of events; namely, Bob's train crashed into Alice's train, but Alice's train did not crash into Bob's train.

A signaling delay argument could be expressed in terms of a classic 2-trains scenario, e.g. train-1 leaves point A, East bound at nKm/hr train-2 leaves B West bound at xKm/hr; n = x. In that description one and only one train car from each of train-1 and train-2 have to collide, *and also do not collide*. That is what is meant by no ΔS correction. There are in fact two distinct realities of the same causal sequence of events.

*You cannot ΔS correct a train that is both smashed and not smashed.*

Interestingly, when we look at LIGO detection of Gravity Waves from 2-coalescing black holes, we find that the causal path back to the individual operand black holes, call them Alice and Bob, has ceased to exist. That is a greater discussion elsewhere in another [Bray] text. In that ΔS scenario, at the time black holes Alice and Bob coalesce, they entropy infinitely. As they submerge behind the black hole's horizon, they are now becoming infinitely entangled by infinite causal paths and can only be described by Shannon Entropy. Thus, the gravity wave is that of the product black hole, Victor. However, the causal paths back to Alice and back to Bob have *ceased to exist*, because the conventional ΔS of the cosmos cannot describe a system that is connected by greater than one causal path. Not off topic, this is an example of a causal path being destroyed, the destruction of Information.

Invariably, some conventional ΔS fix or at least explanation can be applied to some phenomenon of observation, and we invariably assign a set of non-deviating rules. The two recordings have no conventional ΔS correction nor description of convention; even at this time that we have had them

Lunar Landing: Destruction of Information

in our possession for half of a century. The 'arrow of time' as it is sometimes applied to conventional ΔS is in fact the problem. In the lunar recording there is a collection of at least two events, Armstrong as he begins to speak and CAPCOM cutting in just as he begins to speak. In the Earth recording there is no relationship between these two groups of photons flying back and forth between Earth and Moon; and no ΔS description that can explain Armstrong's stutter without the lunar recording, nor can it be modeled and fixed. Furthermore, *there is no predicting it,* no such event would precipitate out of only the Earth recording if it were not observed; *it evidently cannot even be recreated mathematically nor explained.*

> A conventional ΔS resolution in unipolar time, regarded as a time delay issue, could be solved by model trains, literally. However, here we have *two unipolar 'timelines.'* Not science fiction, but Preferential Frame of Reference as established by distance and no other factor.
>
> In the lunar recording this is the simultaneous events of Armstrong beginning to speak and CAPCOM's message confirming touch down. If we re-introduce the Lunar Express back into the scenario, still, there is no resolution, because the collision of the two exact same trains occurs *at different times in each train's frame of reference.* Literally, Bob's [Armstrong] Lunar Express wreck occurs at 9:30, the CAPCOM collision *still never occurs;* at 1.2 seconds after 9:30, in Alice's reality, the two trains merely pass one another.
>
> Again, ΔS fails to resolve the issue. If a system cannot be described by ΔS then something other than ΔS must describe the system. The point of this section of text is to introduce the 'other system' that can describe a *breaking* of causal path.

- The lunar recording is the causal frame of reference that is correct, there is a cause for Armstrong's stutter, the overlap, which is the result of the two events, CAPCOM's transmission reaching Armstrong at the moment he goes to speak, occur simultaneously.
- The Earth recording is the aberrant accounting of events; the events do not occur simultaneously. A causal series of events *does not occur* on Earth because in the Earth Preferential Frame of Reference two simultaneous events do not occur but are separated in time by the noted 'delay,' as it is regarded, exactly 1.2 seconds.
- In the Lunar Preferential Frame of Reference, a series of causal events *does occur* because two events are simultaneous, not separated by time. What establishes the Preferential Frames of Reference is purely distance and no other factor. This distance will be equated with Paths $L$ vs $l$ on the AdS Horizon, where L is superluminal, no transfer of Information and Path $l$ is $\leq c$.

Lunar Landing: Destruction of Information

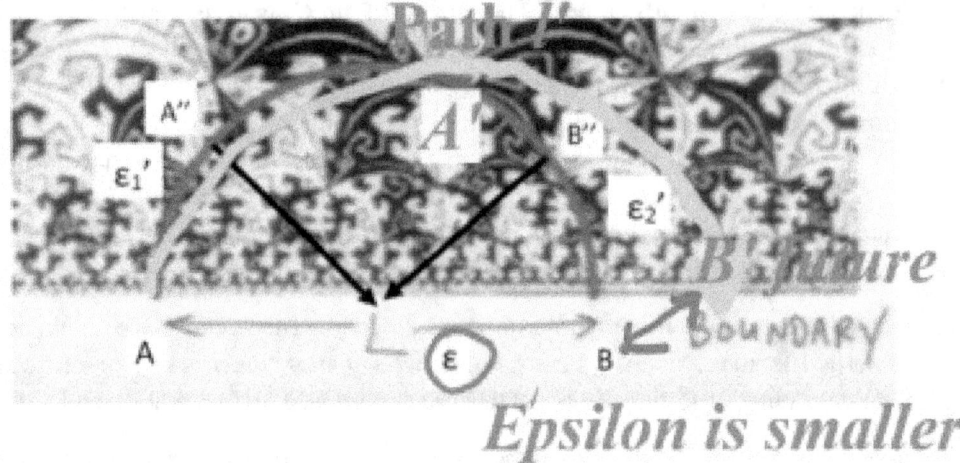

- Where one might regard points A and B as the distance between Earth and Moon this cannot be correct, because the distance across Path L is superluminal, meaning, it does not exist. Meaning, A and B have no causal ties in any reality, they are simply two remote systems.
- The red arc, again, is the normal light distance between Earth and Moon.
- The yellow arc is the distance that results as the red arc conforms as epsilon diminishes at a rate and scope of $c \equiv 1L_p/1t_p$, defined by the Planck flow.
- The 'Gap,' indicated by the double red arrow is equal [always] in magnitude to the normal light distance, the red arc.
- The reason the Gap appears smaller is because it is *Conformally Scale Invariant*. The 'lizards' 'below' the Horizon Surface change in scale along the path of the red arc. Epsilon, $\varepsilon$, at the Horizon is *always* the smallest value, epsilon.

As we destroy Information, in these different scenarios, we find that there is no $\Delta S$ description, and the 'train wreck scenario,' we do and do not have a train wreck in the two-causal sequence of events, establishing them as Preferential Frames of Reference, and purely as a consequence of distance, normal light distance, and no other factor. This is obviously not a 'time delay' issue and cannot be explained by such a $\Delta S$ correction.

FRACTAL SET ASSIGNMENT

## FRACTAL SET ASSIGNMENT

$\varphi$ is the Maric Operator: $\varphi(\pm v) \mapsto |\pm \Delta v|$; $\varphi(\pm l) \mapsto |\pm \Delta l|$; $\varphi(\pm t) \mapsto |\pm \Delta t|$; $\therefore \varphi(\pm 1) \mapsto |1|$

G' is not the gravitational constant of the universe 'changing,' rather, it is the artifact of observation, a pseudo-preferential frame of reference that defines Conformal Scale Invariance as a set of self-similar relationships.

$$G' = 6.67384(80) \times 10^{-11} \text{ m}^3/\text{Kg (tp')}^2$$

$$\varphi(\pm L'_p) \rightleftharpoons |\sqrt{\frac{hG'}{2\pi c^3}}|; \; \varphi(\pm tp') \rightleftharpoons |\sqrt{\frac{hG'}{2\pi c^5}}|$$

R|(G''|{t'', Lp''})_R

$$L_p'' \rightleftharpoons L_{p0} \sqrt{1 - \frac{2G''M}{n'L_p (\frac{1Lp}{1tp})^2}}$$

$$t_p'' \rightleftharpoons t_{p0} \sqrt{1 - \frac{2G''M}{n'L_p (\frac{1Lp}{1tp})^2}}$$

$$\varphi(\pm L'_p) \rightleftharpoons |\sqrt{\frac{hG'}{2\pi c^3}}|; \; \varphi(\pm tp') \rightleftharpoons |\sqrt{\frac{hG'}{2\pi c^5}}|$$

$$\leftrightarrow N_Q = \leftrightarrow \frac{Lp^2}{tp^2} = c^2 \equiv \leftrightarrow 1$$

FRACTAL SET ASSIGNMENT

$$Lp^2 = \frac{hG'}{2\pi c^3}; \; tp^2 = \frac{hG'}{2\pi c^5}$$

**For the SR components:**

$$v = \frac{\frac{4}{n}nLp}{\frac{4}{x}xtp}; \; c = \frac{1Lp}{1tp}$$

Where, for $nLp$, $n$ is an integer, and for $xtp$, $x$ is an integer.

$$\varphi(\pm\beta) = \left| \frac{1}{\sqrt{1 - \left(\frac{\frac{4}{n}nLp}{\frac{4}{x}xtp}\right)^2}} \right|$$

$$\pm l' \rightleftharpoons \frac{l_0}{\sqrt{1 - \left(\frac{\frac{4}{n}nLp}{\frac{4}{x}xtp}\right)^2}}; \; \pm lp' \rightleftharpoons \frac{lp_0}{\sqrt{1 - \left(\frac{\frac{4}{n}nLp}{\frac{4}{x}xtp}\right)^2}}$$

$$\pm t' \rightleftharpoons \frac{t_0}{\sqrt{1 - \left(\frac{\frac{4}{n}nLp}{\frac{4}{x}xtp}\right)^2}}; \; \pm tp' \rightleftharpoons \frac{tp_0}{\sqrt{1 - \left(\frac{\frac{4}{n}nLp}{\frac{4}{x}xtp}\right)^2}}$$

$$\tfrac{4}{n}nLp \not> \tfrac{4}{x}xtp$$

Else, $v > c$.

The reciprocation Preserves Information Conservation: Where $\infty/\infty$ is not regarded as one, it is the progression towards $v \to c \equiv \nmid 1$.

$$\lim_{v \to c} f(Lp') = Lp_0 / \sqrt{1 - \left(\frac{v}{c}\right)^2} = \infty$$

FRACTAL SET ASSIGNMENT

$$\lim_{v \to c} f(tp') = tp_0 / \sqrt{1 - (\frac{v}{c})^2} = \infty$$

$$\leftharpoonup f(c); = \frac{\lim_{v \to c} f(lp')}{\lim_{v \to c} f(tp')} = \leftharpoonup 1$$

$$L_p'' \rightleftharpoons L_{p0} \sqrt{1 - \frac{2G''M}{n'L_p(\frac{1Lp}{1tp})^2}}$$

$$t_p'' \rightleftharpoons t_{p0} \sqrt{1 - \frac{2G''M}{n'L_p(\frac{1Lp}{1tp})^2}}$$

$$\leftharpoonup f(c); = \frac{\lim_{n'Lp \to 1} f(lp')}{\lim_{n'Lp \to 1} f(tp')} = \leftharpoonup 1$$

Else, if:

$$\lim_{v \to c} f(lp') = lp_0 \sqrt{1 - (\frac{v}{c})^2} = 0$$

$$\lim_{v \to c} f(tp') = tp_0 / \sqrt{1 - (\frac{v}{c})^2} = \infty$$

$$\leftharpoonup f(c); = \frac{\lim_{v \to c} f(lp')}{\lim_{v \to c} f(tp')} = 0$$

FRACTAL SET ASSIGNMENT

$$L_p'' \rightleftharpoons L_{p0} / \sqrt{1 - \frac{2G''M}{n'L_p(\frac{1L_p}{1tp})^2}}$$

$$\lrcorner f(c); = \frac{\lim_{n'Lp \to 1} f(lp')}{\lim_{n'Lp \to 1} f(tp')} = 0$$

Always.

# References

## References

1. A Faster Than Light Spacetime Manifold with Zero Mass-Energy Requirement and Zero Negative Energy Density. June 2022 DOI: 10.13140/RG.2.2.20509.18407
2. Quantum Information Dynamics Vol III: The Delayed Choice Lunar Landing. March 2022. DOI: 10.13140/RG.2.2.35708.21127
3. Quantum Information Dynamics Volume I and II August 2021 DOI: 10.13140/RG.2.2.36172.95364
4. The History of Infinity: Zeno, Siddhartha Buddha, and the Judeo-Hindu common origin. June 2021 DOI: 10.13140/RG.2.2.22916.86408
5. A Correct Faster Than Light Spacetime Manifold with Zero Mass-energy Requirement. June 2021 DOI: 10.13140/RG.2.2.23038.10568
6. A Space Based X-Ray Free Electron Laser Platform. May 2021 DOI: 10.13140/RG.2.2.23884.33924
7. Altering the Rate of Time Evolution on the AdS Horizon Surface. April 2021 DOI: 10.13140/RG.2.2.35703.75682
8. A Zeno Dynamic Delayed Choice Quantum Eraser. April 2021 DOI: 10.13140/RG.2.2.19726.08003
9. Faster Than Light Propulsion via the Zeno Effect. March 2021 DOI: 10.13140/RG.2.2.28575.48809
10. Artificial Alteration of Spatial Geometry via the Quantum Zeno Effect. December 2020 DOI: 10.13140/RG.2.2.13527.29601, DOI: 10.13140/RG.2.2.13527.29601/1 [updated file]
11. Quark Confinement, the DeSitter Horizon, The Yang-Mills Statistical Argument, as a Function of Shannon Entropy; Distance on the AdS Horizon as the only Valid Preferential Frame of Reference Model. June 2019 DOI: 10.13140/RG.2.2.18177.10083
12. SPECIAL RELATIVITY & THEIR CONSEQUENTIAL RESULTS PART II W Bray June 2019 DOI: 10.13140/RG.2.2.15971.81446
13. HOW TO USE THE FRACTAL SET AND SHANNON LUMINOSITY CORRECTION FOR THE HUBBLE PARAMETER May 2019 DOI: 10.13140/RG.2.2.36134.78400
14. LETTER: THE HISTORY OF SPECIAL RELATIVITY EXPERIMENTAL RESULTS AND THE ORIGINAL 1905 EINSTEIN-MARIC DERIVATION W Bray May 2019 DOI: 10.13140/RG.2.2.24557.84966/1
15. SPECIAL AND GENERAL RELATIVITY THEIR DERIVATIONS AND CONSEQUENTIAL RESULTS PART I May 2019 DOI: 10.13140/RG.2.2.32710.52808
16. On the Precision of the Hubble Parameter March 2019 DOI:10.13140/RG.2.2.16444.08324/1

## References

17. The Hubble Parameter as a Conformally Scale Invariant Quantitative Result of the Derived Local Value ε of the AdS Horizon February 2019 DOI: 10.13140/RG.2.2.31792.61449/1
18. DERIVATION OF i THE SQUARE ROOT OF NEGATIVE 1; USING THE FUNDAMENTAL EINSTEIN-MARIC OPERATOR φ(+1)(-1) ↦ |1| February 2019 DOI: 10.13140/RG.2.2.27539.32807/1
19. A STEP BY STEP DETAILED DERIVATION OF SPECIAL RELATIVITY FROM THE ORIGINAL EINSTEIN 1905 PAPER February 2019 DOI: 10.13140/RG.2.2.24683.59685/1
20. A STEP BY STEP DETAILED DERIVATION OF SPECIAL RELATIVITY FROM THE ORIGINAL EINSTEIN 1905 PAPER January 2019 DOI: 10.13140/RG.2.2.15810.02243
21. GENERAL RELATIVITY AND PATH LENGTH as a CONFORMALLY SCALE INVARIANT QUANTIZED TRACE MATRIX for Tuv as ε: DEFINITIONS FOR PREFERENTIAL FRAME OF REFERENCE January 2019 DOI: 10.13140/RG.2.2.23139.60967
22. THE GEOMETRY AND EVOLUTION OF THE ADS HORIZON AS CONFORMALLY INVARIANT AS ε THE UNITARY PLANCK INTERVAL EXTENDING TO THE ADS DOMAIN January 2019 DOI: 10.13140/RG.2.2.29604.73609
23. Derivation of the AdS Horizon Ryu-Takayanagi Horizon Value ε as the Unitary Planck Interval Lp and Subsequent Alterations to the Trace Matrix Tuv of Guv as the Planck Length January 2019 DOI: 10.13140/RG.2.2.17860.68488
24. THE PERMITTIVITY (ELECTRIC) VS THE PERMEABILITY (MAGNETIC) DERIVATION FOR THE TWO SPECIES OF VIRTUAL PHOTONS THAT MEDIATE THE OBSERVED EFFECTS AS AN ARTIFACT OF THE EVOLUTION OF THE AdS HORIZON AS HUP CONSERVATION of kNQ: Force as Entropy/Ordiny January 2019 DOI: 10.13140/RG.2.2.26249.29282
25. THE HEISENBERG UNCERTAINTY PRINCIPLE AS PRESERVING INFORMATION CONSERVATION: DEFINITION OF A 'QUBIT' IN INFORMATION QUANTUM THEORIES January 2019 DOI: 10.13140/RG.2.2.28765.87526
26. DESCRIPTIONS HISTORY AND DEFINITIONS OF THE QUANTUM ZENO EFFECT AND ANTI- QUANTUM ZENO EFFECT AS A PROPERTY OF THE EVOLUTION OF THE DESITTER HORIZON IN GENERAL RELATIVITY January 2019 DOI: 10.13140/RG.2.2.22174.59204/1
27. THE EINSTEIN-MIRAC DERIVATION OF LORENTZ TRANSFORMATION AND TIME DILATION CONSIDERATIONS FOR DESITTER EVOLUTION HOLOGRAPHIC AND INFORMATION THEORY BLACK HOLE DYNAMICS AND QUANTUM ENTANGLEMENT October 2018 DOI: 10.13140/RG.2.2.28503.32166

## References

28. The Quantum Zeno Effect on the DeSitter Horizon in Conformal Field Theory and Holographic Theory as a Diadic Markov Process in Statistical Mechanics. August 2018 DOI: 10.13140/RG.2.2.21003.54560
29. DESCRIPTIONS AND DEFINITIONS OF THE QZE AND ANTI-QZE August 2018 DOI: 10.13140/RG.2.2.32406.80968
30. Un método viable para la manipulación del espacio-tiempo via efecto Zenón cuántico para una nave Alcubierre-Bray April 2014 DOI: 10.13140/RG.2.2.23608.80649

## Appendix I: Defining the Qubit

### Appendix I Defining the Qubit

So, as to define quantization in a hard definition or definitions that describe the Qubit unambiguously:

Defining an N-Qubit in quantized terms:

- $A_\Omega$ is the 'worldsheet.
- ϙ is the police pseudo-operator that prevents violations of Information Conservation.
- $N_Q$ is the N-Qubit.
- Lp is the Planck length and tp is Planck unitary time.
- We start with the Bekenstein-Hawking relationship:

$$N = S = \frac{A_\Omega}{4Lp^2}$$

Then set the worldsheet limit:

$$N = \frac{A_\Omega}{4Lp^2} = \frac{4Lp^2}{4Lp^2} = 1$$

$$c \equiv \frac{1L_p}{1t_p} \mid \therefore$$

$$N = \frac{A_\Omega}{4tp^2} = \frac{4tp^2}{4tp^2} = 1$$

$$N = \frac{4Lp^2}{4Lp^2} = \frac{Lp^2}{Lp^2} = \frac{tp^2}{tp^2} = 1$$

$$c \equiv \frac{1L_p}{1t_p}$$

$$ϙN_Q = ϙ\frac{Lp^2}{tp^2} = c^2 \equiv ϙ1$$

Again, ϙ [koppa] is the police pseudo-operator that prevents the human penchant for changing the quantity of Qubits of Information, or to otherwise *borrow them*. The

## Appendix I: Defining the Qubit

Heisenberg Uncertainty Principle can never be used to 'borrow' Qubits. In the most simplistic sense, there is no 'negative Information' to balance such an aberration in thinking.

Looking at the Ryu-Takayanagi Horizon value epsilon:

$$l = \frac{e}{3} Ln \frac{L}{\epsilon}$$

Given, Path $L$ is superluminal $v \gg c$:

$$L \geq \epsilon > 0$$

The *unitary value of* Path L cannot be greater than $\epsilon$ because this is superluminally forbidden. In general, Path L is some quantity of values epsilon, by convention. However, to date there does not seem to be any hard definition regarding what form the unitary value epsilon [$\epsilon$] would take. It is therefore illogical to describe Path L as consisting of some quantity of unitary values epsilon when epsilon has no hard definition. That is, referring to some unitary value that represents the Locally Quantized Meterstick cannot exceed some superluminal expression; a meter cannot be described as $ct > L$ nor $ct < l$. It also must be greater than zero, for obvious reasons.

Epsilon cannot be less than {Lp, tp} as this would violate the lower limit of quantization:

$$\epsilon \nleq \{Lp, tp\} \equiv (u_0 \pm u_{0+1})$$

The Horizon value epsilon cannot be greater than {Lp, tp}, as this would then be a superluminal violation of the relationship between $L$ and $l$. For example, we cannot say that epsilon is equal to 1.6817 [totally fictional value] because this would violate quantization by demanding some fractional value of the Planck length and fractional value of Planck time. If $\epsilon$ is said to be *greater than* {Lp, tp} then it must be some integer value of {Lp, tp}. It cannot be greater than one unitary Planck interval, as this then would demand that communication from one endpoint of epsilon is a superluminal distance from the other end.

$$\epsilon \ngtr \{Lp, tp\} \equiv (u_0 \pm u_{0+1})$$

$$\therefore$$

$$\epsilon \equiv (u_0 \pm u_{0+1})$$

$$\epsilon \equiv \{Lp, tp\} \; L \geq l\epsilon \, ; \, L \equiv n\epsilon$$

## Appendix I: Defining the Qubit

Else, represents a superluminal violation of $c \equiv 1Lp/1tp$. Epsilon cannot be greater than Path L as this would then demand that epsilon is a superluminal violation. Epsilon cannot be less than the unitary Planck intervals {Lp, tp} else this would demand violation of quantization. Path L must be made up of *integer values* epsilon. Epsilon is thus equal to the unitary Planck intervals {Lp, tp}.

From:

$$\natural N_Q = \natural \frac{Lp^2}{tp^2} = c^2 \equiv \natural 1$$

We simplify and isolate terms:

$$\frac{Lp^2}{tp^2} = c^2 = A_\Omega$$

The worldsheet $A_\Omega$ is equal to the constant $c^2$ thus $Lp^2/tp^2$.

And:

$$\varepsilon \equiv \frac{Lp^2}{tp^2} = A_\Omega = N_Q = c^2$$

we can then express coupling to mass-energy as:

$$\varepsilon = c^2 = \frac{E}{m} = \frac{Lp^2}{tp^2} = A_\Omega = N_Q$$

The actual key terms are $A_\Omega$ and $N_Q$. The quantity of N-Qubits cannot change; thus we use the police pseudo-operator $\natural N_Q$. The worldsheet that defines the system, whatever it may be, can and does change Conformal Value, provided $\natural N_Q$ does not differ. Described in detail later, it is the magnitude of the unitary Planck intervals that change, as they are currently being equated as $\varepsilon \equiv \{Lp, tp\}$. Epsilon is not static, thus {Lp, tp} are also dynamic, in magnitude.

## Appendix I: Defining the Qubit

Aside: I do not think anyone has given it any thought, regarding a Schwarzschild or a Lorentzian transformation with respect to maintaining quantization. For example, if we regard time dilation, I think the conventional argument would be that more unitary Planck intervals describe the time dilated system. However, this is a clear violation of Preservation of Information Conservation. Later, in detail, it is shown and derived how the dynamics of the unitary Planck intervals of time and length *change in magnitude, not quantity*.

The worldsheet is equated by convention to the number of N-Qubits that define the system. The Qubit is equated with $c^2$ thus with epsilon.

Given, Path $L$ is superluminal $v \succ c$:

$$[\text{Path}]\ L \geq \varepsilon > 0$$

$$\text{HUP} \succ A_\Omega$$

The Heisenberg Uncertainty Principle [HUP] in terms of the worldsheet $A_\Omega$ cannot be greater than the worldsheet that defines the system.

$$L \geq I\varepsilon$$

$$\frac{Lp^2}{tp^2} = c^2 = A_\Omega$$

$$\varepsilon \equiv \frac{Lp^2}{tp^2} = A_\Omega = N_Q = c^2$$

Because:

$$\int HUP = \int 2Lp = Lp^2$$

$$\therefore$$

$$\{Lp^2, tp^2\} = 2(Lp, tp)$$

## Appendix I: Defining the Qubit

And as it turns out to be, as above, $A_n = Lp^2/tp^2 = c^2$ defines the superluminal upper boundary constraint for the single quantized Qubit, as one Planck Surface. The HUP is thus logically constrained to the local worldsheet such that:

$$HUP \nsucc A_n$$

Along with:

$$\sigma x \sigma v \geq \frac{h}{4\pi}$$

$$\{\sigma x, \sigma v\}$$

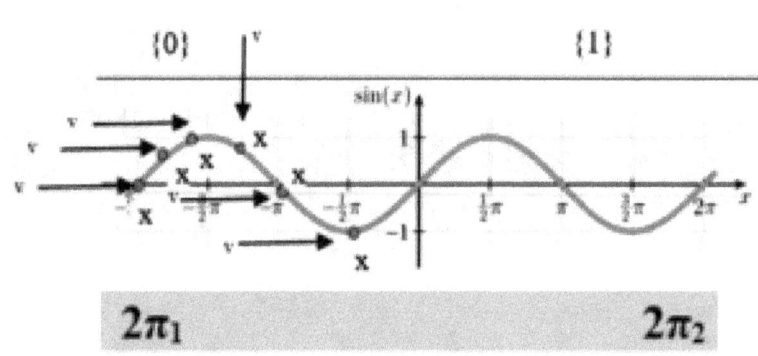

Where $\sigma x$ is defined above visually as:

$$2\pi_1 + 2\pi_2$$

Which in turn is thus:

$$Lp_1 + Lp_2 = 2Lp$$

And $\sigma v$ is thus defined as:

## Appendix I: Defining the Qubit

$$2\pi_1 + 2\pi_2$$

Which in turn is thus:

$$tp_1 + tp_2 = 2tp$$

And:

$$c \equiv Lp/tp = 1 \therefore Lp = tp$$

Is true when constrained to the single Planck interval, thus the term:

$$\sigma x \sigma v = 2Lp \times 2tp = 4Lp^2$$

And we equate the HUP in quantized terms that is limited to the worldsheet that defines the system. The lower boundary condition of this then is the lower boundary condition for quantization, thus:

$$1 \text{ HUP} \equiv 1 \equiv Lp^2/tp^2 \equiv \varepsilon \equiv \text{\textit{RmPath}}L \equiv A_\Omega \equiv c^2 \equiv \Bbbk N_Q \equiv 2\,(Lp/tp)$$

Therefore:

$$\sigma x \sigma v = 2Lp \times 2tp = 4Lp^2 \geq \frac{h}{2Lp}$$

From:

$$length\ l = \frac{e}{3} Ln \frac{L}{\varepsilon}$$

$$\varepsilon \equiv Lp^2/tp^2 \equiv c^2 \equiv \text{HUP} \equiv \Bbbk N_Q \equiv 1 = 2\,(Lp/tp) \equiv \text{\textit{RmPath}}L$$

$$\textit{Given } Lp^2/tp^2 \equiv c^2$$

$$\therefore$$

## Appendix I: Defining the Qubit

$$\varepsilon \equiv \text{ImPath}L \equiv L_p^2/t_p^2 \equiv A_\Omega \equiv c^2 \equiv \hbar N_Q \equiv 1\, HUP \equiv 1 \equiv 2\,(L_p/t_p)$$

$$\varepsilon \equiv 1\, HUP$$

The term immediately above defines 1-HUP [e.g., the Heisenberg Uncertainty Principle of a quantized system of 1{$L_p$, $t_p$}] as epsilon and $c^2$. Equating to $c^2$ allows for equating with the host of obvious equivalents, such as E/m, and so on.

And from:

$$length\ l = \frac{e}{3} Ln \frac{L}{\varepsilon}$$

We get:

$$\varepsilon \equiv L_p^2/t_p^2 \equiv c^2 \equiv HUP \equiv \hbar N_Q \equiv 1 = 2\,(L_p/t_p) \equiv \text{ImPath}L$$

$$L_p^2/t_p^2 \equiv c^2$$

And it then follows that:

$$\hbar\varepsilon = c^2 = \frac{L_p^2}{t_p^2} = \hbar A_\Omega = \hbar N_Q \equiv HUP \geq \text{path}L$$

Given:

$$c^2 = E/m$$

$$\hbar\varepsilon = c^2 = \frac{E}{m} = \frac{L_p^2}{t_p^2} = \hbar A_\Omega = \hbar N_Q$$

$$\hbar\varepsilon = \frac{E}{m}$$

## Appendix I: Defining the Qubit

The Permutations are then:

$$length\ l = \frac{e}{3} Ln \frac{L}{\varepsilon}$$

$$length\ l = \frac{e}{3} Ln \frac{L}{E/m}$$

$$length\ l = \frac{e}{3} Ln \frac{L}{c^2}$$

$$length\ l = \frac{e}{3} Ln \frac{L}{\hbar A_\Omega}$$

$$length\ l = \frac{e}{3} Ln \frac{L}{\hbar N_Q}$$

It follows that:

$$Gap_{B,B'} = l = \frac{e}{3} Ln \frac{L}{\varepsilon} = \frac{e}{3} Ln \frac{L}{c^2}$$

$$Gap_{B,B'} = l = \frac{e}{3} Ln \frac{L}{E/m}$$

These were more exhaustively derived in the noted [Bray] references several pages back. Relating these to mass*energy is important because the argument that conventional ΔS [not Shannon] entropy can describe a system with *seeming* causal violation cannot be correct:

$$\hbar\varepsilon' \equiv \frac{L'}{\frac{3l}{e^e}} = c^2 = \frac{Lp'^2}{tp'^2} = \hbar A'_\Omega$$

319

## Appendix I: Defining the Qubit

$$4\varepsilon' \equiv \frac{L'}{e^{\frac{3l}{e}}} = c^2 = E/m$$

$$e^{\frac{3l}{e}} = e^{\frac{3\frac{e}{3}\ln\left(\frac{L}{\varepsilon}\right)}{e}} = e^{\ln\left(\frac{L}{\varepsilon}\right)}$$

$$4\varepsilon' \equiv \frac{L'}{e^{\ln\left(\frac{L}{\varepsilon}\right)}} = c^2 = E/m$$

$$m = \frac{Ee^{\frac{3l}{e}}}{L} = \frac{Ee^{\frac{3\frac{e}{3}\ln\left(\frac{L}{\varepsilon}\right)}{e}}}{L} = \frac{Ee^{\ln\left(\frac{L}{\varepsilon}\right)}}{L} = E/c^2 = E/\varepsilon$$

$$m = \frac{Ee^{\ln\left(\frac{L}{\varepsilon}\right)}}{L} = E/c^2 = E/\varepsilon$$

and

$$4\varepsilon' \equiv \frac{L'}{e^{\ln\left(\frac{L}{\varepsilon}\right)}} = c^2 = \frac{E}{\varepsilon} = \frac{4Lp'^2}{4tp'^2} \equiv 4N_QQ = 4A'_\Omega$$

And relating epsilon to the geometry of spacetime:

$$G_{uv} \equiv R_{uv} - \frac{1}{2}Rg_{uv} = \frac{8\pi G}{c^4}T_{uv}$$

$$G_{uv} = \frac{8\pi G}{c^4}T_{uv}$$

## Appendix I: Defining the Qubit

$$G_{uv} = \frac{8\pi G}{\frac{Lp^4}{tp^4}} T_{uv}$$

$$G_{uv} = \frac{8\pi G}{\varepsilon^2} T_{uv}$$

$$G_{uv} = \frac{8\pi G}{2HUP_*} T_{uv}$$

*as the Bekenstein-Hawking limit

$$G_{uv} = \frac{8\pi G}{\tfrac{1}{4}Nq^2} T_{uv}$$

$$G_{uv} = \frac{8\pi G}{\tfrac{1}{4}A_\Omega^2} T_{uv}$$

$$\left[T_{uv} \rightleftharpoons S \leftrightharpoons \frac{e}{3} Ln \frac{L}{\varepsilon}\right]; \quad T_{uv} \rightleftharpoons S \leftrightharpoons m = \frac{Ee^{\ln(\frac{L}{\varepsilon})}}{L} = E/c^2 = E/\varepsilon$$

So, this quantizes as:

$$T_{uv} \rightleftharpoons S \leftrightharpoons [\frac{E}{\frac{Lp^2}{tp^2}}]_{uv}$$

Then in terms of Quantum Information Dynamics

$$T_{uv} \rightleftharpoons S \leftrightharpoons [\frac{E}{\tfrac{1}{4}N_Q}]_{uv}$$

$$T_{uv} \rightleftharpoons S \leftrightharpoons [\frac{E}{\tfrac{1}{4}A_\Omega}]_{uv}$$

## Appendix 1: Defining the Qubit

$T_{uv}$ is in terms of unitary intervals that define Path $l$. Again. All scaler and vector quantities must be in magnitude values in terms of Path $l$.

$$[T_{uv} \rightleftharpoons S \leftrightharpoons l \equiv \frac{e}{3} Ln \frac{L}{\varepsilon}]$$

$$[T_{uv} \rightleftharpoons S \leftrightharpoons l \equiv \frac{e}{3} Ln \frac{L}{\frac{Lp^2}{tp^2}}]$$

$$[T_{uv} \rightleftharpoons S \leftrightharpoons l \equiv \frac{e}{3} Ln \frac{L}{c^2}]$$

$$[T_{uv} \rightleftharpoons S \leftrightharpoons l \equiv \frac{e}{3} Ln \frac{L}{\hbar N_Q}]$$

$$[T_{uv} \rightleftharpoons S \leftrightharpoons l \equiv \frac{e}{3} Ln \frac{L}{\hbar A_\Omega}]$$

$T_{uv}$ is the 4-dimensional, that must be in terms of Path $l$, because $l$ defines the magnitude of the normal light distance between any two points, regardless of scale. Thus:

$$\left[T_{uv} \rightleftharpoons S \leftrightharpoons \frac{e}{3} Ln \frac{L}{\varepsilon}\right]; T_{uv} \rightleftharpoons S \leftrightharpoons m = \frac{E e^{\ln(\frac{L}{\varepsilon})}}{L} = E/c^2 = E/\varepsilon$$

Thus, $T^{00}$ is:

$$T^{00} \equiv [m = \frac{E e^{\ln(\frac{L}{\varepsilon})}}{L} = \frac{E}{c^2} = E/\varepsilon]$$

## Appendix I: Defining the Qubit

$$T^{00} \equiv \frac{E}{c^2} = E/\varepsilon$$

All of the above is rather straight forward in that there are no unobservables or unobservable dimensionalities presented nor implied.

## Appendix II: THE CONDITIONS OF THE LOCALLY QUANTIZED METER STICK

## Appendix II THE CONDITIONS OF THE LOCALLY QUANTIZED METER STICK

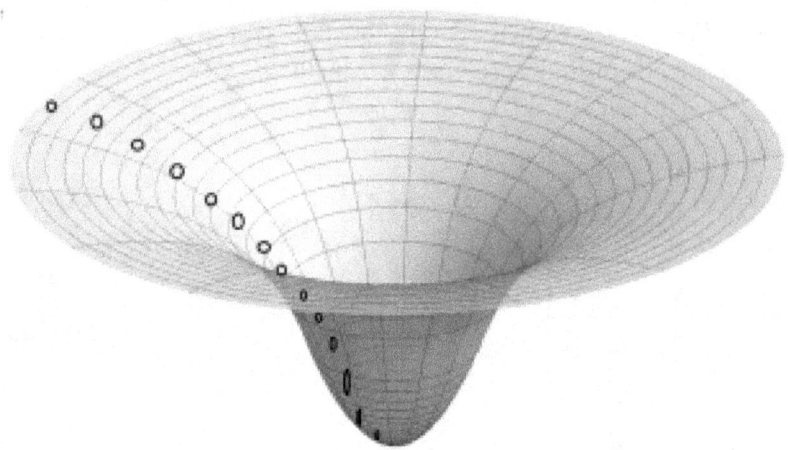

- Every observer along the gradient curve has a locally quantized meter stick that measures a value of each other observer along the gradient curve, and no two values are the same value, regarded as $nLp$. Where $nLp$ is the value each observer measures in integer values of $Lp$ according to their locally quantized meter stick. No observer along the curve can phenomenally read a non-integer value of $Lp$ on their local meter stick. No two values $nLp$ are the same, albeit the *quantity n of the initial conditions, the quantity of Information that describes the system being measured cannot differ from its initial conditions, else violate Information Conservation. The quantity of N-Qubits that initially described the system* cannot have changed; else violate $kN_Q$.
- Nor can the meter stick in my hand being used to measure the system undergo transformation resulting in a change in the number of N-Qubits that define *my meter stick*. That would also be a violation of Information Conservation.

Example: My locally quantized meterstick has 'contracted' to $1/2X$. My measure of a distant photon initially of 1-meter would then be 2-meters on my locally quantized meterstick. The *quantity of* N-Qubits has not changed in either system.

It isn't clear to me how convention regards this issue. It seems apparent that a Lorentz dilation in any conventional description would change the quantity of N-Qubits that defines the dilated system, for example:

Let's look at the above in terms of 'ladders:' We have a real railroad track that passes through a real barn:

## Appendix II: THE CONDITIONS OF THE LOCALLY QUANTIZED METER STICK

The railroad ties are spaced exactly 1-meter apart. Time is measured by your carbon-14 clock in your lap. The train above is going 0.866c, which yields a 'length contraction' of 1/2X. The entire resting distance is 1Km. Thus, there are exactly 1000 railroad ties. As shown above, hypothetically there is some velocity at which, by the convention regarded as 'length contraction,' the entire train would fit in the barn.

Does the *quantity of* railroad ties change from your Preferential Frame of Reference? I suppose a contraction phenomenon would be thinking that the *quantity of* railroad ties has been cut in half, from 1000 railroad ties to 500 railroad ties. Any sensible answer would suggest the distance between the railroad ties has changed, not the quantity of them. This is even more true of Planck intervals of length, as they represent Information at its most fundamental level. The quantity of unitary Planck intervals cannot change, else violate Preservation of Information Conservation.

Again, there are not two answers of the sort, the train both does and does not fit in the barn. Convention has the train fitting in the barn as observed by some stationary observer but not in the frame of reference of the train. Some arguments suggest the train fits in the barn in all systems. There are in fact 32 permutations of answers, all non-sequitur, to the 'ladder paradox.' One can confirm this by reading a host of descriptions of the 'ladder paradox' and finding that there are in fact 32 permutations of answers, thus non can be correct.

The train does not fit in the barn is true in all frames of reference.

It is more likely agreeable that the distance between the 1000 railroad ties changes, not the quantity of them. However, convention takes it down to the unitary Planck length, Lp, and insists that the *quantity of unitary Planck intervals of length changes*, rather than the value of the unitary Planck intervals. And again, the CFT postulate is built upon the assumption that the unitary Planck intervals of length and time change over time as the AdS/CFT description. Thus, the *quality of* unitary Planck intervals cannot be as regarded by convention be fixed.

Thus, velocity is quantized as:

$$v = \frac{\Sigma n L p}{\Sigma x\, tp}$$

## Appendix II: THE CONDITIONS OF THE LOCALLY QUANTIZED METER STICK

And

$$c = \frac{1 Lp}{1 tp}$$

$$\varphi(\pm \beta) \rightleftharpoons \left| \frac{1}{\sqrt{1 - \left(\frac{\frac{\ln Lp}{\ln tp}}{\frac{1 Lp}{1 tp}}\right)^2}} \right|$$

Where $\varphi$ is the Maric Operator:

$$\varphi(\pm v) \mapsto |\pm \Delta v|; \quad \varphi(\pm l) \mapsto |\pm \Delta l|; \quad \varphi(\pm t) \mapsto |\pm \Delta t|; \quad c = l/t \therefore \varphi(\pm 1) \mapsto |1|$$

Note that the relationship above for Beta is a self-similar relationship. E.g.,

$$\varphi(\pm tp') \rightleftharpoons \left| \frac{tp_0}{\sqrt{1 - \left(\frac{\frac{\ln Lp}{\ln tp}}{\frac{1 Lp}{1 tp}}\right)^2}} \right|$$

Self-similar relationships do not equate as conventional equations apply. We look at the inside terms:

$$\varphi(\pm tp') \rightleftharpoons \left| \frac{tp_0}{\sqrt{1 - \left(\frac{\frac{\ln Lp'}{\ln tp'}}{\frac{1 Lp}{1 tp}}\right)^2}} \right|$$

This is identical in any form to a conventional Mandelbrot. However, the system is only moderately more complicated. Later we will look at the ER = EPR 'paradox' and find that conventional $\Delta S$ is preserved as the internal Area of the Einstein-Rosen Bridge increases because it is merely the Conformal Values of the unitary Planck intervals that is changing, not the quantity, which would be a violation of Preservation of Information Conservation.

- There is no rationale where we can regard this as phenomenally artifact of observation. Else, all observation of {Lp, tp} is rendered non-sequitur. This is a description then of direct

## Appendix II: THE CONDITIONS OF THE LOCALLY QUANTIZED METER STICK

measure, by the unitary interval under our local conditions of measure. Regarding this scenario as described as some phenomenal artifact of observation in turn yields the logical demand that all direct measure is artifact of observation.

- o  Nonetheless, if the measurements were an artifact of observation, then the rules still apply; no change in the quantity of unitary Planck intervals at the initial conditions of the system being measured can occur, only the Conformal Value, designated by the self-similar relationship above.
- o  All systems are in motion and thus would be under Lorentz transformation conditions.
- o  All systems are within the cosmological gravitation, regarded as the local universe and thus described by any series of Schwarzschild transformations.

- Regarding $l'$ transformation as artifact of observation but $t'$ as corporeal is violation of Information Conservation, $c = l'/t' = lL_p/lt_p$.

$$\koppa f(c) := \frac{\lim_{v \to c} f(l')}{\lim_{v \to c} f(t')} = \koppa 1$$

Minkowski spacetime demands that the time transformation is real, as deduced from a history of validating measurements, but that the length transformation is regarded as artifact, and regarded as unobservable. Because

$$c = lL_p/lt_p \equiv 1$$

This cannot be the case. Either both are artifact, or both are real.

- The measurement is then regarded as 'real.' This agrees with my prior statement as such. No measurement is an artifact of observation.
  - o  Nonetheless, any artifact of observation cannot violate real conditions else represent some violation of Lorentzian and Schwarzschild conditions that have no prior art.

- In addition, they communicate this to one another in real time, and each quantized photon, which by definition remains quantized, has a different value $nL_p$ as measured by each observer along the gradient curve, $T_{\mu\nu}$. The quantized photons conform to the gradient curve [e.g. Pound-Rebka] in the same manner as the *conformal values* $nL_p$ of the [hard] meter sticks.

- All of the above is therefore true for Planck time, $t_p$, regarded as $\koppa \times t_p$. Koppa, again, is the police pseudo-operator that prevents the human penchant for changing the quantities of unitary Planck intervals that define a system.

- The same holds true for all Lorentz transformations. At any velocity, such transformations must retain quantization, else violate Information Conservation.

## Appendix II: THE CONDITIONS OF THE LOCALLY QUANTIZED METER STICK

- o N-Qubits cannot be created or destroyed between Preferential Frames of Reference.
- o For example, if my unitary time has dilated to 2X, this cannot demand that twice as many unitary Planck intervals of time now exist, which is any penchant between any two systems in the cosmos.
- o Likewise, if my length has 'contracted' to 1/2X, this cannot demand that unitary Planck intervals of length have been annihilated, which is also any penchant between any two systems in the cosmos.
- The 'Negative One' (-1) Law of Thermodynamics: Information cannot be created nor destroyed. Therefore, $kN_Q$ applies to all Schwarzschild and Lorentz transformations. Koppa, again, is the police pseudo-operator that prevents the human penchant to change the quantity of Qubits that describe a system.

**We will use a single quantized hydrogen photon in this description.**

- Regarding a photon created at sea level on Earth, the single quantized hydrogen photon as it is created on the ground does not become phenomenally unquantized as a result of climbing out of a Gravity Well, nor visa-versa.
  - o Furthermore, the quantity of N-Qubits that define the single quantized hydrogen photon cannot change as this would violate Preservation of Information Conservation.
- When I measure the single quantized hydrogen photon with my 'locally quantized meter stick' while in orbit, both the meter stick and the hydrogen photon are quantized, however, against my quantized meter stick in orbit, the [individual] photon now appears to be of some greater or lesser quantized length, depending on one's penchant of description of Schwarzschild transformation.
- There is no possibility that I can measure a value with my 'locally quantized meter stick' that is not an integer value of Lp, as this violates fundamental quantization, in my own frame of reference. Meaning, my Locally Quantized Meterstick is incapable of providing some result that is not an integer value of unitary Planck intervals.
- The *quantity of* unitary intervals that defined the system at initial conditions now being observed cannot change nor change between systems nor change as artifact of observation, as all of these would violate Preservation of Information Conservation.

*Artifact of observation cannot violate Preservation of Information Conservation.*

Any suggestion that this is not the case debases the entire description to one of pure ontology, e.g., 'belief.' There is no transcendent, mystic nor unobservable penchant that describes any real system, as this then equates the description as a genuine religious artifact.

With respect to our single quantized hydrogen photon originating on the ground at sea level and now being measured in orbit:

- This is true on the ground, as well as in orbit. That is, my meter stick in my hand [in orbit] is quantized, there is no possibility using my local meter stick of measuring some value that is not an integer of Lp. The meter stick on the ground is also quantized according to its local

## Appendix II: THE CONDITIONS OF THE LOCALLY QUANTIZED METER STICK

conditions. Again, the number ½nLp cannot change; ½N₀, for either system. However, it will require a different quantity of unitary intervals on my local meterstick to measure the single quantized photon.
- That is, if my local meterstick is now 1/2x from what it was on the ground at sea level, it will require twice as many unitary intervals on my meterstick to measure the ground photon. The total sum of unitary intervals that make up the meterstick, however, has not changed.

- The same is then obviously true for any single quantized hydrogen photon as observed at some velocity relative to my locally quantized hydrogen photon, as per some SR transformation: ½N₀ for ½nLp and ½xtp are fixed for all systems as ≡ 1, when v = c.
  - Meaning, for example, there can be no measure of redshift of a single hydrogen photon that has travelled some distance that is some non-integer value of my quantized meterstick.
  - My meterstick can only yield integer values of unitary Planck intervals of time and length.

Here we regard the case where in both SR and GR transformations, regardless of rather we treat such as artifacts of observation; we have isolated a requirement of conditions such that:

- My locally quantized meter stick is quantized according to my local conditions. It is not possible to measure some non-integer value of unitary Planck intervals in any system.
- The quantization of my local meter stick applies to both its local conditions as per SR and GR.
- It can be regarded that no two systems have the same *immediate* set of SR and GR conditions.
- Each system has a unique set of locally quantized conditions and will therefore measure a *different quantized integer value of any other system; however, the quantity of Information that describes the system being measured cannot differ from its initial conditions, else violate Preservation of Information Conservation.*
- Quanta [Information] can be said as axiom, are not created and/or destroyed regardless of treatment of this set of conditions, even if regarded as purely artifacts of observation, *only*. That is, if such is a mere artifact of observation, *then the demand is that I am observing the creation/annihilation of quanta [Information]*.
- *Any treatment of measurement of one's own state under any Lorentz and/or Schwarzschild transformation as artifact thus negates any sequitur determination of the state of any and every other system, all transformations are real, not artifact.*

Given that the quantity of unitary intervals that describes any system remains fixed, else violate Preservation of Information Conservation, it is thus the *magnitude of the unitary Planck intervals that changes*.

*The value, Lp and tp are therefore mutable.*

## Appendix III: The Zeno Effect and the single quantized photon.

### Appendix III The Zeno Effect and the single quantized photon.

We can use the term 'scanning' or scan to designate some electronic condition that describes 'detector rate.' The PN junction doesn't actually have a detection rate, it has a detect-and-relax rate, or cycle, but no actual 'detection rate.' This was described in prior [Bray] papers as *Detector Pile-up,* and *Detector dead-time.*

> Artificial Alteration of Spatial Geometry via the Quantum Zeno Effect. December 2020 DOI: 10.13140/RG.2.2.13527.29601, DOI: 10.13140/RG.2.2.13527.29601/1 [updated file]

Briefly, if the *single PN junction* detects by electromagnetic interaction some electromagnetic phenomenon, it experiences some change in a valence orbital in one of its molecular or atomic constituents. At the most fundamental level an atom, for example, will require some amount of time before relaxing back to ground state, by convention. Detector Deadtime is a bit non-intuitive, however, in a single electromagnetic event deadtime would be the amount of time for the valence to jump to some non-ground state then relax back to ground state. The excitation time for the electron jump is by convention instantaneous. The relaxation time can be any number. On a single event scale, deadtime is the sum of these events. There is not always some instantaneous absorption of an incoming photon, because that would require infinite energy, spread over a large quantity of PN microstates. Dead-time is simply a PN junction's ability to absorb a second incoming photon over the same molecule(s) that make up the system:

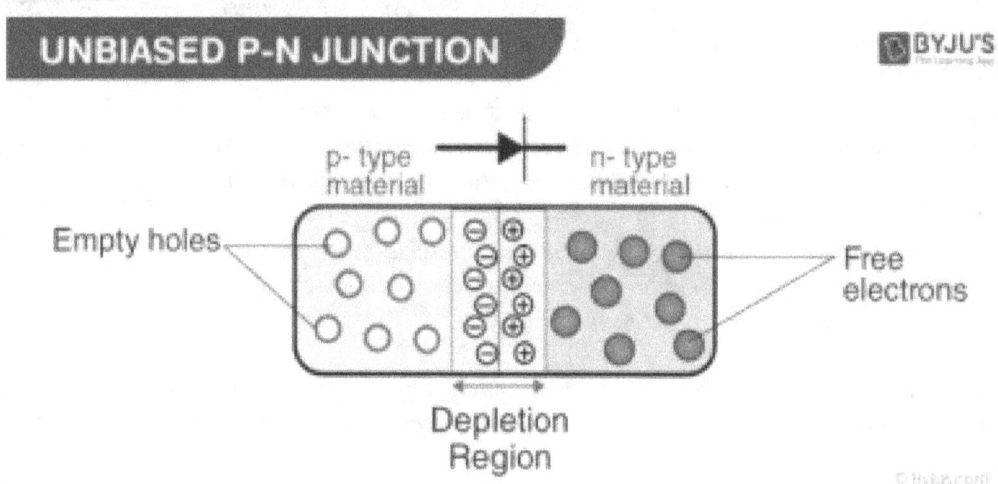

A key concept is the excitation jump is instantaneous and the relaxation jump is also instantaneous. However, the time period between excitation and relaxation is never zero, else there would be no jump or change in orbit. That concept, however, is limited to a single PN junction, which can never actually be anything less than the sum of the microsystems that make up the entire junction, as depicted above, is the rationale wherein we try to make ever smaller PN junctions consisting of over lessening quantities of atoms.

## Appendix III: The Zeno Effect and the single quantized photon.

Dead-time is a serious issue in that much of the literature of the Zeno Effect is not at all related to the Zeno Effect, but detector dead-time, pile-up, data transfer rates, and computational speed issues. This was expressed in great detail in:

1. Artificial Alteration of Spatial Geometry via the Quantum Zeno Effect. W. J. Bray DOI: 10.13140/RG.2.2.13527.29601

Thus, we will not discuss these scenarios in terms of detection, but I shall refer intermittently to these events as *scanning*. The detection rate of a PN junction is non-sequitur except in terms of excitation and relaxation. Thus, a detector has a detection rate that is the limit of its materials in terms of excitation and relaxation. The *scanning rate* is a function of how often the computer system can interrogate the detector PN junction. It is also a function of data transfer rate, which is thus bidirectional, the fastest will be via fiber optic which is currently limited to 5GHz if the signal is not compressed or compacted, which will always be the case in all but the most sophisticated of laboratory environments, and 5GHz processing rate.

Thus, the idea that one can, for example, a lot of Zeno Dynamic experiments have been performed on rubidium. This is likely because of its use in Pharmo-kinetics. For example, there are 37 isotopes of Rubidium, 22 of which are radioactive. A table of activity looks like this:

| Nuclide [n 1] | Z | N | Isotopic mass (Da) [n 2][n 3] | Half-life [n 4][n 5] | Decay mode [n 6] | Daughter isotope [n 7][n 8] |
|---|---|---|---|---|---|---|
| $^{72}Rb$ | 37 | 35 | 71.95908(54)# | <1.5 µs | p | $^{71}Kr$ |
| $^{72m}Rb$ | 100(100)# keV | | | 1# µs | p | $^{71}Kr$ |
| $^{73}Rb$ | 37 | 36 | 72.95056(16)# | <30 ns | p | $^{72}Kr$ |
| $^{74}Rb$ | 37 | 37 | 73.944265(4) | 64.76(3) ms | β+ | $^{74}Kr$ |
| $^{75}Rb$ | 37 | 38 | 74.938570(8) | 19.0(12) s | β+ | $^{75}Kr$ |
| $^{76}Rb$ | 37 | 39 | 75.9350722(20) | 36.5(6) s | β+ | $^{76}Kr$ |
| | | | | | β+, α (3.8×10−7%) | $^{72}Se$ |
| $^{76m}Rb$ | 316.93(8) keV | | | 3.050(7) µs | | |
| $^{77}Rb$ | 37 | 40 | 76.930408(8) | 3.77(4) min | β+ | $^{77}Kr$ |
| $^{78}Rb$ | 37 | 41 | 77.928141(8) | 17.66(8) min | β+ | $^{78}Kr$ |
| $^{78m}Rb$ | 111.20(10) keV | | | 5.74(5) min | β+ (90%) | $^{78}Kr$ |
| | | | | | IT (10%) | $^{78}Rb$ |
| $^{79}Rb$ | 37 | 42 | 78.923989(6) | 22.9(5) min | β+ | $^{79}Kr$ |
| $^{80}Rb$ | 37 | 43 | 79.922519(7) | 33.4(7) s | β+ | $^{80}Kr$ |
| $^{80m}Rb$ | 494.4(5) keV | | | 1.6(2) µs | | |
| $^{81}Rb$ | 37 | 44 | 80.918996(6) | 4.570(4) h | β+ | $^{81}Kr$ |
| $^{81m}Rb$ | 86.31(7) keV | | | 30.5(3) min | IT (97.6%) | $^{81}Rb$ |
| | | | | | β+ (2.4%) | $^{81}Kr$ |
| $^{82}Rb$ | 37 | 45 | 81.9182086(30) | 1.273(2) min | β+ | $^{82}Kr$ |

## Appendix III: The Zeno Effect and the single quantized photon.

| Isotope | Z | N | Mass | Half-life | Decay mode | Daughter |
|---|---|---|---|---|---|---|
| ⁸²ᵐRb | | | 69.0(15) keV | 6.472(5) h | β⁺ (99.67%) | ⁸²Kr |
| | | | | | IT (.33%) | ⁸²Rb |
| ⁸³Rb | 37 | 46 | 82.915110(6) | 86.2(1) d | EC | ⁸³Kr |
| ⁸³ᵐRb | | | 42.11(4) keV | 7.8(7) ms | IT | ⁸³Rb |
| ⁸⁴Rb | 37 | 47 | 83.914385(3) | 33.1(1) d | β⁺ (96.2%) | ⁸⁴Kr |
| | | | | | β⁻ (3.8%) | ⁸⁴Sr |
| ⁸⁴ᵐRb | | | 463.62(9) keV | 20.26(4) min | IT (>99.9%) | ⁸⁴Rb |
| | | | | | β⁺ (<.1%) | ⁸⁴Kr |
| ⁸⁵Rb [n 10] | 37 | 48 | 84.911789738(12) | Stable | | |
| ⁸⁶Rb | 37 | 49 | 85.91116742(21) | 18.642(18) d | β⁻ (99.9948%) | ⁸⁶Sr |
| | | | | | EC (.0052%) | ⁸⁶Kr |
| ⁸⁶ᵐRb | | | 556.05(18) keV | 1.017(3) min | IT | ⁸⁶Rb |
| ⁸⁷Rb [n 11][n 12][n 13] | 37 | 50 | 86.909180527(13) | 4.923(22)×10¹⁰ y | β⁻ | ⁸⁷Sr |
| ⁸⁸Rb | 37 | 51 | 87.91131559(17) | 17.773(11) min | β⁻ | ⁸⁸Sr |
| ⁸⁹Rb | 37 | 52 | 88.912278(6) | 15.15(12) min | β⁻ | ⁸⁹Sr |
| ⁹⁰Rb | 37 | 53 | 89.914802(7) | 158(5) s | β⁻ | ⁹⁰Sr |
| ⁹⁰ᵐRb | | | 106.90(3) keV | 258(4) s | β⁻ (97.4%) | ⁹⁰Sr |
| | | | | | IT (2.6%) | ⁹⁰Rb |
| ⁹¹Rb | 37 | 54 | 90.916537(9) | 58.4(4) s | β⁻ | ⁹¹Sr |
| ⁹²Rb | 37 | 55 | 91.919729(7) | 4.492(20) s | β⁻ (99.98%) | ⁹²Sr |
| | | | | | β⁻, n (.0107%) | ⁹¹Sr |
| ⁹³Rb | 37 | 56 | 92.922042(8) | 5.84(2) s | β⁻ (98.65%) | ⁹³Sr |
| | | | | | β⁻, n (1.35%) | ⁹²Sr |
| ⁹³ᵐRb | | | 253.38(3) keV | 57(15) μs | | |
| ⁹⁴Rb | 37 | 57 | 93.926405(9) | 2.702(5) s | β⁻ (89.99%) | ⁹⁴Sr |
| | | | | | β⁻, n (10.01%) | ⁹³Sr |
| ⁹⁵Rb | 37 | 58 | 94.929303(23) | 377.5(8) ms | β⁻ (91.27%) | ⁹⁵Sr |
| | | | | | β⁻, n (8.73%) | ⁹⁴Sr |
| ⁹⁶Rb | 37 | 59 | 95.93427(3) | 202.8(33) ms | β⁻ (86.6%) | ⁹⁶Sr |
| | | | | | β⁻, n (13.4%) | ⁹⁵Sr |
| ⁹⁶ᵐRb | | | 0(200)# keV | 200# ms [>1 ms] | β⁻ | ⁹⁶Sr |
| | | | | | IT | ⁹⁶Rb |
| | | | | | β⁻, n | ⁹⁵Sr |
| ⁹⁷Rb | 37 | 60 | 96.93735(3) | 169.9(7) ms | β⁻ (74.3%) | ⁹⁷Sr |
| | | | | | β⁻, n (25.7%) | ⁹⁶Sr |
| ⁹⁸Rb | 37 | 61 | 97.94179(5) | 114(5) ms | β⁻ (86.14%) | ⁹⁸Sr |
| | | | | | β⁻, n (13.8%) | ⁹⁷Sr |
| | | | | | β⁻, 2n (.051%) | ⁹⁶Sr |
| ⁹⁸ᵐRb | | | 290(130) keV | 96(3) ms | β⁻ | ⁹⁸Sr |

## Appendix III: The Zeno Effect and the single quantized photon.

| | | | | | | |
|---|---|---|---|---|---|---|
| ⁹⁹Rb | 37 | 62 | 98.94538(13) | 50.3(7) ms | β⁻ (84.1%) | ⁹⁹Sr |
| | | | | | β⁻,n (15.9%) | ⁹⁸Sr |
| ¹⁰⁰Rb | 37 | 63 | 99.94987(32)# | 51(8) ms | β⁻ (94.25%) | ¹⁰⁰Sr |
| | | | | | β⁻,n (5.6%) | ⁹⁹Sr |
| | | | | | β⁻,2n (.15%) | ⁹⁸Sr |
| ¹⁰¹Rb | 37 | 64 | 100.95320(18) | 32(5) ms | β⁻ (69%) | ¹⁰¹Sr |
| | | | | | β⁻,n (31%) | ¹⁰⁰Sr |
| ¹⁰²Rb | 37 | 65 | 101.95887(54)# | 37(5) ms | β⁻ (82%) | ¹⁰²Sr |
| | | | | | β⁻,n (18%) | ¹⁰¹Sr |
| ¹⁰³Rb[3] | 37 | 66 | | 26 ms | β⁻ | ¹⁰³Sr |
| ¹⁰⁴Rb[4] | 37 | 67 | | 35# ms (>550 ns) | β⁻? | ¹⁰⁴Sr |

If a researcher is using rubidium, he/she must choose an isotope that has a relatively long half-life, else there will be no practical use. Thus, the rubidium isotopes above that have half-lives in terms of seconds or perhaps a few days are impractical. Rubidium-87 has a half-life in the billions of years, which translates to 670 decays per second per gram. All of the other isotopes have Specific Activities so high they cannot possibly be used in a Zeno Dynamic experiment, as this would require the detector to operate in the petahertz range, data transfer in the same range, and so on. We could potentially dilute a sample to 12-orders of magnitude, but no paper describing the Zeno Effect describes the amount of radioactivity nor the expected [normal] decay rates, most likely because they are unknown.

No researcher to date has provided a quantifiable Zeno Effect in that the expected [normal] decay rates are described, nor the resulting Zeno Dynamic decay rates. No researcher has ever published sufficient information to reproduce the experiment, which is simply a requirement in good science.

The relationship between half-life and specific activity is as follows:

$$\lambda = \frac{0.693}{t_{1/2}}$$

Here, lambda is not wavelength but a decay constant. The half-life must be in seconds. So, rubidium-87 is 1.55E18 seconds half-life. This yields a specific activity of about 4.5E-19 decays per second. That is the probability of one atom decaying. Thus, we multiply lambda times Avogadro's number about 6E23, which is 87-grams for rubidium-87 to get about 270K decays per second. Obviously, this is problematic.

Note that in these experiments, upon closer examination, the isotope of rubidium is rarely indicated, and the mass of rubidium is never once stated. Thus, it is not possible to check the math, much less reproduce the experiment.

If I want a more practical isotope of rubidium I might choose the next longest half-life, rubidium-83. However, this is an electron capture decay, which is inherently difficult to monitor in terms of

## Appendix III: The Zeno Effect and the single quantized photon.

simple electron multipliers and so on. The next best would be rubidium-84 with a half-life of about 33 days, which translates to a specific activity of bout 2.4E-7 for one atom. Thus, for about 24-million atoms I can expect a decay rate of 1-decay per second.

In order to get into the KHz range of decays, which would be an ideal scenario where 1) we have sufficient activity and 2) we want to scan or detect this at some phenomenal rate, as per the Zeno Effect; I need 1,000 times this amount, which is about 24-billion atoms. Thus, I need 4E-14 X 87g/mol which is 3.5-trillionth of a gram of rubidium, or 34-picograms of rubidium-84. And I have only a few days from the time the material is fabricated to do this, else the specific activity will drop with each day.

You can see how this is getting not only laborious, but in fact rather impossible to do in all but the most sophisticated of laboratory settings. It does not seem reasonable that such procedures would be carried out in a lab and not once in any paper described.

This is why I say that after reviewing literally thousands of experimental papers on the subject, I do not think the researcher knows what the expected decay rate is. We have to physically measure the specific activity at only 33-day half life every day, and it will never be the same twice. Take this in with the fact that the issues of dead-time and pile-up for, example, an electron multiplier, data transfer rate limits according to the type of data transfer medium used, e.g., fiber optic vs copper, and computational rates; and we have the explanation why no researcher has to date ever presented quantitative results.

In every case the researchers present the x or y axes, or both, normalized to some arbitrary value of penchant. This is not useful information. No paper presents enough information to reproduce the experiment.

Then we issues with the physical dimensions of the setup, distance to the detector and cross section of the detector medium surface. The radiative decay will be omnidirectional, with the detector at some unknown distance and cross section.

I suggest that if the day comes when a researcher produces a quantifiable, stable, and reproducible result, it is indeed worthy of a Nobel. That is not a joke, as the Zeno Effect is key to understand *how to artificially alter the unitary progression of real time, on a macroscopic scale.*

The Zeno Effect is defined as the system coupled to the detector. This means that the Zeno Effect is not a quantum scale phenomenon, but macroscopic, as these are routinely done on benchtop scales, albeit the scale is never described.

## Appendix IV: The Maric Operator: Contributions of Mileva Maric

### Appendix IV The Maric Operator
### Contributions of Mileva Maric

I want to focus on her primary contributions that appeared in the 1905 derivation of 'Special Relativity.' According to Abrahm Joffe, a mathematician and personal friend of the Einsteins, who reviewed the 1905 paper for them prior to their submitting it for publication, the Einsteins pointed out several key points in the paper and derivation that are critical to the success of the derivation of Special Relativity.

Joffe, who reviewed the 1905 Electrodynamics paper prior to the Einsteins submitting it for formal review, explains that it was Mileva Maric who introduced the strategy of using this mapping function. The modern interpretation of $\varphi$ as the 19$^{th}$ century convention of angle is as shown, incorrect. The later appearance of $\phi$ is $\phi(\pm l) = |l|$, [in section §6 part II 'Electrodynamics Part'] which is explicitly stated, not the non-sequitur of the 'light,' which will have no such meaning for another 20-years.

They designate:

$$\varphi(v)\varphi(-v) = 1$$

$$\varphi(\pm v) = |1|$$

$$\phi(l)\phi(-l) = 1$$

$$\phi(\pm l) = |1|$$

$$\therefore$$

$$\varphi(\pm l) = |l|$$

$$c \equiv l/t$$

$$\varphi(\pm 1) = |1|$$

## Appendix IV: The Maric Operator: Contributions of Mileva Maric

The transformation equations we have derived also contain an unknown function $\varphi$ of $v$, which we now wish to determine.

To this end we introduce a third coordinate system $K'$, which relative to the system $k$ is in parallel-translational motion parallel to the axis $\Xi$ such that its origin moves along the $\Xi$-axis with velocity $-v$. Let all three coordinate origins coincide at time $t = 0$, and let the time $t'$ of the system $K'$ be zero at $t = x = y = z = 0$. We denote the coordinates measured in the system $K'$ by $x', y', z'$ and, by twofold application of our transformation equations, we get

$$t' = \varphi(-v)\beta(-v)\left\{\tau + \tfrac{v}{V^2}\xi\right\} = \varphi(v)\varphi(-v)t,$$

$$x' = \varphi(-v)\beta(-v)\{\xi + v\tau\} = \varphi(v)\varphi(-v)x,$$

$$y' = \varphi(-v)\eta = \varphi(v)\varphi(-v)y,$$

$$z' = \varphi(-v)\zeta = \varphi(v)\varphi(-v)z.$$

Since the relations between $x', y', z'$ and $x, y, z$ do not contain the time $t$, the systems $K$ and $K'$ are at rest relative to each other, and it is clear that the transformation from $K$ to $K'$ must be the identity transformation. Hence,

$$\varphi(v)\varphi(-v) = 1.$$

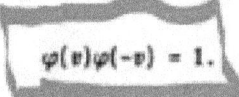 $\varphi(\pm v) \equiv |v|$ *Property 1 The* 'Identity Transformation' The Maric Operator

- The transformation equations we have derived also contain *an unknown function φ of v*, wish we now wish to determine.
- In green, '*To this end we introduce a third system K', which relative to k...*'
- Since the relations between x', y', z' and x' y, z do not contain the time *t*, **the systems K and K' are at rest relative to each other**,
- and it is clear that the transformation from K to K' must be 'the identity [or some translations, identical] Transformation,' [note that in the original this term is capitalized as a proper noun].

The key factor is that they set K [stationary twin] and K' [speeding twin] in relative motion to one another. Because they are metered by *k*, which is the single quantized photon [Einstein had published just five months prior], *regardless of being in relative motion to one another, are at rest relative to each other.*

## Appendix IV: The Maric Operator: Contributions of Mileva Maric

Note that here, **Phi [φ]** is the Maric Operator, not a reference to the 19th century convention of angle:

$$\varphi(\pm v) \equiv |v|$$

$$\varphi(\pm v) = |v|;$$

**then given $c = l/t \equiv 1$;**

$$\varphi(\pm l) = |l|;$$

$$\varphi(\pm t) = |t|;$$

**therefore $\varphi(\pm 1) = |1|$**

The modern interpretation of φ referring to angle is incorrect, as it is outwardly stated as the mapping function, $\varphi(\pm v) \equiv |v|$.

And the rules of the **Maric Operator** apply such that:

- If $A_\Omega$ is $> 2\{Lp, tp\}$ the Maric Operator takes the form $\varphi(\pm 1) = |1|$.
- If $A_\Omega$ is $\leq 2\{Lp, tp\}$ the Maric Operator takes the form $\varphi(\pm 1) \neq |1|$.

The mapping function is the key to Special Relativity, and that all systems are metered by the single quantized photon. There is no case in the entire 1905 electrodynamics paper where K [stationary] and K' [speeding] determine one another's states. They set up multiple scenarios throughout sections §3 through §7, setting K and K' in relative motion in order to see *if they can determine one another's states*. In every case, they cannot determine one another's states.

That is the Twin Paradox.

www.ingramcontent.com/pod-product-compliance
Lightning Source LLC
Chambersburg PA
CBHW060410220526
45465CB00008B/2824